Real-Time PDE-Constrained Optimization

COMPUTATIONAL SCIENCE & ENGINEERING

Computational Science and Engineering (CS&E) is widely accepted, along with theory and experiment, as a crucial third mode of scientific investigation and engineering design. This series publishes research monographs, advanced undergraduate- and graduate-level textbooks, and other volumes of interest to a wide segment of the community of computational scientists and engineers. The series also includes volumes addressed to users of CS&E methods by targeting specific groups of professionals whose work relies extensively on computational science and engineering.

Series Volumes

Biegler, Lorenz T., Omar Ghattas, Matthias Heinkenschloss, David Keyes, and Bart van Bloemen Waanders, Editors, *Real-Time PDE-Constrained Optimization*

Chen, Zhangxin, Guanren Huan, and Yuanle Ma, *Computational Methods for Multiphase Flows in Porous Media*

Shapira, Yair, *Solving PDEs in C++: Numerical Methods in a Unified Object-Oriented Approach*

Real-Time PDE-Constrained Optimization

Edited by

Lorenz T. Biegler
Carnegie Mellon University
Pittsburgh, Pennsylvania

Omar Ghattas
The University of Texas at Austin
Austin, Texas

Matthias Heinkenschloss
Rice University
Houston, Texas

David Keyes
Columbia University
New York, New York

Bart van Bloemen Waanders
Sandia National Laboratories
Albuquerque, New Mexico

siam

Society for Industrial and Applied Mathematics
Philadelphia

10 9 8 7 6 5 4 3 2 1

This project was financially supported by the Computer Science Research Institute at Sandia National Laboratories; National Science Foundation grants 0121360, 0121667, 0314647, 0326449, 0352334, 0427464, 0427985, 0438279, 0457379, 0511624, 0520280, and 0540372; and Department of Energy grant DE-FC02-01ER25477.

Library of Congress Cataloging-in-Publication Data

Real-time PDE-constrained optimization / edited by Lorenz Biegler.
p. cm. — (Computational science and engineering ; 3)
Includes bibliographical references and index.
ISBN: 978-0-898716-21-4 (alk. paper)
1. Mathematical optimization. 2. Differential equations, Partial—Numerical solutions.
I. Biegler, Lorenz T.

QA402.5.R41 2007
519.6—dc22

2006051292

List of Contributors

Frank Allgöwer
Institute for Systems Theory in Engineering (IST)
University of Stuttgart
70049 Stuttgart, Germany
allgower@ist.uni-stuttgart.de

Lorenz T. Biegler
Department of Chemical Engineering
Carnegie Mellon University
Pittsburgh, PA 15213 USA
lb01@andrew.cmu.edu

Pavel B. Bochev
Computational Mathematics and Algorithms Department
Sandia National Laboratories
Albuquerque, NM 87185-1110 USA
pbboche@sandia.gov

Hans Georg Bock
Interdisciplinary Center for Scientific Computing (IWR)
University of Heidelberg
Im Neuenheimer Feld 368
69120 Heidelberg, Germany
bock@iwr.uni-heidelberg.de

Alfio Borzì
Institute für Mathematik und Wissenschaftliches Rechnen
Karl-Franzens-Universität Graz
A-8010 Graz, Austria
alfio.borzi@uni-graz.at

Julien Cortial
Ecole Nationale des Ponts et Chaussées
Champs-sur-Marne, France
jcortial@stanford.edu

Moritz Diehl
Interdisciplinary Center for Scientific Computing (IWR)
University of Heidelberg
Im Neuenheimer Feld 368
69120 Heidelberg, Germany
m.diehl@iwr.uni-heidelberg.de

Hana El-Samad
Department of Mechanical Engineering
University of California
Santa Barbara, CA 93106 USA
helsamad@biochem.ucsf.edu

Charbel Farhat
Department of Aerospace Engineering Sciences and Center for Aerospace Structures
University of Colorado at Boulder
Boulder, CO 80309-0429 USA
cfarhat@stanford.edu

Rolf Findeisen
Institute for Systems Theory in Engineering (IST)
University of Stuttgart
70049 Stuttgart, Germany
findeise@ist.uni-stuttgart.de

Omar Ghattas
Institute for Computational Engineering and Sciences
University of Texas at Austin
Austin, TX 78712 USA
omar@ices.utexas.edu

Martin A. Grepl
Massachusetts Institute of Technology
Cambridge, MA 02139 USA
magrepl@mit.edu

Max D. Gunzburger
School of Computational Science and Information Technology
Florida State University
Tallahassee, FL 32306-4120 USA
gunzburg@csit.fsu.edu

Eldad Haber
Department of Mathematics and Computer Science
Emory University
Atlanta, GA 30322 USA
haber@mathcs.emory.edu

Subhendu B. Hazra
Department of Mathematics
University of Trier
D-54286 Trier, Germany
harzra@uni-trier.de

Matthias Heinkenschloss
Department of Computational and Applied Mathematics
Rice University
Houston, TX 77005 USA
heinken@rice.edu

Michael Hintermüller
Department of Mathematics and Scientific Computing
University of Graz
A-8010 Graz, Austria
michael.hintermueller@uni-graz.at

Chris Homescu
Department of Computer Science and Department of Mechanical Engineering
University of California
Santa Barbara, CA 93106 USA
chome@math.fsu.edu

David E. Keyes
Department of Applied Physics and Applied Mathematics
Columbia University
New York, NY 10027 USA
david.keyes@columbia.edu

Mustafa Khammash
Department of Mechanical Engineering
University of California
Santa Barbara, CA 93106 USA
khammash@engineering.ucsb.edu

Ekaterina Kostina
Interdisciplinary Center for Scientific Computing (IWR)
University of Heidelberg
Im Neuenheimer Feld 368
69120 Heidelberg, Germany
ekaterina.kostina@iwr.uni-heidelberg.de

Karl Kunisch
Department of Mathematics and Scientific Computing
University of Graz
A-8010 Graz, Austria
karl.kunisch@kfunigraz.ac.at

Carl D. Laird
Department of Chemical Engineering
Carnegie Mellon University
Pittsburgh, PA 15213 USA
claird@andrew.cmu.edu

Friedemann Leibfritz
University of Trier
FB-IV Department of Mathematics
D-54286 Trier, Germany
leibfr@uni-trier.de

Gui R. Liu
National University of Singapore
Singapore 117576
mpeliugr@nus.edu.sg

Kirsten Meeker
Department of Computer Science
University of California
Santa Barbara, CA 93106 USA
kmeeker@cs.ucsb.edu

Jan Modersitzki
Institute für Mathematik
Universität Lübeck
D-23560 Lübeck, Germany
modersit@math.uni-luebeck.de

Ngoc C. Nguyen
National University of Singapore
Singapore 117576
cuongng@mit.edu

Anthony T. Patera
National University of Singapore
Singapore 117576
patera@mit.edu

Linda Petzold
Department of Computer Science and
Department of Mechanical Engineering
University of California
Santa Barbara, CA 93106 USA
petzold@engineering.ucsb.edu

Johannes P. Schlöder
Interdisciplinary Center for Scientific Computing (IWR)
University of Heidelberg
Im Neuenheimer Feld 368
69120 Heidelberg, Germany
j.schloeder@iwr.uni-heidelberg.de

Volker Schulz
Department of Mathematics
University of Trier
D-54286 Trier, Germany
volker.schulz@uni-trier.de

Gustaf Söderlind
Centre for Mathematical Sciences
Lund University
Lund, Sweden
gustaf.soderlind@na.lu.se

Stefan Ulbrich
Fachbereich Mathematik
AG10
Technishe Universität Darmstadt
64289 Darmstadt, Germany
ulbrich@mathematik.tu-darmstadt.de

Bart G. van Bloemen Waanders
Optimization and Uncertainty Estimation Department
Sandia National Laboratories
Albuquerque, NM 87185-1110 USA
bartv@sandia.gov

Karen Veroy
National University of Singapore
Singapore 117576
veroy@alum.mit.edu

Stefan Volkwein
Institute of Mathematics and Scientific Computing
University of Graz
A-8010 Graz, Austria
stefan.volkwein@uni-graz.at

K. Willcox
Department of Aeronautics and Astronautics
Massachusetts Institute of Technology
Cambridge, MA 02139 USA
kwillcox@mit.edu

Lei Xie
Department of Mathematics and Scientific Computing
University of Graz
A-8010 Graz, Austria
lei.xie@uni-graz.at

Contents

Preface

Many engineering and scientific problems in design, control, and parameter estimation can be formulated as optimization problems that are governed by partial differential equations (PDEs). The size and complexity of the PDE simulations often present significant optimization challenges. Recent years have seen sustained progress in PDE solvers and optimization algorithms, and the rapid rise in computing capability. Accompanying these advances is a growing interest in real-time and online simulation-based optimization in such diverse areas as aerodynamics, atmospheric and geosciences, chemical process industry, environment, homeland security, infrastructure, manufacturing, and medicine.

Real-time, online optimization arises in the form of (a) *inverse problems*, in which sensor data is assimilated repeatedly into simulations of dynamic processes, (b) *control problems*, in which optimal strategies are repeatedly generated based on new data, and (c) *design problems*, in which algorithms need to be configured to aid in the solution, as well as coordination, of components (a) and (b). An important underlying aspect to online optimization is the availability of powerful large-scale optimization algorithms that can target time-dependent objectives and constraints on challenging engineering systems.

The rapid solution of an optimization problem, however, is only one component needed for the successful realization of real-time optimization. The challenges for real-time and online optimization methods include the ability to do the following:

- *Run sufficiently quickly for decision making at relevant time scales.* Optimization algorithms must return acceptable approximations of the solution in time for decision making. The time scales for decision making vary by application and can range from milliseconds to days. Reliability and efficiency of the optimization algorithms are crucial. Their implementations must be capable of bootstrapping current solutions and yielding meaningful results when terminated prematurely.

- *Adjust to requirement and properties of the process.* Implementations of online optimization algorithms must adjust to different solution accuracy requirements in the process model they feed into. They must be robust and deal with ill-posedness of the problems and tolerate incomplete, uncertain, or errant data.

- *Scale to large problem sizes.* Applications for which real-time and online optimizations are required or desired increase both in number and in complexity. In particular, many of these applications require modeling by PDEs instead of lumped parameter models. Consequently, online optimization algorithms must be able to handle increasingly large and complex PDE models and they have to be able to effectively utilize parallel computing environments.

This volume addresses these issues for real-time optimization of systems governed by PDEs. With 15 contributions, it complements two recent volumes in related areas:

- *Online Optimization of Large Scale Systems*, M. Grötschel, S. Krumke, and J. Rambau (eds.), Springer-Verlag, Berlin (2001). Comprising the work of 25 research projects over six years, this volume deals with a broad array of real-time optimization strategies and applications. In addition to online optimization of systems of differential equations, this volume also considers concepts related to nonlinear and mixed integer optimization, along with stochastic programming and online planning.
- *Large-Scale PDE-Constrained Optimization*, Lecture Notes in Computational Science and Engineering, Vol. 30, L.T. Biegler, O. Ghattas, M. Heinkenschloss, and B. van Bloemen Waanders (eds.), Springer-Verlag, Berlin (2003). This volume deals with a variety of algorithms and strategies for optimization of systems governed by PDEs. The range of PDE applications is broad but largely confined to design and offline optimization.

This volume focuses on the interface of these two areas by considering inversion, control, and design problems that are governed by PDEs and have online or real-time requirements. To this end, the 15 contributions in this volume address the following essential components for real-time optimization:

- Part I: Concepts and properties of real-time, online strategies.
- Part II: Fast PDE-constrained optimization solvers.
- Part III: Reduced-order modeling within the context of design and control.
- Part IV: Applications for real-time decision making.

Concepts and Properties of Real-Time, Online Strategies

The chapters in Part I of this volume address core problems of online optimization of systems governed by differential equations. Key issues in all of these chapters are the formulation of online problems as well as the interaction of the model, solver, and the "plant" (i.e., the real-world system described by the PDEs). Important considerations are online stability properties of the closed-loop system, robustness to uncertainties introduced by the plant as well as the solution algorithm, and efficient implementation of the computations. These issues are considered for differential equation models in the following chapters:

1. "Constrained Optimal Feedback Control of Systems Governed by Large Differential Algebraic Equations," H.G. Bock, M. Diehl, E. Kostina, and J.P. Schlöder.
2. "A Stabilizing Real-Time Implementation of Nonlinear Model Predictive Control," M. Diehl, R. Findeisen, and F. Allgöwer.
3. "Numerical Feedback Controller Design for PDE Systems Using Model Reduction: Techniques and Case Studies," F. Leibfritz and S. Volkwein.

4. "A Least-Squares Finite Element Method for Optimization and Control Problems," P.B. Bochev and M.D. Gunzburger.

Chapter 1 presents an overview of recent work in nonlinear model predictive control (NMPC) by Bock and coworkers. Here the authors derive NMPC controllers based on a multiple shooting formulation. However, a number of approximations are made so that they do not require the online calculation of (expensive) sensitivity matrices. The authors develop four variants of the NMPC controller and also provide convergence proofs for their optimization strategies.

Chapter 2 by Diehl, Findeisen, and Allgöwer presents an overview of recent work in NMPC based on the dual mode control concept and stability analysis. The authors then discuss a novel real-time iteration scheme that allows incomplete solution of the dynamic optimization problem. The stability analysis is modified by coupling the dual mode approach with contractive properties of Newton's method. An experimental case study on the NMPC control of a distillation column, operating at the University of Stuttgart, is presented.

Chapter 3 by Leibfritz and Volkwein presents a set of controller designs based on static output feedback (SOF) along with linear matrix inequality (LMI) constraints to enforce stability. The controller design is then applied to reduced-order models of distributed parameter systems derived using proper orthogonal decomposition (POD). The resulting controller design problems are nonconvex semidefinite programs and are solved using an interior point trust-region algorithm. Three systems are considered and alternative controller designs are presented.

Chapter 4 by Bochev and Gunzburger studies problem formulation and discretization for a class of PDE-constrained optimization problems. Unlike penalty methods, which enforce the PDE constraints by using well-posed least-squares penalty terms added to the original cost functional, this chapter discusses a new approach in which the cost functional is constrained by the least-squares formulation of the PDE constraints. This leads to a bilevel optimization problem. Characterization of solutions and convergence properties of finite element approximations are discussed. The approach is illustrated on a linear quadratic optimal control problem governed by the Stokes equations.

Fast PDE-Constrained Optimization Solvers

Fast optimization solvers remain at the core of online optimization, and recent advances in this area are represented in the five chapters of Part II of this volume. In particular, optimization strategies for time-dependent PDE applications are considered. The first three chapters discuss the development of solution strategies that extend multigrid and domain decomposition methods to time-dependent PDE optimization problems, and also leverage parallel architectures.

5. "Space-Time Multigrid Methods for Solving Unsteady Optimal Control Problems," A. Borzì.

6. "A Time-Parallel Implicit Methodology for the Near-Real-Time Solution of Systems of Linear Oscillators," J. Cortial and C. Farhat.

7. "Generalized SQP Methods with "Parareal" Time Domain Decomposition for Time-Dependent PDE-Constrained Optimization," S. Ulbrich.

Chapter 5 by Borzì gives an overview of space-time multigrid methods for real-time optimal control of parabolic systems, with an application to reaction-diffusion problems. Here reaction-diffusion optimal control problems are approximated by finite differences along with backward Euler schemes, and space-time multigrid schemes are formulated to solve the resulting discrete optimality systems, using a full approximation storage framework. For long time intervals this approach is enhanced with the development of a multigrid receding horizon algorithm, and numerical experiments validate the numerical performance of the space-time multigrid algorithms. Finally, the optimal control of cardiac arrhythmia is presented as an example.

Chapter 6 by Cortial and Farhat treats new methods for time parallelism in the solution of PDEs. Here time slices are generated with a domain decomposition in the time direction. The solution is first approximated on this coarse time grid to provide an initial condition to each time slice. Then, the ODE solver is applied independently and therefore concurrently in each time slice to advance the solution from the starting point of this time slice to its end point. Finally, an iterative process is invoked to improve the solution and eliminate the jumps of the solution on the coarse time grid. This approach led to development of two related approaches, the *parareal* (parallel real-time) algorithm and the *PITA* (parallel implicit time-integrator) method. Here a new extension is presented to improve the stability properties of both methods. Favorable numerical results are demonstrated for linear hyperbolic systems.

Chapter 7 by Ulbrich extends the recent *parareal* time-domain decomposition concept to PDE-constrained optimization problems. The underlying optimization algorithm is a generalized sequential quadratic programming (SQP) method. Unlike traditional SQP methods that require the solution of the linearized state equation, the generalized SQP method studied in this chapter allows the incorporation of approximate state equation solvers. Here parareal time-domain decomposition is applied to approximately solve the state as well as the adjoint equation system. Global convergence of the generalized SQP method is shown, and the method is applied to the optimal control of a semilinear parabolic equation. Numerical tests demonstrate the efficiency of the approach.

The following two chapters deal with essential improvements to accelerate the optimization of PDE-constrained problems:

8. "Simultaneous Pseudo-Timestepping for State-Constrained Optimization Problems in Aerodynamics," S.B. Hazra and V. Schulz.

9. "Digital Filter Stepsize Control in DASPK and Its Effect on Control Optimization Performance," K. Meeker, C. Homescu, L. Petzold, H. El-Samad, M. Khammash, and G. Söderlind.

Chapter 8 by Hazra and Schulz presents a method for solving aerodynamic shape optimization problems involving state constraints, based on simultaneous pseudo-timestepping. This approach solves the KKT system by evolving an embedded nonstationary system of equations in pseudo-time. Advantages of this method include no additional globalization techniques in the design space, and direct extensions to a previously existing pseudo-timestepping method for the states only. Moreover, a block preconditioner that stems from reduced SQP methods can be used for convergence acceleration. The approach is demonstrated on two airfoil application examples to reduced drag with constant lift. Results show a significant reduction of the computational cost compared to usual gradient methods that solve the state and adjoint PDEs exactly at each optimization iteration.

Finally, Chapter 9 by Meeker et al. deals with the issue of nonsmoothness of adaptive ODE and DAE solvers, with respect to perturbations in initial conditions or other problem parameters. Here a recently developed digital filter stepsize controller leads to a much smoother dependence of the solution on problem parameters along with better convergence characteristics for the control and optimization of dynamical systems. This chapter discusses the implementation of the digital filter stepsize controller in the DAE solver *DASPK*3.1. The improved performance of the optimizer, due to the new stepsize controller, is demonstrated on a biological problem regarding the heat shock response of *E. coli*.

Reduced-Order Modeling

An important consideration for online optimization is an accurate representation of the real-world system. Often, the online solution of a detailed plant model, even with efficient state-of-the-art optimization algorithms, is precluded by the computational effort engendered by the governing PDEs. This barrier can often be overcome by the application of reduced-order models. Here a key factor is the development of reduced-order models that capture the accuracy needed for online purposes over a range of input parameters. Clearly, a trade-off occurs between the accuracy achieved by the reduced-order model and the effort needed to solve it. The following chapters deal with the generation of accurate reduced-order models and demonstrate their potential for online optimization:

10. "Certified Rapid Solution of Partial Differential Equations for Real-Time Parameter Estimation and Optimization," M.A. Grepl, N.C. Nguyen, K. Veroy, A.T. Patera, and G.R. Liu.

11. "Model Reduction for Large-Scale Applications in Computational Fluid Dynamics," K. Willcox.

12. "Suboptimal Feedback Control of Flow Separation by POD Model Reduction," K. Kunisch and L. Xie.

Chapter 10 by Grepl et al. derives reduced-order approximations to input-output functions (representing PDE-based systems) with certified accuracy. Here the focus is on processes operating with associated *Assess-Act* scenarios. In the *Assess* stage, robust parameter estimation (inverse) procedures are pursued that map measured-observable outputs to (all) possible system-characteristic and environment-state inputs. In the subsequent *Act* stage, adaptive design and optimization procedures are considered that map mission-objective outputs to best control-variable inputs. The chapter describes a method for real-time certified evaluation of PDE input-output relations. The two ingredients are reduced-basis approximation as well as a posteriori error estimation. Both linear elliptic second-order PDEs as well as extensions to certain nonlinear and parabolic equations are considered.

Chapter 11 by Willcox reviews model reduction of computational fluid dynamics systems and compares two reduction methods, proper orthogonal decomposition (POD) and Fourier model reduction (FMR) in the context of an aerodynamic application, i.e., active flow control for a supersonic diffuser. In particular for linear systems, FMR has advantages in the computation of rigorous, though noncomputable error bounds, and the development of an efficient reduction process when FMR is combined with balanced truncation. Numerical results show significant improvements in approximation over the POD approach. Open questions related to applications of FMR for nonlinear systems are also discussed.

Finally, Chapter 12 by Kunisch and Xie addresses optimal control of fluid dynamical systems using POD applied to the solution of the Hamilton–Jacobi–Bellman (HJB) equations. Both optimal control and boundary feedback control are developed for control problems of Navier–Stokes flows. The approach is implemented in parallel on a cluster, by using a message passing interface to exchange data between MATLAB sessions. This parallel implementation reduces the computing time to an acceptable level for online use.

Applications

The chapters in this final section focus on applications related to online optimization and decision making. While all chapters in this volume incorporate applications to demonstrate the effectiveness of their approaches, the following chapters emphasize problem formulations and solution strategies that are tailored to a particular application. All three chapters deal with inverse problems, which is a core element of real-time optimization and control. The first two chapters are concerned with imaging for medical applications, while the third deals with source detection in water networks.

13. "A Combined Shape-Newton and Topology Optimization Technique in Real-Time Image Segmentation," M. Hintermüller.

14. "COFIR: Coarse and Fine Image Registration," J. Modersitzki and E. Haber.

15. "Real-Time, Large-Scale Optimization of Water Network Systems Using a Subdomain Approach," C. D. Laird, L. T. Biegler, and B. G. van Bloemen Waanders.

Chapter 13 by Hintermüller addresses the determination of tumor boundaries from high resolution magnetic resonance images (MRIs). For this purpose, an edge-detector-based image segmentation algorithm is presented that utilizes both shape as well as topological sensitivity information. Here shape sensitivity and topological derivatives are employed in a serial fashion and no synchronous blending of information takes place. Typically, the algorithm is initiated by the topological phase for computing initial guesses and, if necessary, shape sensitivity is used to drive the contour. A benefit of this approach is that the segmentation becomes fully automatic, without manual selection of an initial contour. This reduction of required user interaction is an important feature in real-time applications. The approach is demonstrated on a number of MRI examples.

Chapter 14 by Modersitzki and Haber presents a general framework for image registration based on a variational principle. Image registration seeks to find a transformation between two related images. Compared to other variational registration schemes, the approach has two new features. First, the transformation space is separated into coarse and fine spaces, orthogonal complements, and only the fine part of the transformation is regularized. Second, a simultaneous solution scheme is developed for both parts. Compared to a multilevel sequential scheme, this can lead to the avoidance of local minima without added computational cost. These features are significant for real-time clinical applications. The approach is applied to registration of an MRI for a human knee.

Chapter 15 by Laird, Biegler, and van Bloemen Vaanders describes a subdomain approach to the problem of source detection to detect contaminants in municipal water networks. Previous work led to efficient and reliable optimization formulations for this problem. The chapter adapts this approach to online implementations and considers a, possibly moving, window of sensor measurements along with problem formulations that can be modified over time. Because the solution of these online optimization problems is time-critical, the moving window approach allows the consideration of very large networks while still maintaining a manageable problem size. The approach is demonstrated on a water distribution network characteristic of a large city.

Conclusions and Acknowledgments

This volume summarizes recent advances toward the solution of time-critical PDE-constrained optimization problems that arise in online implementations. Addressing key issues of online formulations, fast algorithms, reduced-order models, and applications, these chapters deal with essential ingredients for realization of real-time optimization for challenging applications. They also pose a number of open questions that will spark future work in this exciting area.

Finally, we wish to acknowledge the Computer Science Research Institute at Sandia National Laboratories; National Science Foundation grants 0121360, 0121667, 0314647, 0326449, 0352334, 0427464, 0427985, 0438279, 0457379, 0511624, 0520280, 0540372; and Department of Energy grant DE-FC02-01ER25477, for financial support of this project.

L.T. Biegler
O. Ghattas
M. Heinkenschloss
D. Keyes
B. van Bloemen Waanders

Part I

Concepts and Properties of Real-Time, Online Strategies

Chapter 1

Constrained Optimal Feedback Control of Systems Governed by Large Differential Algebraic Equations

Hans Georg Bock, Moritz Diehl*, Ekaterina Kostina*, and Johannes P. Schlöder**

1.1 Introduction

Feedback control based on an online optimization of nonlinear dynamic process models subject to constraints, and its special case, nonlinear model predictive control (NMPC) [1, 20], is an emerging control technique, in particular in chemical engineering [21]. Among its advantages are the capability to directly handle equality and inequality constraints as well as the flexibility provided in formulating the objective function and the process model. It is in particular the increasing availability of detailed nonlinear process models, usually in the form of differential algebraic equations (DAEs) or partial differential algebraic equations (PDAEs), which makes NMPC a practicable control technique with major economic benefits.

Recent years have seen significant advances both in the theory of NMPC and the fast and reliable numerical online computation of constrained feedback control laws for large-scale systems of DAEs and PDAEs. A major aim is to develop algorithms that are able to treat large-scale nonlinear first-principle models without further need of modeling or model reduction.

In this chapter we present and investigate several variants of the "real-time iteration" approach to online computation of constrained optimal feedback control laws. The present realization of this approach is based on the direct multiple shooting method [4] for DAE models [18]. Some of the ideas in this chapter and details on the standard variant of the real-time iteration approach can, e.g., be found in [2, 8, 10, 13]. The main emphasis in this

*Interdisciplinary Center for Scientific Computing (IWR), University of Heidelberg, Im Neuenheimer Feld 368, 69120 Heidelberg, Germany.

chapter is on real-time iteration variants that do not need exact Jacobians and lead to cheap feasibility or optimality refinements. In particular, the presented refinement strategies work in the presence of inequality constraints and active set changes.

1.1.1 Differential Algebraic Equation Systems

Throughout this chapter, we consider DAE models of index one in the following form:

$$B(x(t), z(t), u(t), p)\, \dot{x}(t) = f(x(t), z(t), u(t), p), \tag{1.1}$$

$$0 = g(x(t), z(t), u(t), p). \tag{1.2}$$

Here, x and z denote the differential and the algebraic state vectors, respectively, u is the vector-valued control function, whereas p is a vector of system parameters. This equation type covers many problems in practical engineering applications, from systems of ODEs to reactive flow problems, e.g., the Navier–Stokes equation with chemical reactions. For the sake of simplicity we restrict ourselves in this chapter to the DAE of index 1; however, the generalization to higher-index problems by reduction to index 1 problems with invariants can be derived following well-known techniques [22].

For notational simplicity, we will omit the parameters p in the following. Those parameters which are expected to change during the runtime of the process due to perturbations, and should therefore be observed in the online context, can conceptually be regarded as—nominally—constant differential system states.

1.1.2 Nonlinear Model Predictive Control

Given an observed system state x_0, a *nonlinear model predictive control* (NMPC) scheme obtains a feedback control $\bar{u}(x_0)$ from the solution of an optimal control problem on a prediction horizon $[0, T_p]$ with length T_p:

$$\min_{u(\cdot),x(\cdot),z(\cdot)} \int_0^{T_p} L(x(t), z(t), u(t))\, dt + E(x(T_p)) \tag{1.3a}$$

subject to the initial value constraint

$$x(0) = x_0 \tag{1.3b}$$

and the DAE system

$$B(\cdot)\dot{x}(t) = f(x(t), z(t), u(t)) \quad \forall t \in [0, T_p], \tag{1.3c}$$

$$0 = g(x(t), z(t), u(t)) \quad \forall t \in [0, T_p]. \tag{1.3d}$$

In addition, state and control inequality constraints

$$h(x(t), z(t), u(t)) \geq 0 \quad \forall t \in [0, T_p] \tag{1.3e}$$

as well as terminal constraints

$$r\left(x\left(T_p\right)\right) \geq 0 \tag{1.3f}$$

have to be satisfied.

Solving this problem for a given initial value x_0, we obtain an open-loop optimal control $u^*(t; x_0)$ and corresponding state trajectories $x^*(t; x_0)$, $z^*(t; x_0)$. Based on this solution, a constrained nonlinear feedback control law is given by

$$\bar{u}(x_0) := u^*(0; x_0). \tag{1.4}$$

Due to its origin from an optimal control formulation, the NMPC feedback law has several appealing properties: among them are the possibility to base the feedback on economic criteria, to make use of important process knowledge in the form of nonlinear first-principle models, and to include constraints (1.3e) in a straightforward way. Given suitable choices of the objective function and the final state constraint (1.3f), stability of the nominal NMPC dynamics can be proven [5, 6, 7, 19].

The present chapter is concerned with efficient ways to calculate the feedback control $\bar{u}(x_0)$ or a suitable approximation in *real time*, during the process development.

1.1.3 Overview

In Section 1.2 we give some background material on the classical "direct" multiple shooting method for optimization problems subject to instationary differential equations and outline in Section 1.3 the major ideas of the standard "real-time iteration" scheme. In Section 1.4 four real-time iteration variants are presented that largely avoid costly approximations of Jacobians and Hessians as well as decompositions during the runtime of the online algorithm, and therefore are particularly suitable for large-scale models. Instead, in these variants only few calculations of increasing complexity are needed during each online feedback computation.

- In the linearized optimal feedback control variant presented in Section 1.4.2, only a matrix-vector multiplication as a solution of a small-scale quadratic program (QP) is necessary online. This variant can be interpreted as a constrained linear MPC law and is based on possibly infeasible and suboptimal reference trajectories.

- In the "feasibility improving" suboptimal feedback control variants presented in Sections 1.4.3 and 1.4.4 (for least-squares problems), one additional forward simulation of the nonlinear DAE system is necessary. In the limit, both variants yield feasible but only approximately optimal controls.

- In the online approach in Section 1.4.5, in addition also a "backwards sweep" for computation of the gradient of the Lagrangian is needed. This computation requires only a small factor of the cost of one forward simulation (typically 5–6). In the limit this method yields both feasible and optimal controls.

- Finally, in Section 1.4.6 we propose a multilevel real-time algorithm that simultaneously performs computations of different complexity that correspond to the presented real-time iteration variants.

In Section 1.5 we investigate the convergence properties of the proposed real-time iteration variants when applied to an offline optimal control problem. This convergence analysis may be seen as a first step towards an analysis of the resulting closed-loop behavior. Section 1.6 finishes the chapter with conclusions.

1.2 Direct Multiple Shooting for DAE

Our approaches to the online solution of the optimal control problem (1.3a)–(1.3f)—the real-time iteration schemes—are based on the direct multiple shooting method [4] for DAE models [18], which is briefly reviewed in this section.

1.2.1 Parameterization of the Infinite Optimization Problem

The parameterization of the infinite optimization problem consists of two steps. For a suitable partition of the time horizon $[0, T_p]$ into N subintervals $[t_i, t_{i+1}]$, not necessarily equidistant,

$$0 = t_0 < t_1 < \cdots < t_N = T_p$$

we first parameterize the control function u

$$u(t) = \varphi_i(t, u_i) \quad \text{for } t \in [t_i, t_{i+1}].$$

Note that any parameterization φ_i with local support can be used without changing the structure of the problem as analyzed in the next sections.

In a second step, the DAE solutions are parameterized by *multiple shooting*. For simplicity of presentation we choose the same grid points here as for the controls. The DAE solution is decoupled on the N intervals $[t_i, t_{i+1}]$ by introducing the initial values s_i^x and s_i^z of differential and algebraic states at times t_i as additional optimization variables.

On each subinterval $[t_i, t_{i+1}]$ independently, the trajectories $x_i(t)$ and $z_i(t)$ can be computed as solutions of an initial value problem:

$$B(\cdot)\dot{x}_i(t) = f(x_i(t), z_i(t), \varphi_i(t, u_i)), \tag{1.5a}$$

$$0 = g(x_i(t), z_i(t), \varphi_i(t, u_i)) - \alpha_i(t) g(s_i^x, s_i^z, \varphi_i(t, u_i)), \tag{1.5b}$$

$$x_i(t_i) = s_i^x, \quad z_i(t_i) = s_i^z. \tag{1.5c}$$

Here the subtrahend in (1.5b) is deliberately introduced to relax the DAE and allow an efficient solution for initial values and controls s_i^x, s_i^z, u_i that may violate temporarily the consistency conditions (1.3d). The scalar damping factor $\alpha_i(t)$ is chosen such that $\alpha_i(t_i) = 1$, and $\alpha_i(t) > 0$ is nonincreasing on $t \in [t_i, t_{i+1}]$. The simplest choice is $\alpha_i(t) \equiv 1$, but in most applications an exponential decay down to about one percent at the end of the interval proved to be most efficient. For more details on the relaxation of the DAE the reader is referred, e.g., to [3, 18, 22]. More traditionally, we could also avoid relaxation and delete s_i^z from the variable set, but then we must ensure that the constraints $g(s_i^x, s_i^z, \varphi_i(t_i, u_i)) = 0$ remain feasible.

Since the trajectories $x_i(t)$ and $z_i(t)$ on the interval $[t_i, t_{i+1}]$ are functions of the initial values $s_i := (s_i^x, s_i^z)$ and control parameters u_i only, they will be referred to as $x_i(t; s_i, u_i)$ and $z_i(t; s_i, u_i)$ in the following. The integral part of the cost function is evaluated on each interval independently:

$$L_i(s_i, u_i) := \int_{t_i}^{t_{i+1}} L(x_i(t), z_i(t), \varphi_i(t, u_i))\, dt. \tag{1.6}$$

Note that up to now the multiple shooting parameterization does not involve any discretization of differential operators f, g but is exact.

1.2.2 Structured Nonlinear Programming Problem

The parameterization of problem (1.3a)–(1.3f) using multiple shooting and a suitable control representation leads to the following structured nonlinear programming (NLP) problem:

$$\min_{u,s} \sum_{i=0}^{N-1} L_i(s_i, u_i) \quad + \quad E(s_N^x) \tag{1.7a}$$

subject to the initial condition

$$s_0^x = x_0, \tag{1.7b}$$

the consistency conditions

$$0 = g(s_i^x, s_i^z, \varphi_i(t_i, u_i)), \qquad i = 0, 1, \ldots, N-1, \tag{1.7c}$$

and the continuity conditions

$$s_{i+1}^x = x_i(t_{i+1}; s_i, u_i), \qquad i = 0, 1, \ldots, N-1. \tag{1.7d}$$

Control and path constraints are supposed to be imposed pointwise for a suitable discretization (at n_i points τ_{ij} on each interval, $\tau_{ij} \in [t_i, t_{i+1})$, $j = 0, \ldots, n_i - 1$)

$$h(x_i(\tau_{ij}; s_i, u_i), z_i(\tau_{ij}; s_i, u_i), u_i) \geq 0, \quad j = 0, \ldots, n_i - 1, \quad i = 0, \ldots, N-1. \tag{1.7e}$$

We also have the terminal constraint

$$r(s_N^x) \geq 0. \tag{1.7f}$$

The NLP (1.7a)–(1.7f) can be summarized as

$$P(x_0): \qquad \min_{w} \quad a(w) \quad \text{subject to} \quad \left\{ \begin{array}{rcl} b_{x_0}(w) & = & 0, \\ c(w) & \geq & 0, \end{array} \right. \tag{1.8}$$

where w contains all the multiple shooting state variables and controls:

$$w = (s_0^x, s_0^z, u_0, s_1^x, s_1^z, u_1, \ldots, u_{N-1}, s_N^x) \in \mathbb{R}^{n_w}.$$

The function $a(w)$ is the objective (1.7a), the vector-valued equation $b_{x_0}(w) = 0$ summarizes all equalities from (1.7b)–(1.7d), and the vector-valued $c(w) \geq 0$ contains the inequality constraints (1.7e) and (1.7f).

It is important to note that the initial condition (1.7b) is a linear constraint among the equality constraints, with the varying parameter x_0 entering linearly only in this constraint, so that

$$b_{x_0}(w) = \begin{bmatrix} s_0^x - x_0 \\ g(s_0^x, s_0^z, \varphi_0(t_0, u_0)) \\ s_1^x - x_0(t_1; s_0^x, s_0^z, u_0) \\ \vdots \end{bmatrix} = b_0(w) + Lx_0 \quad \text{with} \quad L := \begin{bmatrix} -\mathbb{I}_{n_x} \\ 0 \\ 0 \\ \vdots \end{bmatrix}. \tag{1.9}$$

1.2.3 Structure of the NLP

Due to the deliberate choice of state and control parameterizations [4] the NLP problem (1.8) has a particular structure: its Lagrangian

$$\mathcal{L}_{x_0}(w, \lambda, \mu) = a(w) - \lambda^T b_{x_0}(w) - \mu^T c(w)$$

(with Lagrange multipliers λ and μ) is *separable* in the sense of [4] so that its Hessian

$$\nabla_w^2 \mathcal{L}(w, \lambda, \mu) = \begin{pmatrix} A_0^{xzu} & & & & \\ & A_1^{xzu} & & & \\ & & \ddots & & \\ & & & A_{N-1}^{xzu} & \\ & & & & A_N^x \end{pmatrix}$$

is *block diagonal*, and obviously independent of x_0 (such that we drop the index x_0 in $\nabla_w^2 \mathcal{L}_{x_0}$). Similarly, the multiple shooting parameterization introduces a characteristic *block sparse structure* of the constraint Jacobian that is also independent of x_0, e.g.,[1]

$$\nabla_w b(w)^T = \begin{pmatrix} \mathbb{I} & & & & & & \\ Z_0^x & Z_0^z & Z_0^u & & & & \\ -X_0^x & -X_0^z & -X_0^u & \mathbb{I} & & & \\ & & & \ddots & \ddots & \ddots & \ddots \end{pmatrix}. \tag{1.10}$$

Furthermore, if the variables w are split into the *state trajectory* $s := (s_0^x, s_0^z, s_1^x, s_1^z, \dots, s_N^x)$ and the *control trajectory* $u := (u_0, u_1, \dots, u_{N-1})$, it is easily seen that $\nabla_s b(w)^T$ is nonsingular, a property to be exploited in the QP linear algebra.

1.2.4 A Newton-Type Method Solution Framework

Throughout the chapter, we will work within a *Newton-type method* framework for the solution of the NLP (1.8). Starting with an initial guess (w_0, λ_0, μ_0), a standard full step iteration for the NLP is

$$w_{k+1} = w_k + \Delta w_k, \tag{1.11}$$

$$\lambda_{k+1} = \lambda_k^{\mathrm{QP}}, \quad \mu_{k+1} = \mu_k^{\mathrm{QP}}, \tag{1.12}$$

where $(\Delta w_k, \lambda_k^{\mathrm{QP}}, \mu_k^{\mathrm{QP}})$ is the solution of a QP. Other than in the classical Gauss–Newton or SQP approaches, we will rather use the more convenient form of the QP

$$\begin{array}{ll} \min\limits_{\Delta w \in \mathbb{R}^{n_w}} & \frac{1}{2} \Delta w^T A_k \Delta w + \left(\nabla_w \mathcal{L}(w_k, \lambda_k, \mu_k)^T + \lambda_k^T B_k + \mu_k^T C_k\right) \Delta w \\ \text{subject to} & \begin{cases} b_{x_0}(w_k) + B_k \Delta w = 0, \\ c(w_k) + C_k \Delta w \geq 0, \end{cases} \end{array} \tag{1.13}$$

[1]We use the definition $\{\nabla_x f\}_{ij} := \frac{\partial f_j}{\partial x_i}$ throughout the chapter.

where A_k is an approximation of the Hessian of the Lagrangian,

$$A_k \approx \nabla_w^2 \mathcal{L}(w_k, \lambda_k, \mu_k),$$

and B_k and C_k are approximations of the constraint Jacobians. Depending on the errors of these approximations we may expect linear or even superlinear convergence. These errors, however, do not influence the accuracy of the solution of the NLP which depends only on the (discretization) errors made in the evaluation of $\nabla_w \mathcal{L}$, b_{x_0}, and c.

1.3 Initial Value Embedding and Real-Time Iterations

In the theoretical approach to constrained feedback control including NMPC, optimal control problems have to be solved online to convergence for varying initial values x_0. We therefore denote the problem (1.7a)–(1.7f), respectively (1.8), by $P(x_0)$. The first question then is to find a strategy to determine an initial guess w_0 for the Newton-type iterations in each problem $P(x_0)$.

From previous optimization steps a solution $w^*(x_0')$ of a neighboring optimization problem $P(x_0')$ is known, including multipliers $\lambda^*(x_0')$ and $\mu^*(x_0')$. A conventional approach would be to use the latest information available, namely to use the old control trajectory, and to compute a new state trajectory by integrating the DAE over the whole horizon using the old control trajectory and the new initial state x_0.

Instead, we propose *not* to make use of x_0 but to use the solution of the previous problem $P(x_0')$ *without any modification*. This initialization for the current problem $P(x_0)$ results, of course, in a violation of the initial value constraint (1.7b) in the NLP (1.7a)–(1.7f), because $s_0^x = x_0' \neq x_0$. However, the constraint is already perfectly satisfied after the first full step Newton-type iteration, due to its linearity. The formulation of the initial value constraint (1.7b) in the NLP (1.7a)–(1.7f) can be considered a linear embedding of each optimization problem into the manifold of perturbed problems. It allows an efficient transition from one optimization problem to the next.

Indeed, in practical applications one observes that the first iteration already yields an excellent approximation of the solution. The effect of this "initial value embedding," or "perturbation embedding," is illustrated in Figure 1.1 for an example problem, an optimal control of a distillation column after a strong perturbation; cf. [13]. In this example the solution of $P(x_0')$ is numerically feasible and optimal, and the Jacobians and Hessians are numerically exact. The left column of Figure 1.1 shows the initialization and the right column shows the solution of $P(x_0)$. One can see that the trajectory of the first iteration (middle column in Figure 1.1) is already very close to the solution in spite of the strong perturbations. Indeed, one can show, as an obvious application of the implicit function theorem (assuming the classical regularity properties) in the case considered (exact Jacobian and Hessian, initialization at solution of $P(x_0')$), that the first QP solution delivers a tangential predictor w_1 to the solution $w^*(x_0)$ of $P(x_0)$,

$$\|w_1 - w^*(x_0)\| = O\left(\|x_0' - x_0\|^2\right).$$

Note, however, that this property is also true even if the change from x_0' to x_0 requires a change of the active set, which can be proven under mild conditions [8]. To sketch the proof

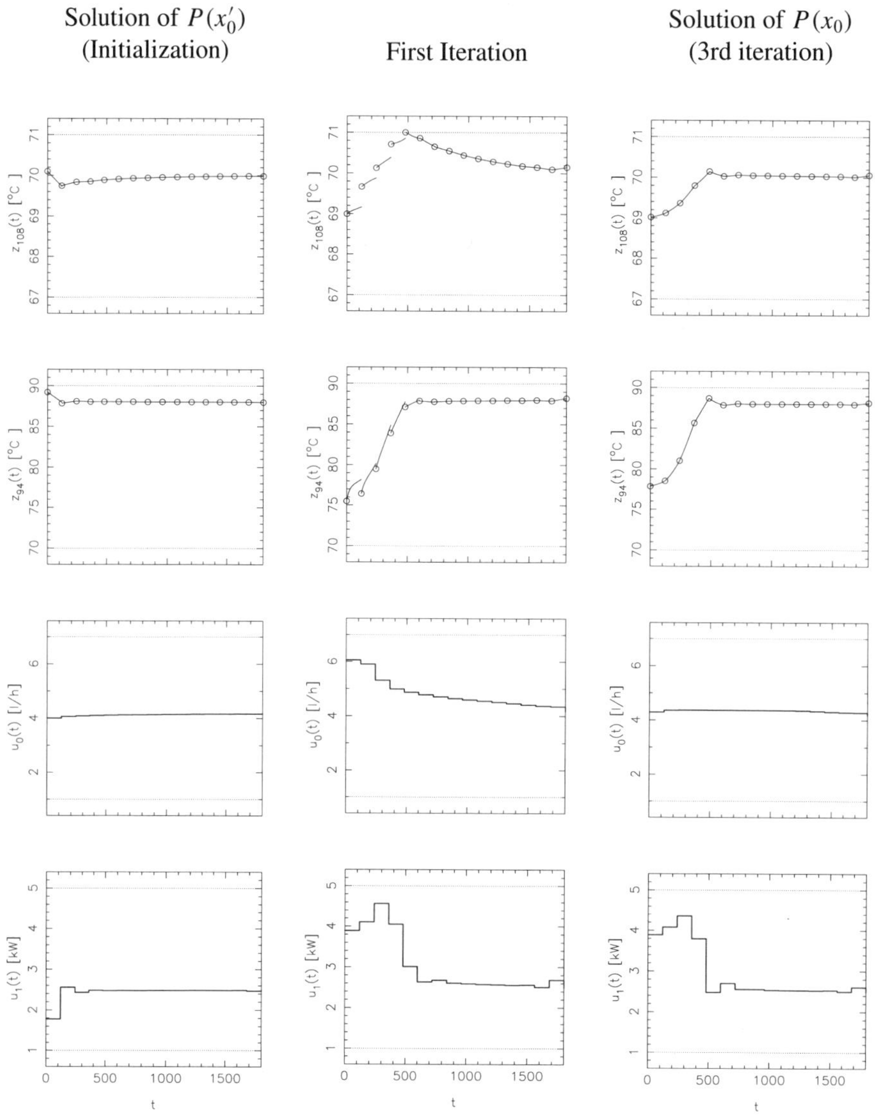

Figure 1.1. *Initial value embedding: after initialization at the solution of $P(x_0')$ (left column), the first iteration (middle column) delivers already a good approximation for the exact solution of $P(x_0)$ (right column); cf. Figure 1.2. The first two graphs of the first iteration (middle column) show discontinuous trajectories, as they are typical in the multiple shooting method.*

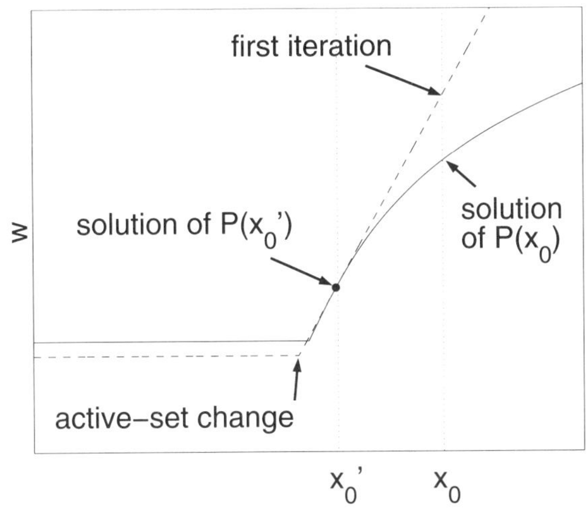

Figure 1.2. *Solution manifold (solid line) and tangential predictor after initial value embedding (dashed line), when initialized with the solution of $P(x_0')$. The first iteration delivers already a good predictor for the exact solution of $P(x_0)$; cf. Figure* 1.1. *This is even true in the presence of active set changes.*

of this counterintuitive property, we assume x_0' to be fixed. An active set change for arbitrarily small $x_0 - x_0'$ occurs only if $P(x_0')$ has some weakly active constraints, and a classical result from parametric optimization [16, 17] states that the solution manifold $w^*(x_0)$ is in this case still directionally differentiable with respect to x_0. Furthermore, the directional derivative $d := \lim_{t\to+0} \frac{w^*(x_0'+t(x_0-x_0'))-w^*(x_0')}{t}$ can be characterized as solution of a QP with the weakly active constraints at x_0' as inequalities. If x_0 enters the problem linearly and if $x_0 - x_0'$ is sufficiently small, then it is possible to show that the optimality conditions of this QP and those of the QP in the first SQP iteration are equivalent; i.e., the first QP solution Δw_0 is equal to d. We conclude that $w_1 = w_0 + \Delta w_0 = w^*(x_0') + d = w^*(x_0) + O(\|x_0' - x_0\|^2)$. For details, we refer the reader to [8, Theorem 3.6]. Note that the solution of the first QP not only gives us directional sensitivity feedback in a small neighborhood of x_0' where the active set does not change anymore but in a larger neighborhood where the linearization is still valid; see the illustration in Figure 1.2. In the case that only approximations of Jacobian and Hessian are used within the QP, we still obtain

$$\|w_1 - w^*(x_0)\| \leq \kappa \left\|x_0' - x_0\right\|,$$

with κ being small if the quality of the approximations is good.

1.3.1 Standard Real-Time Iteration Scheme

Let us now consider the full real-time scenario, where we want to solve a sequence of optimization problems $P(x(t))$ where $x(t)$ is the current system state that changes continuously with time, and which is used as initial value x_0 in problem (1.8).

In the standard real-time iteration scheme [8, 10, 13] we basically proceed as follows: we start with an initial guess (w_0, λ_0, μ_0) and perform the following steps for $k = 0, 1, \ldots$:

1. Preparation: Based on the current solution guess (w_k, λ_k, μ_k), compute all functions and derivatives that are necessary to build the QP (1.13), and prepare the QP solution as much as possible without knowledge of x_0; cf. Section 1.4.1.

2. Feedback Response: At time t_k, obtain the initial value $x_0 := x(t_k)$ from observation of the real system state; solve the QP (1.13) to obtain the step $\Delta w_k = (\Delta s^x_{0k}, \Delta s^z_{0k}, \Delta u_{0k}, \ldots)$, and give the approximation

$$\tilde{u}(x(t_k)) := u_{0k} + \Delta u_{0k}$$

immediately to the real system.

3. Transition: Set the next solution guess as

$$w_{k+1} := w_k + \Delta w_k, \quad \lambda_{k+1} := \lambda_k^{\mathrm{QP}}, \quad \text{and} \quad \mu_{k+1} := \mu_k^{\mathrm{QP}}.$$

1.3.2 Nominal Stability of the Real-Time Iteration Scheme

A central question in NMPC is nominal stability of the closed loop. For the real-time iteration scheme, the state vector of the closed loop consists of the real system state $x(t_k)$ and the content (w_k, λ_k, μ_k) of the prediction horizon in the optimizer. Due to the close connection of system and optimizer, stability of the closed-loop system can only be addressed by combining concepts from both NMPC stability theory and convergence analysis of Newton-type optimization methods. For the standard real-time iteration scheme this analysis has been carried out in [11], and for a related scheme with shift in [12]. In these papers, proofs of nominal stability of the closed loop are given under reasonable assumptions. The class of feedback controls for shrinking horizon problems is treated in [9].

1.4 Real-Time Iteration Variants with Inexact Jacobians

In the standard version of the real-time iteration scheme the time for each cycle corresponds to the time of one SQP iteration. In this chapter, however, we present four variants of the real-time iteration scheme that do not need to evaluate Jacobians online, or at least not in every feedback step. The basic idea for all these variants is to replace the QP (1.13) in the standard real-time iteration scheme by a generic QP that contains only parts of new information but leaves the Hessian A as well as the Jacobians B and C constant; a possible choice is, e.g., $A := \nabla_w^2 \mathcal{L}(\bar{w}, \bar{\lambda}, \bar{\mu})$, $B := \nabla_w b(\bar{w})^T$, and $C := \nabla_w c(\bar{w})^T$ at some reference solution $(\bar{w}, \bar{\lambda}, \bar{\mu})$. In the following, $x_k := x(t_k)$ is the current system state at time t_k:

$$\min_{\Delta w \in \mathbb{R}^{n_w}} \quad \frac{1}{2} \Delta w^T A \, \Delta w + a_k^T \Delta w \tag{1.14a}$$

$$\text{subject to} \quad b_k + L x_k + B \Delta w = 0, \tag{1.14b}$$

$$c_k + C \Delta w \geq 0. \tag{1.14c}$$

The methods, that differ by the choices of a_k, b_k, c_k, proceed by performing the same three steps as in the standard real-time iteration scheme presented in Section 1.3.1, with the only difference that not the QP (1.13) but only the approximated version (1.14) is prepared and solved in each iteration. As the matrices A, B, C are constant, a large share of the computations for preparation and solution of the QP can for all variants already be performed offline, leading to a considerably shorter preparation phase.

We want to point out that for strongly nonlinear processes it might be necessary to update the matrices A, B, C from time to time to ensure sufficient contractivity of the real-time iterates, but we leave this rare updating unconsidered here for simplicity of presentation. For mildly nonlinear systems, however, the matrices A, B, C might really be kept constant without any updates, e.g., evaluated once-for-all at a setpoint solution. While in this case variant 1 is nothing else than linear MPC, variants 2 and 3 converge to nonlinearly feasible (but suboptimal) MPC solutions. Variant 4 will even converge to the true NMPC feedback—without the need to evaluate any derivative matrix online.

1.4.1 Offline Condensing

In all approaches we use fixed approximations of the Jacobians B and C, e.g., by evaluating offline $\nabla_w b(\bar{w})^T$ and $\nabla_w c(\bar{w})^T$ for a reference trajectory $\bar{w}$ that may be an exact or approximate solution of an NLP $P(\bar{x})$ for some state $\bar{x}$. We also use a fixed approximation A of the Hessian, that may be based on the reference solution of $P(\bar{x})$ and computed as $\nabla_w^2 \mathcal{L}(\bar{w}, \bar{\lambda}, \bar{\mu})$, or may be chosen otherwise. Online, we use these fixed components A, B, C to formulate a QP of the form (1.14), where only the vectors a_k, b_k, c_k and the initial value x_k are changing online. It is well known that because $\nabla_s b(\bar{w})$ is invertible, the online QP solution can be prepared by a *condensing* of the QP [4, 8]: We divide Δw into its state and control components Δs and Δu, and presolve the equality constraints (1.14b) to obtain Δs as a linear function of Δu (and x_k), such that we can substitute

$$\Delta w = m(b_k) + \tilde{L} x_k + M \Delta u. \tag{1.15}$$

Note that the matrices $\tilde{L}$ and M are independent of a_k, b_k, c_k, x_k and can in all variants be precomputed offline, exploiting the structure of B in (1.10). In (1.17) below, we show how $m(b_k)$ can be computed efficiently online. We use expression (1.15) to substitute Δw wherever it appears in the QP, to yield the condensed QP:

$$\begin{aligned} \min_{\Delta u} \quad & \frac{1}{2} \Delta u^T A^{\mathrm{c}} \Delta u + \left(a^{\mathrm{c}}(a_k, b_k) + \tilde{A} x_k\right)^T \Delta u \\ \text{subject to} \quad & \left(c^{\mathrm{c}}(c_k, b_k) + \tilde{C} x_k\right) + C^{\mathrm{c}} \Delta u \geq 0. \end{aligned} \tag{1.16}$$

All matrices $A^{\mathrm{c}} := M^T A M$, $\tilde{A} := M^T A \tilde{L}$, $\tilde{C} := C \tilde{L}$, $C^{\mathrm{c}} := CM$ of this condensed QP are precomputed offline. Online, only the vectors $a^{\mathrm{c}}(a_k, b_k) + \tilde{A} x_k$ and $c^{\mathrm{c}}(c_k, b_k) + \tilde{C} x_k$ need to be computed, as shown in the following.

1.4.2 Variant 1: Linear MPC Based on a Reference Trajectory

In the first approach [2, 8], we also compute offline the *fixed* vectors $b := b_0(\bar{w})$, $c := c(\bar{w})$, and $a := \nabla_w a(\bar{w})$, and set $a_k := a$, $b_k := b$, $c_k := c$ in all real-time iterations. We can therefore precompute $m := m(b)$ and also $a^{\mathrm{c}} := a^{\mathrm{c}}(a, b) = M^T(Am + a)$ and $c^{\mathrm{c}} := c^{\mathrm{c}}(c, b) = c + Cm$.

Online, when x_k is known, only two sparse matrix-vector products and two vector additions are needed for computation of $a^c + \tilde{A}x_k$ and $c^c + \tilde{C}x_k$, and the condensed QP (1.16) in variables $\Delta u \in \mathbb{R}^{n_u \times N}$ must be solved. The solution of a QP of this size is standard in linear MPC applications and can usually be achieved quickly. Note that the dimension of the condensed QP (1.16) does *not* depend on the dimensions n_x and n_z of the state vectors, and that the cost of the matrix-vector products grows linearly with n_x and is independent of n_z.

Remark 1.1. It is interesting to consider the special case that a steady state x_{steady} with corresponding algebraic states and controls shall be tracked. By using a least-squares cost function, the choice of $\bar{x} = x_{\text{steady}}$ yields a corresponding steady state solution $\bar{w}$. All multipliers are zero at the solution, and the Hessian $\nabla_w^2 \mathcal{L}(\bar{w}, 0, 0)$ therefore just contains the least-squares penalty matrices. Under these circumstances, the online algorithm results exactly in a linear MPC controller, as observed in [8]. However, when the reference trajectory is updated, e.g., by an upper-level optimization loop, the controller may be called an adaptive linear MPC scheme along reference trajectories that takes future system trajectory behavior into account, in a linearized fashion; cf. Section 1.4.6.

Remark 1.2. It is desirable that the condensed Hessian A^c is positive definite to ensure a unique solution of each online QP, and it might be corrected correspondingly.

Remark 1.3. This online algorithm does not have any memory and basically performs only one feedback step (step 2) at each iteration. Updating of w_k, λ_k, and μ_k in the transition step (step 3) is not necessary, and the preparation step (step 1) is performed only once and offline, before the process starts. However, to speed up the online QP solution, it is desirable to make use of information about the active set in the previous iteration.

1.4.3 Variant 2: Online Feasibility Improvement

In the second variant of the real-time iteration scheme, that was proposed already in [2], we extend the online computational burden by one additional evaluation of $b_0(w_k)$ and $c(w_k)$; i.e., we set $b_k := b_0(w_k)$ and $c_k := c(w_k)$ in the QP (1.14). This allows us to yield a feasibility improvement for nonlinear constraints; see Theorem 1.4 below. Offline, in addition to A, B, C we also compute a fixed objective gradient, e.g., $a = \nabla_w a(\bar{w})$, and then simply set in each iteration $a_k := a + A(w_k - \bar{w})$. This allows us to treat optimality in an approximated fashion.

Recalling the precomputed form of the condensed QP (1.16), only the vectors $a^c(a_k, b_k)$ and $c^c(c_k, b_k)$ have to be computed online, during the preparation phase.

Based on the block sparse structure of B shown in (1.10), the vector $m(b_k) = (m_0^x, m_0^z, m_0^u, \ldots, m_N^x)$ in (1.15) is for given $b_k = (b_0^x, b_0^z, b_1^x, b_1^z, \ldots, b_N^x)$ efficiently computed by a recursion. Starting with $m_0^x := b_0^x$, we compute, for $i = 0, \ldots, N-1$,[2]

$$m_i^u := 0, \quad m_i^z := -\left(Z_i^z\right)^{-1} (b_i^z + Z_i^x m_i^x), \quad m_{i+1}^x := b_{i+1}^x + X_i^x m_i^x + X_i^z m_i^z. \tag{1.17}$$

Based on $m(b_k)$, we can quickly compute $a^c(a_k, b_k) = M^T(Am(b_k) + a_k)$ and $c^c(a_k, b_k) =$

[2]The matrices Z_i^z can be prefactored offline.

$c_k + Cm(b_k)$ (with $a_k = a + A(w_k - \bar{w})$). This computation involves only structured matrix-vector products. This is the end of step 1, the preparation phase. Once x_k is known, the condensed QP (1.16) is solved in the feedback step (step 2), as for the first method.

However, we do have to perform a transition step (step 3), i.e., update $w_{k+1} = w_k + \Delta w_k$, to be able to respect nonlinear constraints. This means that the matrix-vector multiplication $M\Delta u_k$ and the additions $\Delta w_k = m(b_k) + Lx_k + M\Delta u_k$ still need to be done online.

This online algorithm does not make use of λ_k and μ_k and therefore needs not to update them during the online computations.

Remark: It can be shown for the case $x_k = x_0$, i.e., if the optimization problems $P(x_0)$, (1.8) do not change, that the iterates w_k converge under suitable conditions, as will be shown in a forthcoming paper. They converge, however, not to a solution of $P(x_0)$ but to the solution of the following perturbed quadratic problem with nonlinear constraints.

Theorem 1.4 (Limit of Feasibility Improvement). *If, for fixed $x_k = x_0$, the feasibility iterations w_k converge towards a limit w_*, and λ_*^{QP}, μ_*^{QP} are the corresponding multipliers in the QP (1.14) at w_*, then $(w_*, \lambda_*^{\mathrm{QP}}, \mu_*^{\mathrm{QP}})$ is a KKT point of the problem*

$$P_2(x_0): \quad \min_{w} \quad \frac{1}{2}(w - \bar{w})^T A(w - \bar{w}) + (a + e)^T w \quad \textit{subject to} \quad \left\{ \begin{array}{rcl} b_{x_0}(w) & = & 0, \\ c(w) & \geq & 0, \end{array} \right. \tag{1.18}$$

with $e := (\nabla b(w_*) - B^T)\lambda_*^{\mathrm{QP}} + (\nabla c(w_*) - C^T)\mu_*^{\mathrm{QP}}$.

Proof. As $\Delta w = 0$ with λ_*^{QP}, μ_*^{QP} is solution of the QP (1.14) at w_*, the KKT conditions of the QP are satisfied:

$$A(w_* - \bar{w}) + a - B^T\lambda_*^{\mathrm{QP}} - C^T\mu_*^{\mathrm{QP}} = 0, \tag{1.19}$$
$$b_0(w_*) + Lx_0 = 0, \tag{1.20}$$
$$c(w_*) \geq 0, \tag{1.21}$$
$$\mu_*^{\mathrm{QP}} \geq 0, \tag{1.22}$$
$$(\mu_*^{\mathrm{QP}})_i\, c(w_*)_i = 0, \quad i = 1, 2, \ldots, n_c. \tag{1.23}$$

The first equality (1.19) is equivalent to $A(w_* - \bar{w}) + (a+e) - \nabla b(w_*)\lambda_*^{\mathrm{QP}} - \nabla c(w_*)\mu_*^{\mathrm{QP}} = 0$, such that $(w_*, \lambda_*^{\mathrm{QP}}, \mu_*^{\mathrm{QP}})$ satisfies the KKT conditions for the nonlinear problem (1.18). □

1.4.4 Variant 3: Feasibility Improvement for Least-Squares Problems

A slight variation of the above approach for feasibility improvement is applicable to the special case where the objective $a(w)$ in the NLP (1.8) is given by a sum of squares as $a(w) = \frac{1}{2}\|r(w)\|_2^2$ with $r : \mathbb{R}^{n_w} \to \mathbb{R}^{n_r}$. Here the offline choice $A := R^T R$, where R may be given by evaluation of $\nabla_w r(\bar{w})^T$, and the online choice $a_k := R^T r(w_k)$ would be appropriate, and we may call this approach an inexact constrained Gauss–Newton method. All other computations are the same as in variant 2. We can easily establish the following.

Theorem 1.5 (Limit of Feasibility Improvement for Least-Squares Problems). *If, for fixed $x_k = x_0$, the feasibility iterations w_k converge towards a limit w_*, and $\lambda_*^{\mathrm{QP}}, \mu_*^{\mathrm{QP}}$ are the corresponding multipliers in the QP* (1.14) *at w_*, then $(w_*, \lambda_*^{\mathrm{QP}}, \mu_*^{\mathrm{QP}})$ is a KKT point of the problem*

$$P_3(x_0): \quad \min_{w} \quad \frac{1}{2}\|r(w)\|_2^2 + \tilde{e}^T w \quad \textit{subject to} \quad \begin{cases} b_{x_0}(w) &= 0, \\ c(w) &\geq 0, \end{cases} \tag{1.24}$$

with $\tilde{e} := (\nabla b(w_*) - B^T)\lambda_*^{\mathrm{QP}} + (\nabla c(w_*) - C^T)\mu_*^{\mathrm{QP}} - (\nabla r(w_*) - R^T) r(w_*)$.

1.4.5 Variant 4: Online Optimality Improvement

In the fourth variant of the real-time iteration scheme, we further extend the online computational burden by one additional evaluation of the gradient of the Lagrangian $\nabla_w \mathcal{L}(w_k, \lambda_k, \mu_k)$. In an efficient implementation—by means of a reverse sensitivity computation following the principles of automatic differentiation—this will be only a factor of ≈ 5 more expensive than the evaluation of $\mathcal{L}_{x_k}(w_k, \lambda_k, \mu_k)$, which is approximately the cost of variant 2. In the online QP (1.14), we set $a_k := \nabla_w \mathcal{L}(w_k, \lambda_k, \mu_k) + B^T \lambda_k + C^T \mu_k$, as well as $b_k := b_0(w_k)$ and $c_k := c(w_k)$. This approach allows us to yield not only an improvement of feasibility but also of optimality for the original NLP (1.8), as will be shown below.

The remaining online computations are slightly more expensive than for variants 2 and 3, as we need to recover the multipliers $\lambda_k^{\mathrm{QP}}, \mu_k^{\mathrm{QP}}$ of the uncondensed QP (1.14) for the transition step (step 3), as follows.

First, the inequality multipliers μ_k^{QP} are directly obtained as the multipliers μ_k^{cQP} of the condensed QP (1.16): $\mu_k^{\mathrm{QP}} := \mu_k^{\mathrm{cQP}}$. Second, the equality multipliers λ_k^{QP} can be computed as $\lambda_k^{\mathrm{QP}} := (BS^T)^{-T} S(A\Delta w_k + a_k - C^T \mu_k^{\mathrm{QP}})$, where S is a projection matrix that maps w to its subvector s.

The matrix BS^T contains only those columns of B that correspond to the variables s, cf. (1.10), and is thus invertible. Abbreviating $a := S(A\Delta w_k + a_k - C^T \mu_k^{\mathrm{QP}})$, $a = (a_0^x, a_0^z, \ldots, a_N^x)$, we can compute $\lambda_k^{\mathrm{QP}} = (\lambda_0^x, \lambda_0^z, \ldots, \lambda_N^x)$ recursively backwards: Starting with $\lambda_N^x := a_N^x$, we compute, for $i = N-1, N-2, \ldots, 0$,

$$\lambda_i^z = (Z_i^z)^{-T}\left(a_i^z + (X_i^z)^T \lambda_{i+1}^x\right), \quad \lambda_i^x = a_i^x + (X_i^x)^T \lambda_{i+1}^x - (Z_i^x)^T \lambda_i^z,$$

where we employed the submatrix notation of (1.10) for the matrix B, respectively, BS^T. We can show the following.

Theorem 1.6 (Limit of Optimality Improvement). *If, for fixed $x_k = x_0$, the optimality iterations (w_k, λ_k, μ_k) converge towards a limit (w_*, λ_*, μ_*), then this limit is a KKT point of the original NLP* (1.8).

Proof. As $\Delta w = 0$ with λ_*, μ_* is solution of the QP (1.14) at (w_*, λ_*, μ_*), the KKT conditions of the QP are satisfied. In particular, the stationarity (1.19) for the gradient of the Lagrangian of the QP reads

$$\left(\nabla_w \mathcal{L}(w_*, \lambda_*, \mu_*) + B^T \lambda_* + C^T \mu_*\right) - B^T \lambda_* - C^T \mu_* = 0$$

Table 1.1. *Comparison of the four real-time iteration variants.*

Variant	a_k	b_k	c_k	Online computations: Preparation	Online computations: Feedback	Online computations: Transition	Solves NLP
1	fixed	fixed	fixed	-	$\tilde{A}x_k$, $\tilde{C}x_k$, cond. QP	-	-
2	$a+A(w_k-\bar{w})$	$b_0(w_k)$	$c(w_k)$	eval. b, c	as above	Δw_k	(1.18)
3	$R^T r(w_k)$	$b_0(w_k)$	$c(w_k)$	eval. r, b, c	as above	Δw_k	(1.24)
4	$\nabla_w \mathcal{L}(w_k, \lambda_k, \mu_k)$ $+B^T\lambda_k+C^T\mu_k$	$b_0(w_k)$	$c(w_k)$	eval. $\nabla\mathcal{L}$, b, c	as above	Δw_k $\Delta\lambda_k$, $\Delta\mu_k$	(1.8)

which is equivalent to stationarity of the original NLP (1.8). □

An overview of the four real-time iteration variants is given in Table 1.1.

1.4.6 Multilevel Real-Time Iteration Algorithms

Based on the presented variants as well as the standard variant of the real-time iteration scheme presented in Section 1.4, we propose here a multilevel real-time iteration algorithm that employs the variants 1, 2, 4 and the standard real-time iteration scheme in a hierarchical fashion. The algorithm consists of four layers that perform their computations in parallel.

(A) We regard variant 1 as the lowest level, level A, to deliver constrained optimal feedback controls at comparably short sampling times. At each sampling time, a condensed QP of the form (1.16) is solved to obtain feedback to the current system state x_k. This level coincides with the feedback phase of the standard real-time iteration scheme. The constituent vectors and matrices of the QP (1.16) are supplied from the higher levels, as follows.

(B) A modification of variant 2 is used in level B to update the condensed vectors a^c and c^c from time to time, say each n_Ath iterate. This is accomplished by expanding the condensed QP solution of level A at an iterate $k := m \cdot n_A$ to the vector w_k,[3] evaluating $b_0(w_k)$ and $c(w_k)$ as well as $a_k := a + A(w_k - \bar{w})$, and condensing these to the required values a^c and c^c. These computations shall be finished at the time of level A iterate $(m+1)n_A$, so that level A can from then on employ the updated vectors a^c and c^c, while level B starts to work on the next update. The computations of level B need a precomputed approximation a of the gradient of the objective, as well as a reference trajectory $\bar{w}$ that is provided by the next higher level.

(C) On level C, we perform from time to time, say each $(n_A n_B)$th iterate, a new computation of the vector a, the approximation of the gradient of the objective. For this aim, at some iterate $k := m(n_A n_B)$ we determine the multiplier vectors λ_k,

[3]Compute $w_k := w_{(m-1)n_A} + m(b_{(m-1)n_A}) + Lx_k + M\Delta u_k$, where $(m-1)n_A$ is the last iterate in level B.

μ_k from the data of level B, compute the gradient of the Lagrangian, and set $a_k := \nabla_w \mathcal{L}(w_k, \lambda_k, \mu_k) + B^T \lambda_k + C^T \mu_k$. This vector, together with $\bar{w} := w_k$, is given down to level B to be employed until a new update is available.

(D) Finally, on an even higher level D, we perform from time to time, say each $(n_A n_B n_C)$th iterate, a new linearization to compute new matrices A, B, C and the matrices M, $A^c := M^T A M$, $\tilde{A} := M^T A \tilde{L}$, $\tilde{C} := C\tilde{L}$, $C^c := CM$. This allows us to adapt the feedback laws of level A to change process conditions, as well as to improve the contraction rate of the underlying iterations in levels B and C.

Depending on the constants n_A, n_B, n_C, we obtain different algorithms. Choosing $n_A = n_B = n_C = 1$ would recover the standard real-time iteration scheme. The choice $n_A = n_B = 1$ and $n_C = \infty$ results in the new variant 4 (optimality improvement without derivative evaluation), while $n_A = 1$, $n_B = n_C = \infty$ would result in variant 2 (feasibility improvement). Finally, $n_A = n_B = n_C = \infty$ would recover the linear MPC law of variant 1. The performance of the proposed multilevel real-time iteration schemes is currently under investigation.

Remark: Instead of performing an expensive computation of Hessian and Jacobians in level D, it may also be possible to use an update scheme based on secant information from levels B and C in level D, similar to ideas that have been proposed by Griewank and Walther [15] for equality-constrained optimization, and that have recently been adapted to inequalities [14].

1.5 Convergence Analysis

We will briefly investigate under which conditions the real-time iterations w_k converge for variants 2, 3, and 4. Here, we treat only the case $x_k = x_0$, i.e., that the real-time iterates always aim at solving the same problem. This allows us to restrict the attention to the optimization algorithm alone, and interactions with the controlled process need not be considered. A more detailed analysis of the closed-loop behavior of the new real-time iteration variants (similar to the one for the standard real-time iteration scheme [11, 12]) is a subject of future research, as well as an investigation of the multilevel scheme.

As variant 1 has no memory and no iterates w_k, we need only to investigate variants 2, 3, and 4. We will show that they converge under suitable conditions towards the solutions of $P_2(x_0)$, $P_3(x_0)$, and $P(x_0)$, respectively. We will treat all three cases simultaneously, with the focus on variant 4. Let us make the following assumptions.

Assumption 1.7 (Regularity of NLP Solution). *The NLP $P(x_0)$ (respectively, $P_2(x_0)$ or $P_3(x_0)$) has a local optimizer w_* with multipliers λ_*, μ_* that satisfies the following conditions:*

- *The functions $a(w)$, $b_{x_0}(w)$, $c(w)$ are continuously differentiable in a neighborhood of w_*.*
- *The linear independence constraint qualification (LICQ) holds; i.e., the vectors $\nabla_w b_i(w_*), i = 1, \dots, n_b$, and $\nabla_w c_i(w_*), i \in I^* := \{i | c_i(w_*) = 0\}$, are linearly*

independent.

- *The KKT conditions of $P(x_0)$ (respectively, of $P_2(x_0)$ or $P_3(x_0)$) hold. This means, e.g., for $P(x_0)$ that*

$$\nabla_w a(w_*) - \nabla_w b(w_*)\lambda_* - \nabla_w c(w_*)\mu_* = 0, \tag{1.25}$$

$$b_{x_0}(w_*) = 0, \tag{1.26}$$

$$c(w_*) \geq 0, \tag{1.27}$$

$$\mu_* \geq 0, \tag{1.28}$$

$$(\mu_*)_i\, c(w_*)_i = 0, \quad i = 1, 2, \ldots, n_c. \tag{1.29}$$

- *Strict complementarity holds, i.e., $(\mu_*)_i > 0$ for $i \in I_*$.*

Note that we do not need to ensure uniqueness of the solution by a second-order condition yet, but uniqueness will later be implied by Assumption 1.12.

Definition 1.8 (Active Set Functions $\tilde{b}$, $\tilde{B}$, $\tilde{\lambda}$, $\tilde{\mathcal{L}}$ and Projection Π). *For notational simplicity, we will in the following summarize $b_{x_0}(w)$ and $c_i(w)$, $i \in I_*$, in the function $\tilde{b}(w)$, and denote the corresponding Jacobian approximation (composed from B and rows of C) by $\tilde{B}$, $\tilde{B} \approx \nabla_w \tilde{b}(\bar{w})^T$. We also define the corresponding multipliers $\tilde{\lambda}$ and Lagrangian $\tilde{\mathcal{L}}(w, \tilde{\lambda}) := a(w) - \tilde{\lambda}^T \tilde{b}(w)$. By Π we denote a projection such that $\Pi(w, \lambda, \mu) = (w, \tilde{\lambda})$.*

Assumption 1.9 (Matrix Regularity of QP). *The row vectors of $\tilde{B}$ are linearly independent, and the Hessian A is positive definite on the null space of $\tilde{B}$.*

Using Theorem 1.6 (respectively, Theorem 1.4 or 1.5) it is straightforward to show the following.

Lemma 1.10 (Regularity of QP Solution). *If Assumptions* 1.7 *and* 1.9 *hold, the triple $(0, \lambda_*, \mu_*)$ is a strict local minimizer of the QP* (1.14) *formulated at (w_*, λ_*, μ_*) with the following characteristics:*

- *the active set is given by I_*;*
- *LICQ holds;*
- *the KKT conditions of QP* (1.14) *are satisfied;*
- *strict complementarity holds;*
- *the second-order sufficient condition for optimality of the QP is satisfied.*

1.5.1 Stability of the Active Set Near a Solution

Lemma 1.10 opens the way to prove that the QP active set near a solution point is stable, using a standard result of nonlinear optimization.

Lemma 1.11 (Stability of QP Active Set). *If Assumptions 1.7 and 1.9 hold, then there is a neighborhood U of (w_*, λ_*, μ_*) such that the QP (1.14) formulated at any $(w, \lambda, \mu) \in U$ has a unique solution $(\Delta w, \lambda^{\mathrm{QP}}, \mu^{\mathrm{QP}})$ that satisfies strict complementarity, and the active set is given by I_*. Moreover, the QP solution $(\Delta w, \lambda^{\mathrm{QP}}, \mu^{\mathrm{QP}})$ depends continuously on (w, λ, μ).*

1.5.2 Convergence for a Given Active Set

The fact that the active set remains stable near a solution now allows us to establish conditions for local convergence of the SQP variants. For this purpose we first observe that the iterates $(w_k, \tilde{\lambda}_k)$ in variant 4 can be interpreted as inexact Newton iterations towards the solution $y_* := (w_*, \tilde{\lambda}_*)$ of the nonlinear system

$$F(y) = 0, \quad \text{with} \quad y := \begin{bmatrix} w \\ \tilde{\lambda} \end{bmatrix} \quad \text{and} \quad F(y) := \begin{bmatrix} \nabla_w \tilde{\mathcal{L}}(w, \tilde{\lambda}) \\ \tilde{b}(w) \end{bmatrix}.$$

For variants 2 and 3, we accordingly have

$$F_2(y) := \begin{bmatrix} a + A(w - \bar{w}) - \tilde{B}\tilde{\lambda} \\ \tilde{b}(w) \end{bmatrix} \quad \text{and} \quad F_3(y) := \begin{bmatrix} R^T r(w) - \tilde{B}\tilde{\lambda} \\ \tilde{b}(w) \end{bmatrix}.$$

The inexact Newton iterations are performed according to

$$J(y_{k+1} - y_k) = -F(y_k) \tag{1.30}$$

with the fixed matrix

$$J := \begin{bmatrix} A & -\tilde{B}^T \\ \tilde{B} & 0 \end{bmatrix} \neq \frac{\partial F}{\partial y}(y_k) = \begin{bmatrix} \nabla_w^2 \tilde{\mathcal{L}}(w_k, \tilde{\lambda}_k) & -\nabla_w \tilde{b}(w_k) \\ \nabla_w \tilde{b}(w_k)^T & 0 \end{bmatrix}.$$

Note that the matrix J is invertible due to Assumption 1.9. For variants 2 and 3, we accordingly have

$$\frac{\partial F_2}{\partial y}(y_k) = \begin{bmatrix} A & -\tilde{B}^T \\ \nabla_w \tilde{b}(w_k)^T & 0 \end{bmatrix} \quad \text{and} \quad \frac{\partial F_3}{\partial y}(y_k) = \begin{bmatrix} R^T \nabla r(w_k)^T & -\tilde{B}^T \\ \nabla_w \tilde{b}(w_k)^T & 0 \end{bmatrix}.$$

Assumption 1.12. *There exists a closed ball U of radius ρ around (w_*, λ_*, μ_*) such that the following conditions are satisfied:*

- *for each $(w, \lambda, \mu) \in U$ the active set of the QP is I_* (cf. Lemma 1.11);*
- *there is a $\kappa < 1$ such that for all $y \in \Pi U$ it holds that*[4]

$$\left\| \mathbb{I} - J^{-1} \frac{\partial F_{(v)}}{\partial y}(y) \right\| \leq \kappa. \tag{1.31}$$

[4] Depending on the variant number 2, 3, or 4, the index $_{(v)}$ takes the value $_2$ or $_3$ or is empty.

Note that (1.31) implies that $y_* = \Pi(w_*, \lambda_*, \mu_*)$ is the only point in ΠU that satisfies $F_{(v)}(y_*) = 0$. If there was a second point y'_* with $F_{(v)}(y'_*) = 0$, $\Delta y = y'_* - y_* \neq 0$, this would lead to the contradiction

$$\begin{aligned} \|\Delta y\| &= \left\|\Delta y - J^{-1}(F_{(v)}(y'_*) - F_{(v)}(y_*))\right\| \\ &= \left\|\int_0^1 \left(\mathbb{I} - J^{-1}\frac{\partial F_{(v)}}{\partial y}(y_* + t\Delta y)\right)\Delta y dt\right\| \leq \kappa\|\Delta y\| < \|\Delta y\|. \end{aligned}$$

Theorem 1.13 (Convergence towards NLP Solution). *If Assumptions* 1.7, 1.9, *and* 1.12 *hold, the iterations according to variant* 4 *(respectively, variant* 2 *or* 3*) converge towards the optimal solution* (w_*, λ_*, μ_*) *of* $P(x_0)$ *(respectively,* $P_2(x_0)$ *or* $P_3(x_0)$*) for all initial guesses* $(w_0, \lambda_0, \mu_0) \in U$ *that satisfy*

$$\|y_* - y_0\| + \frac{\|J^{-1}F_{(v)}(y_0)\|}{1-\kappa} \leq \rho \quad \textit{with} \quad y_0 := \Pi(w_0, \lambda_0, \mu_0). \tag{1.32}$$

Proof. We first note that the active set is fixed at the first iteration such that it is only necessary to regard the iterates $y_0, y_1, \ldots$ according to (1.30). Note that $y_1 - y_0 = -J^{-1}F_{(v)}(y_0)$. Using a standard trick, we expand

$$\begin{aligned} y_{i+1} - y_i &= -J^{-1}F_{(v)}(y_i) = -J^{-1}\left(F_{(v)}(y_i) - F_{(v)}(y_{i-1}) - J\,(y_i - y_{i-1})\right) \\ &= -\int_0^1 J^{-1}\left(\frac{\partial F_{(v)}}{\partial y}(y_{i-1} + t(y_i - y_{i-1})) - J\right)dt(y_i - y_{i-1}) \end{aligned}$$

such that we can bound

$$\|y_{i+1} - y_i\| \leq \kappa\|y_i - y_{i-1}\|.$$

Therefore, from assumption (1.32), for all $i \in \mathbb{N}$, it holds that $\|y_i - y_0\| \leq \frac{\|y_1 - y_0\|}{1-\kappa} \leq \rho - \|y_* - y_0\|$ and therefore $\|y_i - y_*\| \leq \rho \Rightarrow y_i \in \Pi U$. The Cauchy sequence y_i remains in ΠU and therefore converges to a point $y_* \in \Pi U$ which satisfies $J^{-1}F_{(v)}(y_*) = 0 \Leftrightarrow F_{(v)}(y_*) = 0$ and is therefore the unique local minimizer. □

Remark: Note that the active set is discovered within the first QP and need not be known in advance. However, the convergence result applies only to a neighborhood where the QP active set is always the same. As in the case of classical SQP methods, this assumption of a fixed QP active set is only necessary to obtain the equivalence of the method with a Newton-type method near the solution. In practice, the region of full step convergence of our adjoint based SQP-type method is much larger: as long as the QP data approximate the NLP constraints well enough the method is contracting, even if the correct active set is not determined in the first iteration.

1.6 Conclusions

We have presented a class of methods for online computation of constrained optimal feedback controls in NMPC that are based on the direct multiple shooting method and a "real-time iteration" approach. They use an initial value embedding for efficient initialization of subsequent optimization problems, and treat in each iteration of the optimization process a different optimization problem, with the current system state x_k as initial value.

We proposed four real-time iteration variants that do not need to evaluate Jacobians during the runtime of the online algorithm and are therefore suitable for large-scale DAE models. Particular emphasis is given to inequality constraints, which are crucial for the success of NMPC in practice. In the presented variants online computations with different properties are performed:

- Variant 1 basically requires only the online solution of a condensed QP. It can be interpreted as a linear MPC based on reference trajectories.
- In variants 2 and 3, one additional DAE simulation is needed to evaluate the constraint functions (and a least-squares residual in variant 3). Both approaches are able to achieve feasibility exactly but only optimality with respect to a perturbed objective.
- Variant 4 requires in addition to the forward simulation also a reverse simulation for online computation of the gradient of the Lagrangian, that needs only about 5 times as much effort as the simulation. This variant is able to achieve feasibility as well as optimality for inequality constrained problems, still without evaluating Jacobians online.
- Finally, we proposed a multilevel real-time algorithm that simultaneously performs computations of different complexity that correspond to the presented real-time iteration variants. This algorithm is aimed to combine good local convergence properties with short sampling times and will be the subject of future investigation.

We have shown for the special case of constant initial values, $x_k = x_0$, that the optimization iterations for variant 4 converge locally to the solution of the nonlinear optimization problem, if the approximations of Hessian and constraint Jacobians are sufficiently close to the exact ones. For variants 2 and 3, local convergence towards feasible points was shown, that are, however, only optimal with respect to a perturbed objective. The presented convergence analysis can be seen as a first step towards an analysis of the closed-loop behavior of the real-time iteration variants. Their practical performance in application problems is under investigation.

Acknowledgment

Financial support of the DFG under grant BO864/10-1 within the research collaboration "Optimization Based Control of Chemical Processes" is gratefully acknowledged.

Bibliography

[1] F. Allgöwer, T. Badgwell, J. Qin, J. Rawlings, and S. Wright, *Nonlinear predictive control and moving horizon estimation – An introductory overview*, in Advances in Control, Highlights of ECC'99, P. M. Frank, ed., Springer-Verlag, London, 1999, pp. 391–449.

[2] H. Bock, M. Diehl, D. Leineweber, and J. Schlöder, *A direct multiple shooting method for real-time optimization of nonlinear DAE processes*, in Nonlinear Model Predictive Control, Progr. Systems Control Theory 26, F. Allgöwer and A. Zheng, eds., Birkhäuser, Basel, 2000, pp. 246–267.

[3] H. Bock, E. Eich, and J. Schlöder, *Numerical solution of constrained least squares boundary value problems in differential-algebraic equations*, in Numerical Treatment of Differential Equations, K. Strehmel, ed., Teubner, Leipzig, 1988, pp. 269–280.

[4] H. Bock and K. Plitt, *A multiple shooting algorithm for direct solution of optimal control problems*, in Proceedings of the 9th IFAC World Congress Budapest, Pergamon Press, New York, 1984, pp. 243–247.

[5] H. Chen, *Stability and Robustness Considerations in Nonlinear Model Predictive Control*, Fortschr.-Ber. VDI Reihe 8 Nr. 674, VDI-Verlag, Düsseldorf, 1997.

[6] H. Chen and F. Allgöwer, *A quasi-infinite horizon nonlinear model predictive control scheme with guaranteed stability*, Automatica, 34 (1998), pp. 1205–1218.

[7] G. De Nicolao, L. Magni, and R. Scattolini, *Stability and robustness of nonlinear receding horizon control*, in Nonlinear Model Predictive Control, Progr. Systems Control Theory 26, F. Allgöwer and A. Zheng, eds., Birkhäuser, Basel, 2000, pp. 3–23.

[8] M. Diehl, *Real-Time Optimization for Large Scale Nonlinear Processes*, vol. 920 of Fortschr.-Ber. VDI Reihe 8, Mess-, Steuerungs- und Regelungstechnik, VDI-Verlag, Düsseldorf, 2002.

[9] M. Diehl, H. G. Bock, and J. P. Schlöder, *A real-time iteration scheme for nonlinear optimization in optimal feedback control*, SIAM J. Control Optim., 43 (2005), pp. 1714–1736.

[10] M. Diehl, H. Bock, J. Schlöder, R. Findeisen, Z. Nagy, and F. Allgöwer, *Real-time optimization and nonlinear model predictive control of processes governed by differential-algebraic equations*, J. Proc. Contr., 12 (2002), pp. 577–585.

[11] M. Diehl, R. Findeisen, and F. Allgöwer, *A stabilizing real-time implementation of nonlinear model predictive control*, in Real-Time PDE-Constrained Optimization, L. T. Biegler, O. Ghattas, M. Heinkenschloss, D. Keyes, and B. van Bloemen Waanders, eds., SIAM, Philadlephia, 2007.

[12] M. Diehl, R. Findeisen, F. Allgöwer, H. Bock, and J. Schlöder, *Nominal stability of the real-time iteration scheme for nonlinear model predictive control*, IEE Proc. Control Theory Appl., 152 (2005), pp. 296–308.

[13] M. Diehl, R. Findeisen, S. Schwarzkopf, I. Uslu, F. Allgöwer, H. Bock, E. Gilles, and J. Schlöder, *An efficient algorithm for nonlinear model predictive control of large-scale systems. Part* I: *Description of the method*, Automatisierungstechnik, 50 (2002), pp. 557–567.

[14] M. Diehl, A. Walther, H. Bock, and E. Kostina, *An Adjoint-Based SQP Algorithm with Quasi-Newton Jacobian Updates for Inequality Constrained Optimization*, Technical report MATH-WR-02-2005, Technical University Dresden, Dresden, Germany, 2005.

[15] A. Griewank and A. Walther, *On constrained optimization by adjoint based quasi-Newton methods*, Optim. Methods Softw., 17 (2002), pp. 869–889.

[16] J. Guddat, F. G. Vasquez, and H. T. Jongen, *Parametric Optimization: Singularities, Pathfollowing and Jumps*, B.G.Teubner, Stuttgart, John Wiley & Sons, Chichester, 1990.

[17] K. Jittorntrum, *Solution point differentiability without strict complementarity in nonlinear programming*, Math. Program., 21 (1984), pp. 127–138.

[18] D. Leineweber, *Efficient reduced SQP methods for the optimization of chemical processes described by large sparse DAE models*, vol. 613 of Fortschr.-Ber. VDI Reihe 3, Verfahrenstechnik, VDI-Verlag, Düsseldorf, 1999.

[19] D. Mayne, *Nonlinear model predictive control: Challenges and opportunities*, in Nonlinear Model Predictive Control, Progr. Systems Control Theory 26, F. Allgöwer and A. Zheng, eds., Birkhäuser, Basel, 2000, pp. 23–44.

[20] D. Mayne, J. Rawlings, C. Rao, and P. Scokaert, *Constrained model predictive control: Stability and optimality*, Automatica, 26 (2000), pp. 789–814.

[21] S. Qin and T. Badgwell, *Review of nonlinear model predictive control applications*, in Nonlinear Model Predictive Control: Theory and Application, B. Kouvaritakis and M. Cannon, eds., The Institute of Electrical Engineers, London, 2001, pp. 3–32.

[22] V. H. Schulz, H. G. Bock, and M. C. Steinbach, *Exploiting invariants in the numerical solution of multipoint boundary value problems for DAEs*, SIAM J. Sci. Comput., 19 (1998), pp. 440–467.

Chapter 2

A Stabilizing Real-Time Implementation of Nonlinear Model Predictive Control

Moritz Diehl, Rolf Findeisen*†*,*
and Frank Allgöwer†

2.1 Introduction

Nonlinear model predictive control (NMPC) is a feedback control technique based on the real-time solution of optimal control problems. It has attracted significant interest over the past decades [2, 25, 34, 36], not only from theoreticians but also from practitioners. This interest is mainly motivated by the fact that NMPC allows for a wide flexibility in formulating the objective and the process model, and facilitates directly taking equality and inequality constraints on the states and inputs into account.

One important precondition for the successful application of NMPC is the availability of reliable and efficient numerical dynamic optimization algorithms, since at every sampling time a nonlinear dynamic optimization problem must be solved in real time. Solving such an optimization problem efficiently and quickly is a nontrivial task; see, e.g., [3, 4, 7, 12, 15, 17, 29, 31, 35, 38, 39, 40].

Often classical offline dynamic optimization algorithms are used to solve the optimization problems arising in NMPC. This is done as fast as possible, and the solution is applied to the system once it is available. For small systems using fast computers the feedback delay due to the required computation time is often negligible if compared to the time scale of the system. In this case the computational delay can be considered as a small disturbance, which NMPC, under certain conditions, is able to reject [21, 24, 30, 37].

*Interdisciplinary Center for Scientific Computing (IWR), University of Heidelberg, Im Neuenheimer Feld 368, 69120 Heidelberg, Germany.
†Institute for Systems Theory in Engineering (IST), University of Stuttgart, 70049 Stuttgart, Germany.

However, is it also possible to guarantee stability of the closed loop if the computational delay for an "exact" solution of the optimal control problem is not negligible? Or could one use approximated solutions which are cheaply available, and still achieve stability and good performance, while being able to reject disturbances quickly? These are the questions considered in this chapter.

One possible approach to take account of the computation time is to predict the state at the time we expect the optimization to be finished and carry out the optimization for this prediction [9, 21], allowing us to prove nominal stability. One drawback of this approach is that new information is only incorporated in the feedback once the optimal solution has been found. This can lead to performance decrease if disturbances occurring in between the sampling times drive the system far away from the predicted trajectories.

In contrast to these approaches in the so-called real-time iteration scheme [7, 12, 15, 17] the sampling times are reduced by using approximated solutions of the optimal control problem. The sampling times and feedback delays are reduced by dovetailing the dynamics of the system with the dynamics of the optimization algorithm. In principle only one optimization iteration is performed per sampling instant; i.e., the solution of the optimal control problem and the control of the system are performed in parallel. Applying only approximated solutions reduces the sampling time significantly and thus being able to react to external disturbances significantly faster. The real-time iteration scheme allows us to efficiently treat large-scale systems [18, 20, 22] or systems with short time scales [13] on standard computers and thus pushing forward the frontier of practical applicability of NMPC.

The principal aim of the chapter is to prove that for the real-time iteration scheme the closed loop consisting of the combined system-optimizer dynamics is stable under certain conditions. The investigation combines concepts from both classical stability theory for NMPC as well as from convergence theory for Newton-type optimization methods. Similar results have recently been presented in [16] where a shift of the optimal solutions between iterations is performed to account for the movement of the horizon in time. In contrast to this, we focus in this chapter on a real-time iteration scheme without any shifting between sampling times; i.e., the optimization variables are not modified between iterations. Thus, the real system state that enters the optimization problems in the form of an initial value constraint is the only perturbation to the Newton-type iterations. This simplifies the presentation and proofs significantly, and brings theory and practical implementations of the real-time iteration scheme closer together. The second aim is to illustrate and underpin the obtained theoretical results with numerical experiments considering the control of a high-purity distillation column. The chapter is structured as follows: In Section 2.2 we briefly review the basics of NMPC and present the considered setup. The real-time iteration scheme is outlined in Section 2.3. In Sections 2.4 and 2.5 we derive intermediate results needed for the main contribution, i.e., the proof of stability of the real-time iteration scheme without shift, which is presented in Section 2.6. Section 2.7 contains some numerical experiments for the control of a distillation column, which illustrate the derived theoretical results. We conclude in Section 2.8 with some final remarks.

2.2 Discrete-Time Nonlinear Model Predictive Control

In the following we review the basics of discrete-time NMPC. The presentation is kept short on purpose, since the main objective of the chapter is to derive stability properties of the real-time iteration scheme. More details on NMPC can be found in [1, 10, 23, 32]. Throughout this chapter, we consider the following nonlinear discrete-time system:

$$x^{k+1} = f(x^k, u^k), \quad k = 0, 1, 2, \ldots, \tag{2.1}$$

with system states $x^k \in \mathbb{R}^{n_x}$ and controls $u^k \in \mathbb{R}^{n_u}$. We assume that $f : \mathbb{R}^{n_x} \times \mathbb{R}^{n_u} \to \mathbb{R}^{n_x}$ is twice continuously differentiable, and, without loss of generality, that the origin is a steady state for (2.1), i.e., $f(0, 0) = 0$.

In NMPC the applied controls $u^k = u(x^k)$ depend on the current system state x^k. They are defined such that they are optimal with respect to a specified objective that depends on the predicted system behavior over a horizon of length N. For the derivations in this chapter we assume that the objective minimized at every time instant k is given by

$$\sum_{i=0}^{N-1} L(s_i, q_i) + E(s_N).$$

Here s_i, $i = 0, \ldots, N$, are the predicted states over the fixed prediction horizon $N \in \mathbb{N}$ starting from x^k considering a predicted input sequence $(q_0, q_1, \ldots, q_{N-1})$:

$$s_{i+1} = f(s_i, q_i), \; i = 0, \ldots, N-1, \qquad s_0 = x^k.$$

We use s_i and q_i for the predicted states and controls to distinguish them clearly from the states x^k and controls u^k of the real system (2.1). The functions L and E weigh the system state and input applied to the system. They can be motivated by economical or practical interests and might represent asset values. With respect to the stage cost function L we assume the following.

Assumption 2.1. *The stage cost* $L : \mathbb{R}^{n_x} \times \mathbb{R}^{n_u} \to \mathbb{R}$ *and the terminal penalty* $E : \mathbb{R}^{n_x} \to \mathbb{R}$ *are twice continuously differentiable,* $L(0, 0) = 0$ *and* $E(0) = 0$, *and there exists a positive constant* $m > 0$ *such that*

$$L(x, u) \geq m\|x\|^2 \quad \forall\, x \in \mathbb{R}^{n_x}, u \in \mathbb{R}^{n_u}. \tag{2.2}$$

Often L is quadratic, e.g., $L(x, u) = x^T Q x + u^T R u$, with positive definite matrices Q and R.

The input applied in NMPC is defined as the first input q_0^* of the optimal (optimal values are in the following denoted by a star) predicted input sequence $(q_0^*, \ldots, q_{N-1}^*)$:

$$u(x^k) := q_0^*(x^k). \tag{2.3}$$

Thus the closed-loop system obeys the "ideal NMPC dynamics"

$$x^{k+1} = f(x^k, u(x^k)). \tag{2.4}$$

We denote this as the ideal NMPC dynamics, since no external disturbances nor any numerical errors with respect to the optimality of the applied input are considered.

Summarizing, in an ideal NMPC setting the applied input is given by the solution of a sequence of optimization problems $P(x^k)$ of the following form.

Definition 2.2 ($P(x)$).

$$\min_{\substack{s_0,\ldots,s_N,\\ q_0,\ldots,q_{N-1}}} \quad \sum_{i=0}^{N-1} L(s_i, q_i) + E(s_N) \tag{2.5a}$$

subject to

$$x - s_0 = 0, \tag{2.5b}$$

$$f(s_i, q_i) - s_{i+1} = 0, \quad i = 0, \ldots, N-1. \tag{2.5c}$$

Besides the problem of having to solve this optimal control problem efficiently and online for a practical implementation of NMPC, one central question in NMPC is if the closed-loop system (2.4) is stable. This question has been examined extensively over recent years, and a variety of NMPC schemes exist that can guarantee stability; see, e.g., [1, 10, 23, 33]. For the purpose of this chapter we consider a specific setup of NMPC that ensures stability without enforcing the last predicted state to lie in a certain region (a so-called final region constraint) or even to reach the origin (a so-called zero terminal constraint) [1, 10, 23, 33]. Namely we require that for a certain region of attraction X the end penalty term $E(s_N)$ bounds from above the cost occurring from applying a stabilizing control law $k(x)$ over an infinite time. Since we do not consider any additional path or stability constraints, the setup is close to the setup presented for continuous-time systems in [26].

Before we come to the assumptions ensuring nominal stability of the ideal NMPC setup considered, we state the following assumption, which is required to ensure the existence of unique solutions to problem $P(x)$.

Assumption 2.3. *There exists an open set $X \subset \mathbb{R}^{n_x}$ such that for all initial values $x \in X$ problem $P(x)$ has a unique optimal solution $(s_0^*(x), \ldots, s_N^*(x), q_0^*(x), \ldots, q_{N-1}^*(x))$, the value function $V(x)$ which is defined via the optimal cost for every x by*

$$V(x) := \sum_{i=0}^{N-1} L(s_i^*(x), q_i^*(x)) + E(s_N^*(x)) \tag{2.6}$$

is continuous on the set X, and there exists a (possibly large) $M > 0$ such that $V(x) \leq M\|x\|^2$ for all $x \in X$.

Note that the steady state trajectory $(0, 0, \ldots, 0)$ is the solution of $P(0)$ and has optimal cost $V(0) = 0$, and that because $V(x) \geq L(x, q_0^*(x)) \geq m\|x\|^2$ we also have $V(x) > 0$ for all $x \in X \backslash \{0\}$.

In the remainder of this chapter we are not interested in the set X but rather in a compact level set of V that is contained in X. Thus in the following we consider a fixed $\alpha > 0$ such that

$$X_\alpha := \{x \in X | V(x) \leq \alpha\} \subset X, \tag{2.7}$$

and assume that the set X_α is compact. Clearly, X_α contains a neighborhood of the origin.

We now state the following theorem that ensures stability of the closed loop under ideal NMPC. It is a slight adaptation of the results contained in [1, 10, 32].

Theorem 2.4 (Nominal Stability for Ideal NMPC). *Let Assumptions* 2.1 *and* 2.3 *hold. Then the closed-loop system* $x^{k+1} = f(x^k, u(x^k))$, $k = 0, 1, \ldots$, *defined by the ideal NMPC feedback law* (2.3) *is asymptotically stable with a region of attraction of at least the size of* X_α, *if*

(a) $E(x) > 0$, *for all* $x \in X_\alpha \setminus \{0\}$,

(b) *there exists a local control law* $k(x)$ *defined on* X_α, *such that* $f(x, k(x)) \in X_\alpha$ *for all* $x \in X_\alpha$,

(c) $E(f(x, k(x)) - E(x) \leq -L(x, k(x))$ *for all* $x \in X_\alpha$.

The proof follows from the results presented in [1, 10, 32]. The assumptions required are standard assumptions used in NMPC. The main difference from other approaches is that a strict terminal constraint for the last predicted state (s_N) is avoided, to make the presentations and derivations easier. Note, also, that terminal equality constraints requiring the final state to be fixed are difficult to satisfy in practical applications to large-scale systems.

Assumptions (a) and (c) ensure that E is a local control Lyapunov function in X_α. Assumption (b) ensures that the region X_α is positive invariant under the virtual control law $k(x)$, which is never applied in practice, and guarantees feasibility at the next time instance. A number of methods to determine a terminal penalty term and a terminal region exist; see, e.g., [1, 10, 32].

Results on the "robustness" of Lyapunov functions and results on the stability of NMPC under perturbations suggest that the ideal NMPC controller has some inherent robustness properties with respect to disturbances under the stated assumptions (in particular because V is continuous). This is based on the fact that along solution trajectories of the closed loop under ideal NMPC the following inequality holds:

$$V(f(x, u(x))) \leq V(x) - L(x, u(x)). \tag{2.8}$$

Basically $-L(x, u(x))$ on the right-hand side provides some robustness with respect to disturbances that might lead to a lower decrease—but no increase—of the value function V from time step to time step [23, 24, 27, 37]. Thus, considering the error of an approximate feedback compared to the ideal NMPC input $u(x^k)$ as a disturbance, it can be hoped that under certain conditions the closed loop is stable. We build on somewhat similar arguments in the proof of the main result of this chapter.

Remark: In practical applications, inequality path constraints of the form $h(x_i, q_i) \geq 0$, like bounds on controls or states, are of major interest and should be included in the formulation of the optimization problems $P(x)$. For the derivations of this chapter, however, we leave such constraints unconsidered. Note that in the practical implementation of the real-time iteration scheme they are included.

2.2.1 Online Solution of NMPC: Interconnection of System and Optimizer Dynamics

In ideal NMPC it is assumed that the feedback $u(x^k)$ is available instantaneously at every sampling time k. However, in practice the numerical solution of $P(x^k)$ requires a nonnegligible computation time and involves some numerical errors. Typically the initial value x^k is only known at the time k when the corresponding control u^k is already required for implementation. Thus, instead of implementing the ideal NMPC control $u(x^k)$, often some quickly available but sufficiently good approximation $\tilde{u}(x^k, w^k)$ is used, where the additional argument w^k indicates a data vector $w^k \in \mathbb{R}^n$ (to be defined later) that we use to parameterize the control approximation. These are generated by an online optimization algorithm, and they may be updated from one time step to the next one, according to the law $w^{k+1} = F(x^k, w^k)$, where the argument x^k takes account of the fact that the update shall of course depend on the current system state. To achieve this the computations performed are thus divided into two parts.

1. Preparation: Computation of $w^k = F(x^{k-1}, w^{k-1})$, and generation of the feedback approximation function $\tilde{u}(\cdot, w^k)$, during the transition of the system from state x^{k-1} to x^k.

2. Feedback Response: At time k, give the feedback approximation $u^k := \tilde{u}(x^k, w^k)$ to the system, which then evolves according to $x^{k+1} = f(x^k, u^k)$.

From a system theoretic point of view, instead of the ideal NMPC dynamics (2.4), we have to investigate the combined system-optimizer dynamics

$$x^{k+1} = f(x^k, \tilde{u}(x^k, w^k)), \tag{2.9a}$$
$$w^{k+1} = F(x^k, w^k). \tag{2.9b}$$

The difficulty in the analysis of the closed-loop behavior of this system stems from the fact that the two subsystems mutually depend on each other; see Figure 2.1. The real-time iteration scheme investigated in this chapter is one specific approach for a practical implementation of NMPC, where the data vector w^k is essentially a guess for the optimal solution trajectory of $P(x^k)$. The data update law $w^{k+1} = F(x^k, w^k)$ shall provide iteratively refined solution guesses and is derived from a Newton-type optimization scheme. The approximate feedback law $\tilde{u}(x^k, w^k)$ can be considered as a by-product of this Newton-type iteration scheme.

2.3 The Real-Time Iteration Scheme

In order to characterize the solution of the optimization problem $P(x)$ we introduce the Lagrange multipliers $\lambda_0, \dots, \lambda_N$ for the constraints (2.5b) and (2.5c), and define the Lagrangian function $\mathcal{L}_x(\lambda_0, s_0, q_0, \dots)$ of problem $P(x)$ as

$$\mathcal{L}_x(w) := \sum_{i=0}^{N-1} L(s_i, q_i) + E(s_N) + {\lambda_0}^T(x - s_0) + \sum_{i=0}^{N-1} {\lambda_{i+1}}^T(f(s_i, q_i) - s_{i+1}),$$

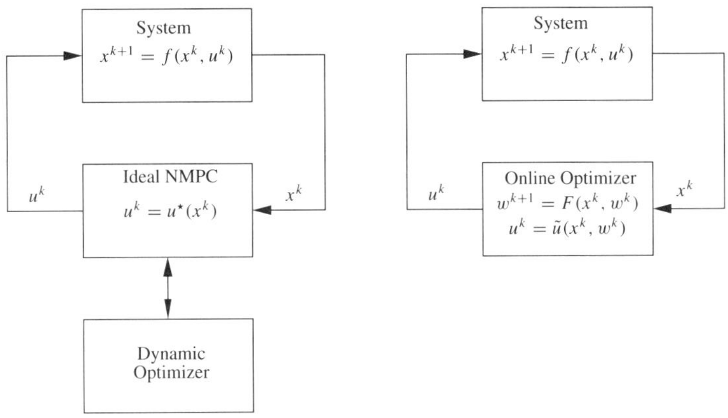

Figure 2.1. *Ideal NMPC (left) and interconnection of system and optimizer dynamics in the real-time iteration scheme (right).*

where all variables are summarized in the (data) vector

$$w := (\lambda_0, s_0, q_0, \ldots, \lambda_N, s_N) \in \mathbb{R}^n. \tag{2.10}$$

Note that under the stated assumptions it follows that $\mathcal{L}_x$ is twice continuously differentiable in its arguments over the considered regions. The necessary optimality conditions of first order for $P(x)$ are

$$\nabla_w \mathcal{L}_x(w) = \begin{bmatrix} x - s_0 \\ \nabla_s L(s_0, q_0) + \frac{\partial f}{\partial s}(s_0, q_0)^T \lambda_1 - \lambda_0 \\ \nabla_q L(s_0, q_0) + \frac{\partial f}{\partial q}(s_0, q_0)^T \lambda_1 \\ \vdots \\ f(s_{N-1}, q_{N-1}) - s_N \\ \nabla_s E(s_N) - \lambda_N \end{bmatrix} = 0. \tag{2.11}$$

One possible solution method for this set of nonlinear equations is to use Newton-type iterations, as outlined in the following.

2.3.1 Review of Newton-Type Iterations

The Newton-type methods investigated in this chapter aim to find the solution w^* of the necessary optimality conditions (2.11) of $P(x)$, $\nabla_w \mathcal{L}_x(w^*) = 0$. Starting at some guess w for w^*, they compute an iterate $w' = w + \Delta w(x, w)$ with

$$\Delta w(x, w) := -J(w)^{-1} \; \nabla_w \mathcal{L}_x(w), \tag{2.12}$$

where $J(w)$ is an approximation of the second derivative $\nabla_w^2 \mathcal{L}_x(w)$. We will use this definition of $\Delta w(x, w)$ throughout the chapter and often also refer to its subvectors as

$$\Delta w(x, w) = (\Delta\lambda_0(x, w), \Delta s_0(x, w), \Delta q_0(x, w), \ldots). \tag{2.13}$$

Note that $\nabla_w^2 \mathcal{L}_x$—often called *Karush–Kuhn–Tucker (KKT)* matrix— is independent of the initial value x, which enters the Lagrangian $\mathcal{L}_x(w)$ only linearly. The index argument x

can therefore be omitted for the KKT matrix; i.e., we will write $\nabla_w^2 \mathcal{L}(w)$ in the following. Of course, its approximation $J(w)$ shall also be independent of x, and we assume in the following that $J(w)$ is continuous over the considered regions. Moreover, the steps shall have the property that $s_0 + \Delta s_0(x, w) = x$, i.e., that the linear initial value constraint (2.5b), $x - s_0 = 0$, is satisfied after one Newton-type iteration. This is easily accomplished by noting that the first n_x rows of $\nabla_w^2 \mathcal{L}(w)$ are constant and choosing them to also be the first n_x rows of $J(w)$.

The Constrained Gauss–Newton Method

An important special case of the Newton-type methods considered in this chapter is the constrained Gauss–Newton method [5], which is applicable for problems with a least-squares form of the objective function, i.e.,

$$L(s_i, q_i) = \frac{1}{2} \|l(s_i, q_i)\|_2^2.$$

For this case, the Hessian blocks $\nabla_{(s_i,q_i)}^2 \mathcal{L}(w)$ within the KKT matrix $\nabla_w^2 \mathcal{L}(w)$ can cheaply be approximated by

$$\left[\frac{\partial l(s_i, q_i)}{\partial(s_i, q_i)}\right]^T \frac{\partial l(s_i, q_i)}{\partial(s_i, q_i)} = \nabla_{(s_i,q_i)}^2 \mathcal{L}(w) + O(\|\lambda_{i+1}\| + \|l(s_i, q_i)\|)$$

neglecting terms that become small when the multipliers and the least-squares residuals are small. In the constrained Gauss–Newton method, this approximation of the Hessian blocks is used within the iteration matrix $J(w)$, while all remaining parts of $J(w)$ are identical to the exact KKT matrix $\nabla_w^2 \mathcal{L}(w)$. It is interesting to note that this iteration matrix $J(w)$ does not depend on the values of the Lagrange multipliers. In the numerical experiments in Section 2.7 we use the constrained Gauss–Newton method. An alternative to the Gauss–Newton method might be to also approximate the constraint Jacobians. In this case adjoint techniques can be used for computing the gradient of the Lagrangian efficiently, and each iteration becomes considerably cheaper [6].

2.3.2 The Real-Time Iteration Algorithm

Assume that during the transition from one sampling instant to the next we have time to perform only one Newton-type iteration. After an initial disturbance it subsequently delivers approximations u^k for the optimal feedback control that allow us to steer the system close to the desired steady state, as will be shown in Section 2.6, under suitable conditions.

The real-time iteration scheme (without shift strategy; cf. [16]) proceeds now as follows:

1. Preparation: Based on the current guess $w^k = (\lambda_0^k, s_0^k, q_0^k, \lambda_1^k, s_1^k, q_1^k, \ldots, \lambda_N^k, s_N^k)$ compute all components of the vector $\nabla_w \mathcal{L}_{x^k}(w^k)$ apart from the first one, and compute the matrix $J(w^k)$. Prepare the computation of $J(w^k)^{-1} \nabla_w \mathcal{L}_{x^k}(w^k)$ as much as possible without knowledge of the value of x^k (a detailed description how this can be achieved is given in [15] or [12]).

2. Feedback Response: At the time k, when x^k measured, compute the feedback approximation $u^k = \tilde{u}(x^k, w^k) := q_0^k + \Delta q_0(x^k, w^k)$ and apply the control u^k immediately to the real system.

3. Transition: Compute the next initial guess w^{k+1} by adding the step vector Δw^k to w^k:
$$w^{k+1} = w^k + \Delta w^k = w^k - J(w^k)^{-1}\ \nabla_w \mathcal{L}_{x^k}(w^k).$$
Continue by setting $k = k + 1$ and going to 1.

In contrast to the ideal NMPC feedback closed loop (2.4), in the real-time iteration scheme we have to regard combined system-optimizer dynamics of the form (2.9), which are given by

$$x^{k+1} = f\left(\ x^k,\ \ q_0^k + \Delta q_0(x^k, w^k)\ \right), \tag{2.14a}$$
$$w^{k+1} = \ w^k + \Delta w(x^k, w^k). \tag{2.14b}$$

In the remainder of the chapter we concentrate on investigating the nominal stability of these system-optimizer dynamics.

2.4 Local Convergence of Newton-Type Optimization

In a first step we review a local convergence result of Newton-type optimization for the solution of one fixed optimization problem; i.e., we regard only the optimizer dynamics (2.14b) for $x^k = x$, with fixed $x \in X_\alpha$. We denote in the following by w_0 an (arbitrary) initial guess for the primal-dual variables of problem $P(x)$. A standard Newton-type scheme for the solution of (2.11), $\nabla_w \mathcal{L}_x(w) = 0$, proceeds by computing iterates $w_1, w_2, \ldots$ according to

$$w_{i+1} = w_i + \Delta w_i, \qquad \Delta w_i := \Delta w(x, w_i).$$

The following theorem states a well-known affinely invariant condition for convergence of Newton-type methods that is due to Bock [5].

Theorem 2.5 (Local Convergence of Newton-Type Optimization [5, 16]). *Assume that $J(w)$ is invertible for all $w \in D$, where $D \subset \mathbb{R}^n$. Furthermore, assume*

(i) *there exist constants $\kappa < 1$, $\omega < \infty$ such that for all $w', w \in D$, $\Delta w = w' - w$, and all $t \in [0, 1]$*

$$\left\| J(w')^{-1} \left(J(w + t\Delta w) - \nabla_w^2 \mathcal{L}(w + t\Delta w)\right) \Delta w \right\| \le \kappa \|\Delta w\|, \tag{2.15a}$$
$$\left\| J(w')^{-1} \left(J(w + t\Delta w) - J(w)\right) \Delta w \right\| \le \omega t \|\Delta w\|^2, \tag{2.15b}$$

(ii) *the first step Δw_0 is sufficiently small such that*

$$\delta_0 := \kappa + \frac{\omega}{2}\|\Delta w_0\| < 1, \tag{2.15c}$$

(iii) *the ball $B_0 := \left\{w \in \mathbb{R}^n \mid \|w - w_0\| \le \frac{\|\Delta w_0\|}{1-\delta_0}\right\}$ is completely contained in D.*

Then the Newton-type iterates $w_0, w_1, \ldots$ are well defined, stay in the ball B_0, and converge towards a point $w^* \in B_0$ satisfying $\nabla_w \mathcal{L}_x(w^*) = 0$. In addition, the iterates satisfy the contraction property

$$\|\Delta w_{i+1}\| \leq \left(\kappa + \frac{\omega}{2}\|\Delta w_i\|\right)\|\Delta w_i\| \leq \delta_0 \|\Delta w_i\|$$

and

$$\|w_i - w^*\| \leq \frac{\|\Delta w_i\|}{1-\delta_0}.$$

Remark 2.6. Equations (2.15a) and (2.15b) are well-known affine invariant assumptions for the convergence of Newton-type methods [5, 11]. Equation (2.15a) is an assumption on the quality of J as an approximation of the second derivative $\nabla_w^2 \mathcal{L}$ (and is satisfied with $\kappa = 0$ for an exact Newton's method), while (2.15b) is an assumption on the Lipschitz continuity of J and is satisfied for any twice differentiable J on a compact set. It is in general difficult to determine κ and ω a priori for a given problem, but a posteriori estimates can be obtained while the Newton-type iterations are carried out.

Remark 2.7. It is interesting to note that the assumptions of the above theorem imply that not only $J(w)$ but also the exact KKT matrix $\nabla_w^2 \mathcal{L}(w)$ is nonsingular, as can be shown by contradiction. Let us for this aim assume that there was, at a point w' in the interior of D, a singular direction $\Delta w \neq 0$ such that $\nabla_w^2 \mathcal{L}(w')\Delta w = 0$. Without loss of generality, assume that Δw is small enough such that $w = w' - \Delta w \in D$. By setting $t = 1$ in (2.15a), we obtain the contradiction $\|\Delta w\| = \left\| J(w')^{-1}\left(J(w') - \nabla_w^2 \mathcal{L}(w')\right)\Delta w\right\| \leq \kappa \|\Delta w\| < \|\Delta w\|$.

2.4.1 Local Convergence of Newton-Type Methods for NMPC

We tailor the above results to the NMPC problem. For this purpose we define two sets $D_C \subset D_{2C}$ which are defined in terms of a fixed $C > 0$:

$$D_C := \left\{ w \in \mathbb{R}^n \,\middle|\, \exists x \in X_\alpha, \quad \|w - w^*(x)\| \leq C \right\}, \tag{2.16}$$

$$D_{2C} := \left\{ w \in \mathbb{R}^n \,\middle|\, \exists x \in X_\alpha, \quad \|w - w^*(x)\| \leq 2C \right\}, \tag{2.17}$$

where $w^*(x)$ is the primal-dual solution of problem $P(x)$, and where X_α is the maximum level set of V in X as introduced in (2.7). Given these sets we can now state the technical assumptions necessary for the following corollary.

Assumption 2.8.

(i) *For each $x \in X_\alpha$ the solution $w^*(x)$ is unique in D_{2C}; i.e., it is the only point $w \in D_{2C}$ satisfying $\nabla_w \mathcal{L}_x(w) = 0$.*

(ii) *For each $w \in D_{2C}$ the matrix $J(w)$ is invertible.*

(iii) *There exist constants $\omega < \infty$, $\kappa < 1$ such that for all $w', w \in D_{2C}$, $\Delta w = w' - w$, and all $t \in [0, 1]$*

$$\left\| J(w')^{-1}\left(J(w + t\Delta w) - \nabla_w^2 \mathcal{L}(w + t\Delta w)\right)\Delta w\right\| \leq \kappa \|\Delta w\|, \tag{2.18a}$$

$$\left\| J(w')^{-1}\left(J(w + t\Delta w) - J(w)\right)\Delta w\right\| \leq \omega t \|\Delta w\|^2. \tag{2.18b}$$

The following two scalars d and δ are used throughout the chapter.

Definition 2.9. *Given a fixed $C > 0$ such that Assumption 2.8 holds, we define the positive scalars*

$$d := \frac{C(1-\kappa)}{1+\frac{\omega}{2}C} \quad \text{and} \quad \delta := \kappa + \frac{\omega}{2}d. \tag{2.19}$$

Note that $\delta = \frac{\kappa+\frac{\omega}{2}C}{1+\frac{\omega}{2}C} < 1$. Now we can state the following corollary that provides conditions for the convergence of Newton-type methods for NMPC.

Corollary 2.10 (Local Convergence of Newton-Type Methods for NMPC). *Suppose Assumption 2.8 holds. If, for some $x \in X_\alpha$ and some $w_0 \in D_C$, it holds that $\|\Delta w(x, w_0)\| \leq d$, then the Newton-type iterates w_i for the solution of $\nabla_w \mathcal{L}_x(w) = 0$, initialized with the initial guess w_0, converge towards the solution $w^*(x)$. Furthermore, the iterates remain in D_C, and satisfy $\|\Delta w_{i+1}\| \leq \delta \|\Delta w_i\|$ as well as $\|w_{i+1} - w^*(x)\| \leq \frac{\delta}{1-\delta}\|\Delta w_i\|$.*

Proof. We start by noting that $C = \frac{d}{1-\delta}$. The ball B_0 defined in Theorem 2.5 is contained in the ball $\{w' \in \mathbb{R}^n \,|\, \|w' - w_0\| \leq C\}$, which itself is contained in the set D_{2C}, as $w_0 \in D_C$. Therefore, there is a solution $w^* \in D_{2C}$ satisfying $\nabla_w \mathcal{L}_x(w^*) = 0$, which must be equal to $w^*(x)$ due to uniqueness. Furthermore, the distance of iterate w_i from $w^*(x)$ is bounded by

$$\|w_i - w^*(x)\| \leq \frac{\|\Delta w_0\|}{1-\delta_0} \leq \frac{d}{1-\delta} = C, \tag{2.20}$$

i.e., $w_i \in D_C$. The remaining two properties immediately follow from Theorem 2.5. □

In the remainder of the chapter we consider fixed values for α and C and assume that Assumption 2.8 is satisfied. Furthermore, we often refer to the set Ξ, which is defined as follows.

Definition 2.11 (Set Ξ). *The set Ξ is defined in the following by*

$$\Xi := \left\{(x, w) \in \mathbb{R}^{n_x} \times \mathbb{R}^n \,|\, x \in X_\alpha, w \in D_C, \|\Delta w(x, w)\| \leq d\right\}. \tag{2.21}$$

This set Ξ contains all pairs (x, w) for which Corollary 2.10 ensures numerical solvability. Note that Ξ is nonempty, as it contains at least the points $(x, w^*(x))$, for all $x \in X_\alpha$, and their neighborhoods.

2.5 Contractivity of the Real-Time Iterations

Before being able to prove stability of the real-time iteration scheme in Section 2.6 we need to establish contraction properties of the Newton-type iterations in the real-time iteration scheme. To investigate the stability of the combined system-optimizer dynamics (2.14) we establish in this section a bound on the size of the steps $\Delta w^k := \Delta w(x^k, w^k)$. For this aim assume the following.

Assumption 2.12. *There exist constants $\tilde{\sigma} > 0$, $\eta > 0$, such that for all $w \in D_C$ and all x, u with $\|u - u(x)\| \leq \frac{\delta d}{1-\delta}$*

$$\left\| J(w)^{-1} \begin{bmatrix} f(x,u)-x \\ 0 \\ \vdots \end{bmatrix} \right\| \leq \eta \|x\| + \tilde{\sigma} \|u - u(x)\|. \tag{2.22}$$

The constant $\tilde{\sigma}$ satisfies $\tilde{\sigma} < \frac{(1-\delta)^2}{\delta}$, and for $\eta > 0$ it holds with $\sigma := \frac{\delta}{1-\delta}\tilde{\sigma}$ that

$$\eta \leq \max \left\{ \frac{1}{2}\sqrt{\frac{m}{\alpha}}(1 - (\delta + \sigma))d,\ \frac{1}{2}\frac{m(1 - (\delta + \sigma))}{\sqrt{32(M + m)\mu}} \right\}. \tag{2.23}$$

Remark 2.13. From this assumption it follows that for all $(x, w) \in \Xi$, $w' = w + \Delta w(x, w)$

$$\left\| J(w')^{-1} \begin{bmatrix} f(s_0',q_0')-s_0' \\ 0 \\ \vdots \end{bmatrix} \right\| \leq \eta \|x\| + \sigma \|\Delta w(x, w)\| \tag{2.24}$$

(with $\sigma = \frac{\delta}{1-\delta}\tilde{\sigma}$), since $s_0' = x$ and $\|q_0' - u(x)\| \leq \|w' - w^*(x)\| \leq \frac{\delta}{1-\delta}\|\Delta w(x, w)\|$. It also follows that $\delta + \sigma < 1$. We will use this and inequality (2.24) in the following.

Remark 2.14. In particular the fact that η might be required to be quite small deserves some discussion. For this aim we first observe that under suitable differentiability assumptions on f and u and because $f(0, 0) = 0$ and $u(0) = 0$ we first have $f(x, u(x)) - x = O(\|x\|)$. We also remark that if $f(x, u)$ is the transition function over time ΔT of a continuous time system, cf. Section 2.7, then $f(x, u) = x + O(\Delta T)$, and we could conclude that $f(x, u(x)) - x = O(\Delta T \|x\|)$. These considerations can be seen as a cautious motivation for feasibility of Assumption 2.12.

Lemma 2.15 (Stepsize Contraction for Real-Time Iterations). *Suppose Assumptions 2.8 and 2.12 are satisfied. Furthermore, assume that $(x^k, w^k) \in \Xi$ and $x^{k+1} := f(x^k, q_0^k + \Delta q_0(x^k, w^k)) \in X_\alpha$. Then, using $\Delta w^k = \Delta w(x^k, w^k)$ and $w^{k+1} = w^k + \Delta w^k$, the following holds:*

$$\|\Delta w(x^{k+1}, w^{k+1})\| \leq (\delta + \sigma)\|\Delta w^k\| + \eta \|x^k\|. \tag{2.25}$$

In particular, $\|\Delta w(x^{k+1}, w^{k+1})\| \leq d$, i.e., $(x^{k+1}, w^{k+1}) \in \Xi$.

Proof. First note that for any $w = (\lambda_0, s_0, q_0, \dots) \in \mathbb{R}^n$ and any $x_1, x_2 \in \mathbb{R}^{n_x}$

$$\nabla_w \mathcal{L}_{x_1}(w) = \nabla_w \mathcal{L}_{x_2}(w) + \begin{bmatrix} x_1 - x_2 \\ 0 \\ \vdots \end{bmatrix}.$$

Furthermore, note that $x^{k+1} = f(x^k, q_0^k + \Delta q_0^k) = f(s_0^{k+1}, q_0^{k+1})$. Therefore, we can deduce

$$
\begin{aligned}
\|\Delta w(x^{k+1}, w^{k+1})\| &= \|\Delta w(f(s_0^{k+1}, q_0^{k+1}), w^{k+1})\| \\
&= \|J(w^{k+1})^{-1} \cdot \nabla_w \mathcal{L}_{f(s_0^{k+1}, q_0^{k+1})}(w^{k+1})\| \\
&= \left\| J(w^{k+1})^{-1} \cdot \left(\nabla_w \mathcal{L}_{x^k}(w^{k+1}) + \begin{bmatrix} f(s_0^{k+1}, q_0^{k+1}) - s_0^{k+1} \\ 0 \\ \vdots \end{bmatrix} \right) \right\| \\
&\leq \|J(w^{k+1})^{-1} \cdot \nabla_w \mathcal{L}_{x^k}(w^{k+1})\| \quad + \quad \eta\|x^k\| + \sigma\|\Delta w^k\| \\
&\leq \qquad \delta \quad \|\Delta w^k\| \qquad + \quad \eta\|x^k\| + \sigma\|\Delta w^k\| \\
&= (\delta + \sigma)\,\|\Delta w^k\| + \eta\|x^k\|,
\end{aligned}
$$

where we have made use of Assumption 2.12 in the fourth transformation and of Assumption 2.8 and Corollary 2.10 in the fifth one. From $m\|x\|^2 \leq V(x) \leq \alpha$ it follows for all $x \in X_\alpha$ that $\|x\| \leq \sqrt{\frac{\alpha}{m}}$, and from $\|\Delta w^k\| \leq d$ and from the first upper bound of η in inequality (2.23), we can finally deduce that $\|\Delta w(x^{k+1}, w^{k+1})\| \leq (\delta + \sigma)\, d + \eta\sqrt{\frac{\alpha}{m}} \leq (\delta + \sigma)\, d + \frac{1}{2}\sqrt{\frac{m}{\alpha}}(1 - (\delta + \sigma))d\sqrt{\frac{\alpha}{m}} \leq d.$ □

The lemma allows us to conclude the following contraction property for the real-time iterations (x^k, w^k) as defined in (2.14), which we use in the following section.

Corollary 2.16 (Shrinking Stepsize for Real-Time Iterations). *Let us, in addition to Assumptions* 2.3, 2.8, *and* 2.12, *assume that the real-time iterations start with an initialization* $(x^0, w^0) \in \Xi$, *and that for a given* $\alpha_0 \leq \alpha$ *and for some* $k_0 > 0$ *we have for all* $k \leq k_0$ *that* $x^k \in X_{\alpha_0}$. *Then* $(x^k, w^k) \in \Xi$ *for all* $k \leq k_0$ *and*

$$
\|\Delta w^k\| \leq (\delta+\sigma)^k \|\Delta w^0\| + \frac{\rho}{2}\sqrt{\alpha_0} \quad, \forall k \leq k_0, \quad \textit{with} \quad \rho := \frac{2\eta}{\sqrt{m}(1 - (\delta + \sigma))}. \tag{2.26}
$$

Proof. Inductively applying Lemma 2.15 to the iterates (x^{k+1}, w^{k+1}), we immediately obtain that $(x^k, w^k) \in \Xi$ for $k = 1, 2, \dots, k_0$. Similarly, from the contraction inequality (2.25), from $\|\Delta w^{k+1}\| \leq (\delta + \sigma)\|\Delta w^k\| + \eta\|x^k\|$, and from the fact that $\|x^k\| \leq \sqrt{\frac{\alpha_0}{m}}$ one obtains inductively that

$$
\|\Delta w^k\| \leq (\delta + \sigma)^k \|\Delta w^0\| + \eta\sqrt{\frac{\alpha_0}{m}} \sum_{i=0}^{k-1} (\delta + \sigma)^i \leq (\delta + \sigma)^k \|\Delta w^0\| + \frac{\eta\sqrt{\frac{\alpha_0}{m}}}{1 - (\delta + \sigma)}. \qquad \square
$$

We may furthermore ask how many iterations we need to reduce the stepsize such that it becomes smaller than a given level. However, considering Corollary 2.16 we must expect that they will not become smaller than the constant $\frac{\rho}{2}\sqrt{\alpha_0}$ in (2.26). But how many iterations do we need, for example, to push the stepsize under twice that level?

Corollary 2.17. *Let us, in addition to Assumptions* 2.3, 2.8, *and* 2.12, *assume that the real-time iterations start with an initialization* $(x^0, w^0) \in \Xi$, *and that for a given* $\alpha_0 \leq \alpha$ *and for some*

$$
k_0 \geq \log_{\delta+\sigma}\left(\frac{\rho\sqrt{\alpha_0}}{2\|\Delta w^0\|}\right) \tag{2.27}
$$

we have for all $k \le k_0$ that $x^k \in X_{\alpha_0}$. Then

$$\|\Delta w^{k_0}\| \le \rho\sqrt{\alpha_0}. \tag{2.28}$$

The proof of this corollary trivially follows from (2.27), since $(\delta + \sigma)^{k_0}\|\Delta w^0\| \le \frac{\rho}{2}\sqrt{\alpha_0}$. This together with (2.26) yields (2.28).

2.6 Stability of the Real-Time Iteration Scheme

In this section we will finally give a proof of asymptotic stability of the combined system-optimizer dynamics (2.14) of the real-time iteration scheme without shift. The line of proof is very similar to the one used in [16] for the real-time iteration scheme with shift.

So far we have already examined the properties of the optimizer iterates w^k but under the crucial assumption that the system states x^k remain in a certain level set X_{α_0}. In this section we will show how this can be assured, such that we will finally be able to prove convergence of the combined iterates (x^k, w^k) towards the origin.

For ideal NMPC, the optimal value function can only decrease due to (2.8), and positive invariance of a level set X_{α_0} would easily follow. In the real-time iteration scheme, however, instead of applying the ideal NMPC control $u(x) := q_0^*(x)$ to the plant, we employ a feedback approximation $\tilde{u}(x, w) := q_0 + \Delta q_0(x, w)$ that depends not only on the system state x but also on the current optimizer parameter vector $w = (\lambda_0, s_0, q_0, \dots)$. This may result in an increase of the value function.

2.6.1 Bounding the Error of Feedback Approximations

Fortunately, under the assumptions of Theorem 2.4 the ideal NMPC feedback inherits under some additional conditions robustness properties as already outlined in Section 2.2. We do not go into details here and instead refer the reader to [16, 37]. We assume the following.

Assumption 2.18. *There exists a constant $\tilde{\mu}$ such that for all $x \in X_\alpha$ and all u with $\|u - u(x)\| \le \frac{\delta}{1-\delta}d$ the following holds:*

$$V(f(x,u)) \le V(x) - L(x,u) + \tilde{\mu}\|u - u(x)\|^2.$$

Remark: The assumption can be motivated by the following informal considerations. For the optimal value function V the following equality holds:

$$V(x) = \min_u \left(L(x,u) + \tilde{V}(f(x,u)) \right),$$

where $\tilde{V}$ is the optimal value function for a shrunk optimization problem similar to (2.5) but with only $N-1$ time steps in the horizon. The minimizer is given by the ideal NMPC law $u(x)$. Assuming suitable differentiability of L, V, and $\tilde{V}$, and using optimality of $u(x)$ we therefore have for u close to $u(x)$

$$L(x,u) + \tilde{V}(f(x,u)) = V(x) + O(\|u - u(x)\|^2).$$

Moreover, from the assumptions of Theorem 2.4 one could conclude that $V(x) \leq \tilde{V}(x)$ for all $x \in X_\alpha$, so that we obtain

$$L(x,u) + V(f(x,u)) \leq L(x,u) + \tilde{V}(f(x,u)) = V(x) + O(\|u - u(x)\|^2),$$

which is a statement of the form of Assumption 2.18.

The following theorem states that the error with respect to the ideal NMPC descent property (2.8) is small if the Newton-type stepsize $\|\Delta w(x,w)\|$ is small. It trivially follows from Assumption 2.18 by observing that for each $(x,w) \in \Xi$ it holds that $\|q_0 + \Delta q_0(x,w) - u(x)\| \leq \frac{\delta}{1-\delta}\|\Delta w(x,w)\|$.

Theorem 2.19 (Error Bound for Approximate Feedback). *Suppose Assumptions* 2.3, 2.8, *and* 2.18 *hold. Then for each* $(x,w) \in \Xi$

$$V(f(x, q_0 + \Delta q_0(x,w))) \leq V(x) - L(x, q_0 + \Delta q_0(x,w)) + \mu\|\Delta w(x,w)\|^2$$

with $\mu := \frac{\delta^2}{(1-\delta)^2}\tilde{\mu}$.

2.6.2 Combining Error Bound and Contractivity

Equipped with the error bound from Section 2.6.1 and the contractivity of the real-time iterations from Section 2.5 we can finally prove nominal stability of the real-time iteration closed-loop scheme. However, since the error in the decrease in the value function depends on the real-time iteration stepsize $\|\Delta w^k\|$, we have to investigate two competing effects: on the one hand, the feedback errors may allow an increase in $V(x^k)$, instead of the desired decrease that was needed to prove nominal stability for ideal NMPC in Theorem 2.4. On the other hand, we know that the stepsizes $\|\Delta w^k\|$ shrink during the iterations, and thus we also expect the errors to become smaller. Since an increase in the value function might imply that we leave the level set X_α, we will not be able to stabilize with the real-time iteration scheme the whole set X_α (at least not if Δw^0 is too large). Thus, we have to provide a safety margin to allow an increase in the value function without leaving X_α until Δw^k is small enough to guarantee a decrease of the value function. For this reason we will distinguish two phases:

- In the first phase we may have an increase in the value function $V(x^k)$; therefore we must allow for a safety back-off. However, the stepsizes $\|\Delta w^k\|$ can be shown to shrink.

- In the second phase, finally, the numerical errors are small enough to guarantee a decrease of both $V(x^k)$ and $\|\Delta w^k\|$, and we can prove convergence of the iterates (x^k, w^k) towards the origin $(0,0)$.

2.6.3 Phase 1: Increase in Objective but Decrease in Stepsize

Exploiting Corollary 2.17, define the number k_α of iterations that are at maximum needed for reduction of the stepsize under the value $\rho\sqrt{\alpha}$ if all iterates stay in the level set X_α.

Definition 2.20 (k_α and Ξ_{attr}). *We define k_α to be the smallest integer such that*

$$k_\alpha \geq \log_{\delta+\sigma}\left(\frac{\rho\sqrt{\alpha}}{2d}\right). \tag{2.29}$$

Furthermore, we define our safety back-off set as the set

$$\Xi_{\text{attr}} := \left\{(x, w) \in \Xi \,\middle|\, V(x) \leq \alpha - k_\alpha \mu d^2\right\}. \tag{2.30}$$

Lemma 2.21 (Increase in Objective, Decrease in Stepsize). *Assume that Assumptions* 2.1, 2.3, 2.8, 2.12, *and* 2.18 *hold and that* $(x^0, w^0) \in \Xi_{\text{attr}}$. *Then for* $k = 0, \dots, k_\alpha$ *it holds that* $(x^k, w^k) \in \Xi$. *Furthermore,* $\|\Delta w^{k_\alpha}\| \leq \rho\sqrt{\alpha}$.

Proof. We make use of Corollaries 2.16 and 2.17. To apply them, we first observe that $(x^0, w^0) \in \Xi$. It remains to be shown that $x^0, \dots, x^{k_\alpha} \in X_\alpha$. We do this inductively, and show that if for some $k \leq k_\alpha$ it holds that $(x^k, w^k) \in \Xi$ and $V(x^k) \leq \alpha + (k - k_\alpha)\mu d^2$, then also $(x^{k+1}, w^{k+1}) \in \Xi$ and $V(x^{k+1}) \leq \alpha + (k + 1 - k_\alpha)\mu d^2$. To show this we first note that $\|\Delta w^k\| \leq d$ as an immediate consequence of Corollary 2.16. Now from Theorem 2.19 we know that

$$V(x^{k+1}) \leq V(x^k) - L(x^k, u^k) + \mu d^2$$

from which we conclude

$$V(x^{k+1}) \leq V(x^k) + \mu d^2 \leq \alpha + (k - k_\alpha)\mu d^2 + \mu d^2 = \alpha + (k + 1 - k_\alpha)\mu d^2. \quad \square$$

Remark: The restriction of the initial system state x^0 to the level set $\{x \in X \,|\, V(x) \leq \alpha - k_\alpha \mu d^2\}$ is unnecessarily restrictive. On the one hand, we neglected the decrease $-L(x^k, u^k)$ in each step; and on the other hand, an initial stepsize $\|\Delta w(x^0, w^0)\|$ considerably smaller than d would allow the errors in the decrease condition to be considerably smaller than μd^2. Note in particular that an initial iterate (x^0, w^0) where the optimizer is initialized so well that $\|\Delta w(x^0, w^0)\| \leq \rho\sqrt{\alpha}$ directly qualifies for Phase 2, if only $V(x^0) \leq \alpha$, without requiring any safety back-off at all. However, to keep the discussion as simple as possible, we chose to stick to our above definition of the set Ξ_{attr} of states attracted by the origin.

2.6.4 Phase 2: Convergence towards the Origin

We now show that the real-time iterations—once the errors have become small enough—not only remain in their level sets but, moreover, are attracted by even smaller level sets. For convenient formulation of the results of this subsection we first define two constant integers.

Definition 2.22 (k_1 and k_2). *Define the constants k_1 and k_2 to be the smallest integers that satisfy*

$$k_1 \geq \frac{6(M + m)}{m} \quad \textit{and} \quad k_2 \geq \log_{\delta+\sigma}\left(\frac{1}{4}\right).$$

Lemma 2.23 (Objective and Stepsize Reduction). *Let us, in addition to Assumptions* 2.1, 2.3, 2.8, 2.12, *and* 2.18 *assume that for an* $\alpha_0 \leq \alpha$ *and a* $k_0 \geq 0$ *it holds that*

$$V(x^{k_0}) \leq \alpha_0 \quad \textit{and} \quad \|\Delta w^{k_0}\| \leq \rho\sqrt{\alpha_0}.$$

Then all iterates $k \geq k_0$ are well defined and also satisfy $V(x^k) \leq \alpha_0$ and $\|\Delta w^k\| \leq \rho\sqrt{\alpha_0}$. Moreover, for $k \geq k_0 + k_1 + k_2$

$$V(x^k) \leq \frac{1}{4}\alpha_0 \quad \textit{and} \quad \|\Delta w^k\| \leq \rho\sqrt{\frac{1}{4}\alpha_0}.$$

Proof. We prove the lemma in three steps: invariance of level sets, attractivity of a small level set for x^k, and reduction of the Newton steps $\|\Delta w^k\|$.

Step 1: *Well-definedness of all iterates and invariance of the level sets.*

The proof is by induction. We assume that for some $k \geq k_0$ it holds that $V(x^k) \leq \alpha_0$ and $\|\Delta w^k\| \leq \rho\sqrt{\alpha_0}$. We will show that then the next real-time iterate is well defined and remains in the level sets, i.e., $V(x^{k+1}) \leq \alpha_0$ and $\|\Delta w^{k+1}\| \leq \rho\sqrt{\alpha_0}$.

First note that by the definition of ρ in (2.26), by $\alpha_0 \leq \alpha$, and by the first upper bound of η in inequality (2.23)

$$\|\Delta w^k\| \leq \frac{2\eta}{\sqrt{m}(1-(\delta+\sigma))}\sqrt{\alpha} \leq \frac{2\frac{1}{2}\sqrt{\frac{m}{\alpha}}(1-(\delta+\sigma))d}{\sqrt{m}(1-(\delta+\sigma))}\sqrt{\alpha} = d;$$

i.e., $(x^k, w^k) \in \Xi$ and the real-time iterate is well defined. Now, due to Assumption 2.3, $\|x^k\| \leq \sqrt{\frac{\alpha_0}{m}}$. By Lemma 2.15 we know that if $x^{k+1} \in X_\alpha$, then

$$\|\Delta w^{k+1}\| \leq (\delta+\sigma)\|\Delta w^k\| + \eta\|x^k\| \leq (\delta+\sigma)\rho\sqrt{\alpha_0} + \eta\sqrt{\frac{\alpha_0}{m}}$$

and therefore, substituting the last term by use of the definition of ρ in (2.26),

$$\|\Delta w^{k+1}\| \leq \rho\sqrt{\alpha_0}\left((\delta+\sigma) + \frac{1}{2}(1-(\delta+\sigma))\right) = \rho\sqrt{\alpha_0}\frac{1+\delta+\sigma}{2} \leq \rho\sqrt{\alpha_0}.$$

It remains to be shown that $x^{k+1} \in X_{\alpha_0} \subset X_\alpha$. To show this we first observe that due to the second upper bound of η in inequality (2.23) in Assumption 2.12 we have

$$\rho \leq \sqrt{\frac{m}{8(M+m)\mu}}$$

and therefore

$$\mu\|\Delta w^{k_0}\|^2 \leq \frac{m}{8(M+m)}\alpha_0 =: \epsilon_0. \tag{2.31}$$

By Theorem 2.19 we obtain $V(x^{k+1}) \leq V(x^k) + \epsilon_0 - m\|x_k\|^2$. We now easily verify that $V(x^{k+1}) \leq \alpha_0$ (i.e., $x^{k+1} \in X_{\alpha_0}$), by distinguishing the following two cases:

(a) $m\|x^k\|^2 \geq 2\epsilon_0$: we have $V(x^{k+1}) \leq V(x^k) - \epsilon_0 \leq \alpha_0 - \epsilon_0$; $\quad\surd$

(b) $m\|x^k\|^2 \leq 2\epsilon_0$: because $V(x^k) \leq M\|x^k\|^2$ we have that $V(x^k) \leq 2\frac{M}{m}\epsilon_0$ and therefore $V(x^{k+1}) \leq 2\frac{M}{m}\epsilon_0 + \epsilon_0 = \frac{1}{4}\alpha_0$ by the definition of ϵ_0 in (2.31). $\quad\surd$

This completes the first step of the proof.

Step 2: *Attraction of the states x^k for $k \geq k_0 + k_1$ by the level set $X_{\frac{1}{4}\alpha_0}$.*

We already showed that all iterates are well defined and satisfy $V(x^k) \leq \alpha_0$ and $\|\Delta w^k\| \leq \rho\sqrt{\alpha_0}$, and furthermore that $\mu\|\Delta w^k\|^2 \leq \epsilon_0$.

To prove the stronger result that the states x^k with $k \geq k_0 + k_1$ are in the reduced level set $X_{\frac{1}{4}\alpha_0} = \{x \in X | V(x) \leq \frac{1}{4}\alpha_0\}$, we first show that for any $k' \geq k_0$ if $V(x^{k'}) \leq \frac{1}{4}\alpha_0$, then also $V(x^{k'+1}) \leq \frac{1}{4}\alpha_0$. We do this again by checking the two cases:

(a) $m\|x^{k'}\|^2 \geq 2\epsilon_0$: we have $V(x^{k'+1}) \leq V(x^{k'}) - \epsilon_0 \leq V(x^{k'}) \leq \frac{1}{4}\alpha_0$; $\quad\checkmark$

(b) $m\|x^{k'}\|^2 \leq 2\epsilon_0$: as before, we have $V(x^{k'+1}) \leq \frac{1}{4}\alpha_0$. $\quad\checkmark$

So let us see how many state iterates x^k can at maximum remain outside $X_{\frac{1}{4}\alpha_0}$. First note that if $V(x^k) \geq \frac{1}{4}\alpha_0$, we also have $M\|x^k\|^2 \geq \frac{1}{4}\alpha_0 = 2\frac{M+m}{m}\epsilon_0 \geq 2\frac{M}{m}\epsilon_0$, i.e., $m\|x^k\|^2 \geq 2\epsilon_0$. Therefore, for every iterate that remains outside $X_{\frac{1}{4}\alpha_0}$, case (a) holds, and $V(x^{k+1}) \leq V(x^k) - \epsilon_0$. We deduce that for any iterate $k \geq k_0$ holds $V(x^k) \leq \alpha_0 - (k - k_0)\epsilon_0$, and therefore for $k \geq k_0 + \frac{6(M+m)}{m}$ that $V(x^k) \leq \alpha_0 - \frac{6(M+m)}{m}\epsilon_0 = \frac{1}{4}\alpha_0$.

Step 3: *Reduction of the steps $\|\Delta w^k\|$ for $k \geq k_0 + k_1 + k_2$ under the level $\rho\sqrt{\frac{1}{4}\alpha_0}$.*

We already know that all iterates $k \geq k_0 + k_1$ satisfy $V(x^k) \leq \frac{1}{4}\alpha_0$ and $\|\Delta w^k\| \leq \rho\sqrt{\alpha_0}$. We can now use Corollary 2.17 with $\|\Delta w^0\|$ replaced by $\rho\sqrt{\alpha_0}$, α_0 replaced by $\frac{1}{4}\alpha_0$, and k_0 replaced by $k - (k_0 + k_1)$, which yields the proposition:

$$\text{If} \quad k - (k_0 + k_1) \geq \log_{\delta+\sigma}\left(\frac{\rho\sqrt{\frac{1}{4}\alpha_0}}{2\rho\sqrt{\alpha_0}}\right), \quad \text{then} \quad \|\Delta w^k\| \leq \rho\sqrt{\frac{1}{4}\alpha_0}.$$

By definition of k_2 this implies $\|\Delta w^k\| \leq \rho\sqrt{\frac{1}{4}\alpha_0}$ for all $k \geq k_0 + k_1 + k_2$. □

2.6.5 Nominal Stability of Real-Time Iterations without Shift

Lemma 2.23 allows us to conclude that at each $k_1 + k_2$ iterations, the level of the objective is reduced by a factor of $\frac{1}{4}$. This allows us to state the main result of this chapter, nominal stability of the real-time iteration scheme without shift.

Theorem 2.24 (Nominal Stability of the Real-Time Iteration Scheme without Shift). *Suppose Assumptions* 2.1, 2.3, 2.8, 2.12, *and* 2.18 *and assume that* $(x^0, w^0) \in \Xi_{\text{attr}}$. *Then all system-optimizer states generated by the combination* (2.14) *of the nominal system dynamics and the real-time iteration scheme without shift are well defined, i.e., satisfy* $(x^k, w^k) \in \Xi$, *and for all integers* $p \geq 0$ *and* $k \geq k_\alpha + p(k_1 + k_2)$ *it holds that* $V(x^k) \leq \alpha\frac{1}{4^p}$ *and* $\|\Delta w^k\| \leq \rho\sqrt{\alpha}\frac{1}{2^p}$. *Therefore,* $(x^k, w^k) \to (0, 0)$.

Proof. The theorem is an immediate consequence of Lemma 2.21 followed by an inductive application of Lemma 2.23. In Phase 1, Lemma 2.21 ensures that the first k_α

system-optimizer iterates remain in Ξ, and that the optimizer stepsize shrinks under the level $\|\Delta w^{k_\alpha}\| \leq \rho\sqrt{\alpha}$. This allows us to apply, at the start of Phase 2, Lemma 2.23 with the constants α_0 and k_0 set to $\alpha_0 := \alpha$ and $k_0 := k_\alpha$. The lemma ensures that after $(k_1 + k_2)$ iterates more, both the objective and the optimizer step are reduced to levels under $\frac{1}{4}\alpha$ and $\rho\sqrt{\frac{1}{4}\alpha}$, respectively. This allows us to apply the same lemma (Lemma 2.23) again, with the constants chosen to be $\alpha_0 := \frac{1}{4}\alpha$ and $k_0 := k_\alpha + (k_1 + k_2)$. The result is a further decrease of the objective to $\frac{1}{4}$ of its previous value and the corresponding decrease in the optimizer stepsize. Repeating this procedure inductively yields the desired bounds $V(x^k) \leq \alpha\frac{1}{4^p}$ and $\|\Delta w^k\| \leq \rho\sqrt{\alpha}\frac{1}{2^p}$ for all $k \geq k_\alpha + p(k_1 + k_2)$. Convergence of x^k towards the origin follows from $m\|x^k\|^2 \leq V(x^k) \leq \alpha\frac{1}{4^p}$. □

Remark: From a practical point of view, the derived result can be interpreted as follows: whenever the system state is subject to a disturbance, but such that after the disturbance the combined system-optimizer state is in the region Ξ_{attr}, the subsequent closed-loop response will lead the system towards the origin with a linear convergence rate. The proof should not be seen as a construction rule for designing suitable real-time iteration schemes. Instead it gives a theoretical underpinning of the real-time iteration scheme.

Similar convergence results as for the real-time iteration scheme would also hold true for numerical schemes where more than one Newton-type iteration is performed per sampling time, sacrificing, however, the instantaneous feedback of the real-time iteration scheme. In the limit of infinitely many iterations per optimization problem, the set Ξ_{attr} would approach the set Ξ and the whole region of attraction of the ideal NMPC controller would be recovered; cf. Theorem 2.4.

2.7 Numerical Experiments: Distillation Control

In order to illustrate the investigated method, tests with a nontrivial process control example, namely a high-purity distillation column differential algebraic equation (DAE) model , with 82 differential and 122 algebraic states are performed. For details of the distillation model and the control problem we refer the reader to [12, 18], where also experimental tests of the real-time iteration scheme at a pilot plant distillation column in Stuttgart are presented. The column has two inputs $u = (L_{\text{vol}}, Q)^T$, reflux L_{vol} and heating Q, and shall track two tray temperatures, T_{14} and T_{28}, at the values 88 °C and 70 °C (that are part of the algebraic states).

Though the DAE model is in continuous time, it is straightforward to apply the discrete-time setting investigated in the previous sections. For this purpose the time axis is divided into intervals $[t_k, t_{k+1}]$ of length ΔT (in the experiments ΔT is set to the different values 60, 120, 300, 600, and 1200 seconds). On these intervals $[t_k, t_{k+1}]$ we leave the control inputs u^k constant, and compute the trajectories $x_k(t)$ and $z_k(t)$ of differential and algebraic states as the solution of an initial value problem:

$$\dot{x}_k(t) = f_{\text{dae}}(x_k(t), z_k(t), u^k), \tag{2.32a}$$

$$0 = g_{\text{dae}}(x_k(t), z_k(t), u^k), \tag{2.32b}$$

$$x_k(t_k) = x^k. \tag{2.32c}$$

Note that the trajectory $x_k(t)$ on the interval $[t_k, t_{k+1}]$ depends only on the initial value x^k and the control input u^k, so that we can refer to it as $x_k(t; x^k, u^k)$. The discrete-time system function from (2.1) is then simply defined by

$$f(x^k, u^k) := x_k(t_{k+1}; x^k, u^k). \tag{2.33}$$

The distillation column is open loop stable, but takes very long to reach the steady state (more than 10 000 seconds), which corresponds to controls $u_S = (4.18, 2.49)^T$.

2.7.1 Optimal Control Problem Formulation

The control horizon is chosen to $T_c = 1200$ seconds. By dividing it by the interval length, we obtain the number of control intervals in the optimal control problem (2.5) as $N = \frac{T_c}{\Delta T}$.

The terminal penalty term $E(s_N)$ appearing in (2.5) is obtained by a prediction interval $[T_c, T_p]$ on which the controls are fixed to the setpoint values u_S. The objective contribution of the prediction interval provides an upper bound of the neglected future costs that are due after the end of the control horizon, if T_p is chosen sufficiently long. For the numerical experiments we use a length of $T_p - T_c = 36\,000$ seconds that we consider long enough for approximating the infinite horizon. Thus, we practically satisfy assumptions (b) and (c) of Theorem 2.4, with the local control law $k(x)$ being just the constant setpoint controls u_S. The resulting open-loop NMPC optimal control problem to solve at each sampling instant t_k is given by

$$\min_{u(\cdot),x(\cdot)} \int_0^{T_p} \left\{ \left\| \begin{bmatrix} T_{14}(\tau)-88 \\ T_{28}(\tau)-70 \end{bmatrix} \right\|_2^2 + 0.01 \left\| u - u_S \right\|_2^2 \right\} \, d\tau \tag{2.34}$$

subject to the DAE model

$$\begin{aligned} \dot{x}(\tau) &= f_{\text{dae}}(x(\tau), z(\tau), u(\tau)) \\ 0 &= g_{\text{dae}}(x(\tau), z(\tau), u(\tau)) \end{aligned} \qquad \text{for } \tau \in [0, T_p].$$

The initial value is given by $x(0) = x^k$. We furthermore require that

$$u(\tau) = q_i \quad \forall \tau \in [i\Delta T, (i+1)\Delta T], \quad i = 0, \ldots, N-1,$$

and on the long prediction interval the steady state control u_S is applied, i.e., $u(\tau) = u_S$ for $\tau \in [T_c, T_p]$. This implicitly defines the stage costs $L(s_i, q_i)$ and the final penalty term $E(s_N)$ in (2.5). It is easy to argue that Assumption 2.1, $L(s_i, q_i) \geq m\|s_i\|^2$, is satisfied even though we penalize only two of the states in the least-squares integrals (2.34). We only have to assume that the continuous-time system dynamics with constant controls will during one sampling time only be able to keep both penalized temperatures at their setpoint values if all components of the initial state s_i are already at the desired setpoint. This is a reasonable assumption that is strongly related to observability of the system with respect to the two temperatures.

In contrast to the simplified problem (2.5), we also formulate state and control inequality constraints by

$$h(x(i\Delta T), z(i\Delta T), u(i\Delta T)) \geq 0, \quad i = 0, \ldots, N,$$

where

$$h(x, z, u) := \begin{bmatrix} D(x,z,u)-D_{\min} \\ B(x,z,u)-B_{\min} \\ u-u_{\min} \\ u_{\max}-u \end{bmatrix}$$

define the lower bounds for the fluxes $D(x, z, u)$ and $B(x, z, u)$ that are determined according to the model assumptions, which should always maintain small positive values, and lower and upper bounds for the controls. Note that while inequality constraints have not been considered in the theoretical investigations of this chapter, their influence on NMPC stability is well understood and does not require major changes of the principal assumptions.

Overall, we claim that the assumptions of Theorem 2.4 are practically satisfied in this NMPC setup, such that nominal stability of the ideal NMPC scheme would be ensured.

Direct Multiple Shooting for DAE

The practical implementation of the real-time iteration scheme is based on the direct multiple shooting method [8] for *DAE* models (see [28]), which is very briefly—and in a simplified form—reviewed here. The main idea is as follows: to transform the continuous time problem into an NLP of the form (2.5), we introduce the node values s_i at times $i\Delta T$, $i = 0, \ldots, N-1$, and compute on each subinterval $\tau \in [i\Delta T, (i+1)\Delta T]$, and also for $\tau \in [T_c, T_p]$, the trajectories $x(\tau)$ and $z(\tau)$ as the solution of an initial value problem, according to (2.32).

The integral part of the cost function is then evaluated on each interval independently:

$$L_i(s_i, q_i) := \int_{i\Delta T}^{(i+1)\Delta T} \left\| \begin{bmatrix} T_{14}(\tau)-88 \\ T_{28}(\tau)-70 \end{bmatrix} \right\|_2^2 + 0.01 \, \|q_i - u_S\|_2^2 \; d\tau \tag{2.35}$$

as well as

$$E(s_N) := \int_{T_c}^{T_p} \left\| \begin{bmatrix} T_{14}(\tau)-88 \\ T_{28}(\tau)-70 \end{bmatrix} \right\|_2^2 \; d\tau \tag{2.36}$$

because $u(\tau) = u_S$ for $\tau \in [T_c, T_p]$. In Figure 2.2 we show the content of the optimizer for an arbitrary sample iteration.

For more details on direct multiple shooting and the practical implementation of the real-time iteration scheme, we refer the reader to [7, 8, 12, 15, 17, 28].

2.7.2 Simulation Results and Discussion

Several closed-loop simulations with different sampling times ΔT over 3000 seconds are performed. In all scenarios, the initial value x^0 is largely disturbed from steady state (to the value that results after applying a control vector of $u = (2.18, 2.49)^T$ with reduced reflux for 1500 seconds). The real-time iterations are initialized with variables w^0 corresponding

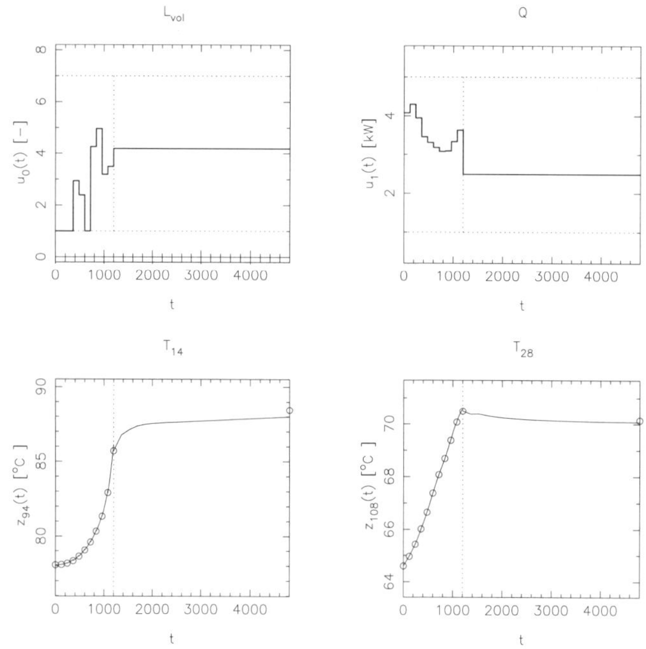

Figure 2.2. *Snapshot of the optimizer content w^k during a simulation scenario for a sampling time of $\Delta T = 120$ seconds. The number of multiple shooting nodes is $N = \frac{T_c}{\Delta T} = 10$. The prediction interval at the end of* 36 000 *seconds length is only partly shown.*

to the constant steady state solution. The result of the simulations are given in Figure 2.3 for $\Delta T = 60$, 120, 300, 600, and 1200 ($=T_c$) seconds sampling time. One can see that convergence of the closed-loop system occurs in every scenario but that the closed-loop performance deteriorates considerably with increasing interval length. This is partly due to the fact that larger intervals mean less freedom for optimization, and partly due to the fact that the nonlinear optimization procedure is more strongly disturbed by larger changes in the initial values; cf. Assumption 2.12. It is important to note that Assumption 2.12 might be violated if the state perturbation $\|x\|$ is large and at the same time the sampling time is too long. We claim, however, that for a sufficiently short choice of sampling time ΔT, Assumption 2.12 can be satisfied even for large variations. This is because the crucial disturbance term in Assumption 2.12, $f(x, u(x)) - x$, is of $O(\Delta T \|x\|)$; cf. the second remark after Assumption 2.12. In the limiting case of infinitesimal sampling times, we would recover the ideal NMPC feedback, as the time needed to converge towards the exact NMPC solutions would become infinitely short.

The necessary CPU time for the preparation phase of each real-time iteration was in all scenarios below 60 seconds on an Intel Pentium 4 processor; cf. Table 2.1. The CPU time is decreasing with the number N of multiple shooting intervals on the control horizon T_c. The feedback phase was always more than two orders of magnitude shorter, with at

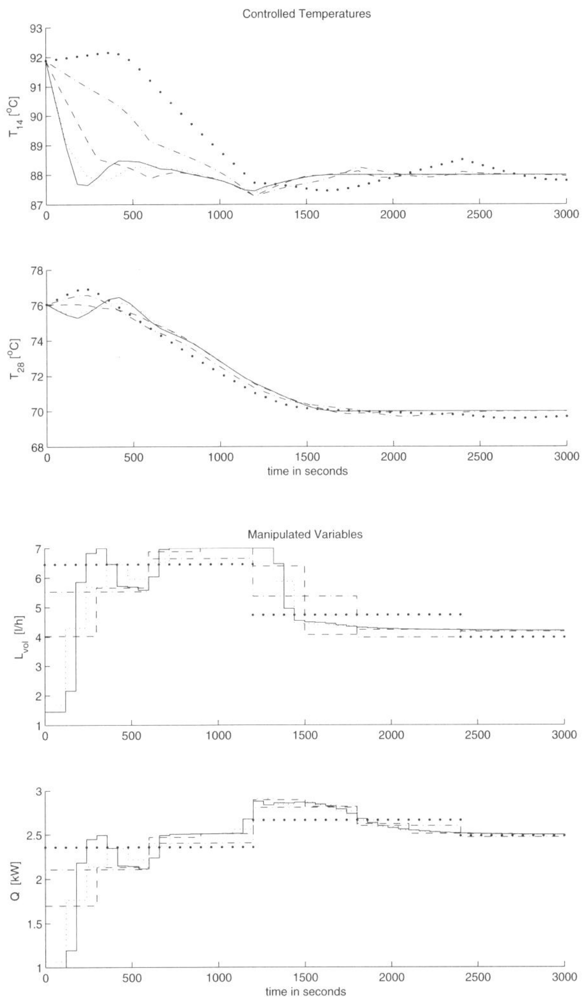

Figure 2.3. *Comparison of simulation runs with sampling times of* $\Delta T = 60$ *(−), 120 (· · ·), 300 (−−), 600 (− · −), and 1200 (·) seconds for application of the real-time iteration scheme to a distillation column model. The optimizer was initialized with the steady state solution.*

maximum 200 milliseconds CPU time (for $\Delta T = 60$s).

2.8 Summary and Conclusions

We have presented a Newton-type method for optimization in NMPC—the real-time iteration scheme without shift—and have proven nominal stability of the resulting system-

Table 2.1. *CPU times for the preparation phase in each real-time iteration. The feedback phase was always more than a factor of* 100 *shorter.*

ΔT [s]	60	120	300	600	1200
N	20	10	4	2	1
CPU [s]	$20 - 45$	$14 - 27$	$11 - 18$	$7 - 14$	$8 - 12$

optimizer dynamics. This scheme is characterized by a dovetailing of the dynamics of the system with those of the optimizer, resulting in an efficient online optimization algorithm which, however, shows intricate dynamics that do not allow us to apply readily available standard stability results from NMPC.

The proof of nominal stability makes use of results from both classical stability theory for NMPC and convergence theory for Newton-type optimization methods.

The derived result gives a theoretical underpinning of the real-time iteration scheme, which has already successfully been applied to several example systems, among them a real pilot-plant distillation column [18, 20]. Our numerical results illustrate the practical experience that the real-time iteration scheme is able to bring the system-optimizer dynamics back into the region of attraction even after rather large disturbances, which even holds in the case of strongly unstable systems (cf. [13, 14, 19]).

Acknowledgments

Financial support by the DFG within priority program 469 and within the research collaboration "Optimization Based Control of Chemical Processes" is gratefully acknowledged, as well as support by the Institute of Mathematics and Its Applications of the University of Minnesota (USA) which hosted the first author while parts of the chapter were being developed.

Bibliography

[1] F. ALLGÖWER, T. BADGWELL, J. QIN, J. RAWLINGS, AND S. WRIGHT, *Nonlinear predictive control and moving horizon estimation – An introductory overview*, in Advances in Control, Highlights of ECC'99, P. M. Frank, ed., Springer-Verlag, London, 1999, pp. 391–449.

[2] F. ALLGÖWER AND A. ZHENG, EDS., *Nonlinear Model Predictive Control*, Progr. Systems Control Theory 26, Birkhäuser, Basel, 2000.

[3] R. BARTLETT, A. WÄCHTER, AND L. BIEGLER, *Active set vs. interior point strategies for model predicitve control*, in Proceedings of the American Control Conference, Chicago, IL, 2000, pp. 4229–4233.

[4] L. Biegler, *Efficient solution of dynamic optimization and NMPC problems*, in Nonlinear Model Predictive Control, Progr. Systems Control Theory 26, F. Allgöwer and A. Zheng, eds., Birkhäuser, Basel, 2000, pp. 219–244.

[5] H. Bock, *Randwertproblemmethoden zur Parameteridentifizierung in Systemen nichtlinearer Differentialgleichungen*, Bonner Mathematische Schriften 183, University of Bonn, Bonn, Germany, 1987.

[6] H. Bock, M. Diehl, E. Kostina, and J. Schlöder, *Constrained optimal feedback control of systems governed by large differential algebraic equations*, in Real-Time PDE-Constrained Optimization, L. T. Biegler, O. Ghattas, M. Heinkenschloss, D. Keyes, and B. van Bloemen Waanders, eds., SIAM, Philadelphia, 2007.

[7] H. Bock, M. Diehl, D. Leineweber, and J. Schlöder, *A direct multiple shooting method for real-time optimization of nonlinear DAE processes*, in Nonlinear Model Predictive Control, Progr. Systems Control Theory 26, F. Allgöwer and A. Zheng, eds., Birkhäuser, Basel, 2000, pp. 246–267.

[8] H. Bock and K. Plitt, *A multiple shooting algorithm for direct solution of optimal control problems*, in Proceedings of the 9th IFAC World Congress, Budapest, Pergamon Press, New York, 1984, pp. 243–247.

[9] W. Chen, D. Ballance, and J. O'Reilly, *Model predictive control of nonlinear systems: Computational delay and stability*, IEE Proc., Part D, 147 (2000), pp. 387–394.

[10] G. De Nicolao, L. Magni, and R. Scattolini, *Stability and robustness of nonlinear receding horizon control*, in Nonlinear Model Predictive Control, Progr. Systems Control Theory 26, F. Allgöwer and A. Zheng, eds., Birkhäuser, Basel, 2000, pp. 3–23.

[11] P. Deuflhard and G. Heindl, *Affine invariant convergence theorems for Newton's method and extensions to related methods*, SIAM J. Numer. Anal., 16 (1979), pp. 1–10.

[12] M. Diehl, *Real-Time Optimization for Large Scale Nonlinear Processes*, vol. 920 of Fortschr.-Ber. VDI Reihe 8, Mess-, Steuerungs- und Regelungstechnik, VDI-Verlag, Düsseldorf, 2002.

[13] M. Diehl, H. Bock, and J. Schlöder, *Newton-type methods for the approximate solution of nonlinear programming problems in real-time*, in High Performance Algorithms and Software for Nonlinear Optimization, G. D. Pillo and A. Murli, eds., Kluwer Academic Publishers, Boston, 2002, pp. 177–200.

[14] M. Diehl, H. G. Bock, and J. P. Schlöder, *A real-time iteration scheme for nonlinear optimization in optimal feedback control*, SIAM J. Control Optim., 43 (2005), pp. 1714–1736.

[15] M. Diehl, H. Bock, J. Schlöder, R. Findeisen, Z. Nagy, and F. Allgöwer, *Real-time optimization and nonlinear model predictive control of processes governed by differential-algebraic equations*, J. Proc. Contr., 12 (2002), pp. 577–585.

[16] M. Diehl, R. Findeisen, F. Allgöwer, H. Bock, and J. Schlöder, *Nominal stability of the real-time iteration scheme for nonlinear model predictive control*, IEE Proc. Control Theory Appl., 152 (2005), pp. 296–308.

[17] M. Diehl, R. Findeisen, S. Schwarzkopf, I. Uslu, F. Allgöwer, H. Bock, E. Gilles, and J. Schlöder, *An efficient algorithm for nonlinear model predictive control of large-scale systems. Part* I: *Description of the method*, Automatisierungstechnik, 50 (2002), pp. 557–567.

[18] M. Diehl, R. Findeisen, S. Schwarzkopf, I. Uslu, F. Allgöwer, H. Bock, E. Gilles, and J. Schlöder, *An efficient algorithm for nonlinear model predictive control of large-scale systems. Part* II: *Application to a distillation column*, Automatisierungstechnik, 51 (2003), pp. 22–29.

[19] M. Diehl, L. Magni, and G. D. Nicolao, *Online NMPC of unstable periodic systems using approximate infinite horizon closed loop costing*, Ann. Rev. Control, 28 (2004), pp. 37–45.

[20] M. Diehl, I. Uslu, R. Findeisen, S. Schwarzkopf, F. Allgöwer, H. Bock, T. Bürner, E. Gilles, A. Kienle, J. Schlöder, and E. Stein, *Real-time optimization for large scale processes: Nonlinear model predictive control of a high purity distillation column*, in Online Optimization of Large Scale Systems: State of the Art, M. Grötschel, S. O. Krumke, and J. Rambau, eds., Springer-Verlag, 2001, pp. 363–384.

[21] R. Findeisen and F. Allgöwer, *Computational delay in nonlinear model predictive control*, in Proceedings of the International Symposium on Advanced Control of Chemical Processes, ADCHEM, 2003, pp. 427–432.

[22] R. Findeisen, M. Diehl, I. Uslu, S. Schwarzkopf, F. Allgöwer, H. Bock, J. Schlöder, and E. Gilles, *Computation and performance assesment of nonlinear model predictive control*, in Proceedings of the 42nd IEEE Conference on Decision and Control, Las Vegas, NV, 2002, pp. 4613–4618.

[23] R. Findeisen, L. Imsland, F. Allgöwer, and B. Foss, *State and output feedback nonlinear model predictive control: An overview*, European J. Control, 9 (2003), pp. 190–207.

[24] R. Findeisen, L. Imsland, F. Allgöwer, and B. Foss, *Towards a sampled-data theory for nonlinear model predictive control*, in New Trends in Nonlinear Dynamics and Control, Lecture Notes in Control and Inform. Sci. 295, W. Kang, C. Borges, and M. Xiao, eds., Springer-Verlag, New York, 2003, pp. 295–313.

[25] M. Henson, *Nonlinear model predictive control: Current status and future directions*, Comp. & Chem. Eng., 23 (1998), pp. 187–202.

[26] A. Jadbabaie, J. Yu, and J. Hauser, *Unconstrained receding horizon control of nonlinear systems*, IEEE Trans. Automat. Control, 46 (2001), pp. 776 –783.

[27] C. Kellett and A. Teel, *On robustness of stability and Lyapunov functions for discontinuous difference equations*, in Proceedings of the 42nd IEEE Conference on Decision and Control, Las Vegas, NV, 2002, pp. 4282–4287.

[28] D. Leineweber, *Efficient reduced SQP methods for the optimization of chemical processes described by large sparse DAE models*, vol. 613 of Fortschr.-Ber. VDI Reihe 3, Verfahrenstechnik, VDI-Verlag, Düsseldorf, 1999.

[29] D. Leineweber, A. Schäfer, H. Bock, and J. Schlöder, *An efficient multiple shooting based reduced SQP strategy for large-scale dynamic process optimization. Part* II: *Software aspects and applications*, Comp. & Chem. Eng., 27 (2003), pp. 167–174.

[30] L. Magni and R. Sepulchre, *Stability margins of nonlinear receding-horizon control via inverse optimality*, Systems Control Lett., 32 (1997), pp. 241–245.

[31] F. Martinsen, L. Biegler, and B. Foss, *Application of optimization algorithms to nonlinear MPC*, in Proceedings of 15th IFAC World Congress, Barcelona, Spain, 2002.

[32] D. Mayne, *Nonlinear model predictive control: Challenges and opportunities*, in Nonlinear Model Predictive Control, Progr. Systems Control Theory 26, F. Allgöwer and A. Zheng, eds., Birkhäuser, Basel, 2000, pp. 23–44.

[33] D. Mayne and H. Michalska, *Receding horizon control of nonlinear systems*, IEEE Trans. Automat. Control, 35 (1990), pp. 814–824.

[34] D. Mayne, J. Rawlings, C. Rao, and P. Scokaert, *Constrained model predictive control: Stability and optimality*, Automatica, 26 (2000), pp. 789–814.

[35] N. Oliveira and L. Biegler, *An extension of Newton-type algorithms for nonlinear process control*, Automatica, 31 (1995), pp. 281–286.

[36] S. Qin and T. Badgwell, *Review of nonlinear model predictive control applications*, in Nonlinear Model Predictive Control: Theory and Application, B. Kouvaritakis and M. Cannon, eds., The Institute of Electrical Engineers, London, 2001, pp. 3–32.

[37] P. Scokaert, J. Rawlings, and E. Meadows, *Discrete-time stability with perturbations: Application to model predictive control*, Automatica, 33 (1997), pp. 463–470.

[38] M. Tenny and J. Rawlings, *Feasible real-time nonlinear model predictive control*, in Proceedings of the 6th International Conference on Chemical Process Control – CPC VI, AIChE Symposium Series, 2001.

[39] M. Tenny, S. Wright, and J. Rawlings, *Nonlinear Model Predictive Control via Feasibility-Perturbed Sequential Quadratic Programming*, Technical report TWMCC-2002-02, Texas-Wisconsin Modeling and Control Consortium, Austin, TX, Madison, WI, 2002.

[40] S. J. WRIGHT, *Applying new optimization algorithms to model predictive control*, in Fifth International Conference on Chemical Process Control – CPC V, J. Kantor, C. Garcia, and B. Carnahan, eds., American Institute of Chemical Engineers, New York, 1996, pp. 147–155.

Chapter 3

Numerical Feedback Controller Design for PDE Systems Using Model Reduction: Techniques and Case Studies

Friedemann Leibfritz and Stefan Volkwein†*

3.1 Introduction

Recently, the application of reduced-order models to optimal control problems for partial differential equations (PDEs) has received an increasing amount of attention. The reduced-order approach is based on projecting the dynamical system onto subspaces consisting of basis elements that contain characteristics of the expected solution. This is in contrast to, e.g., finite element techniques, where the elements of the subspaces are uncorrelated to the physical properties of the system that they approximate. One reduced-basis method is proper orthogonal decomposition (POD). It has been successfully used in a variety of fields; see, for instance, [9, 22] for detailed reference lists.

Output feedback controller synthesis for linear control systems that meet desired performance and/or robustness specifications is an attractive model-based control design tool and has been an active research area of the control community for several decades (see, e.g., [7, 30]). It is not always possible to have full access to the state of the control system, and a controller based on the measurements has to be used. Output feedback synthesis without additional complexity constraints yields in general a controller order equal to n_x, the dimension of the dynamical system. The computation of the controller action becomes more expensive with increasing controller dimension. This is one reason why full-order synthesis control has not been widely used in industry. Recently, linear matrix inequalities (LMIs) have attained much attention in control engineering [4], since many control problems can be formulated in terms of LMIs and thus solved via convex programming approaches.

*University of Trier, FB–IV Department of Mathematics, D-54286 Trier, Germany.
†Institute of Mathematics and Scientific Computing, University of Graz, A-8010 Graz, Austria.

However, the resulting controllers are state feedback or of order n_x equal to the plant. For example, difficulties arise if we want to design a static output feedback (SOF) control law.

In this case, the control problems consist of finding a SOF controller which minimizes a performance measure subject to stability and/or robustness constraints. It is known that they can be rewritten to nonconvex matrix optimization problems (see [10, 12, 18]). Notice that in [17] the first author extends these matrix optimization problems to nonlinear semidefinite programs (NSDPs) by including explicitly the stability condition (modeled by two matrix inequalities) into the problem formulation. Finding a numerical solution to the nonconvex NSDP is a difficult task, particularly, if the dimension of the NSDP is large. Usually this will be the case if the control system dynamics are given by a PDE. Then, the dimension of the discretized counterpart can be very large and the computation of an output feedback controller may be impossible. In particular, the SOF case requires the solution of a large-scale NSDP with several million variables, which is usually not solvable. This is one of our main motivations for considering the SOF problem for PDE-constrained control problems in combination with the POD method for deriving a low-dimensional control system and the interior point trust-region (IPCTR) algorithm for solving the corresponding low-dimensional NSDP. The SOF control law can be constructed from the solution of the low-dimensional NSDP. In our numerical examples, we will demonstrate that this SOF can be used for controlling the large-dimensional PDE system. In particular, we illustrate that the choice of the POD norms as well as the POD snapshots can improve the stability properties of the computed SOF gain.

The chapter is organized in the following manner. In Section 3.2 we review the POD method and recall some prerequisites needed for the numerical experiments. The discussion of SOF control design problems for finite-dimensional systems and a sketch of IPCTR is contained briefly in Section 3.3. Numerical tests are carried out in Section 3.4, and in the last section we draw some conclusions.

Notation. Throughout this chapter, $\mathcal{S}^n$ denotes the linear space of real symmetric $n \times n$ matrices. In the space of real $m \times n$ matrices we define the inner product by $\langle M, Z \rangle = \mathrm{Tr}(M^T Z)$ for $M, Z \in \mathbb{R}^{m\times n}$, where $\mathrm{Tr}(\cdot)$ is the trace operator, and $\|\cdot\|$ denotes the Frobenius norm given by $\|M\| = \langle M, M \rangle^{1/2}$, while other norms and inner products will be specified. For a matrix $M \in \mathcal{S}^n$ we use the notation $M \succ 0$ or $M \succeq 0$ if it is positive definite or positive semidefinite, respectively. For a twice differentiable mapping $G : \mathcal{U} \to \mathcal{W}$ ($\mathcal{U}, \mathcal{W}$ Banach spaces) we denote by G_U, G_{UU} the first and second partial derivatives of G with respect to U. Moreover, $G_U(\cdot)H$ is used when a linear operator $G_U(\cdot)$ is applied to an element $H \in \mathcal{U}$. Furthermore, $G_U^*(\cdot)$ denotes the adjoint of $G_U(\cdot)$ and $\mathcal{L}(\mathcal{U}, \mathcal{V})$ refers to the space of linear, bounded operators endowed with the common norm.

3.2 Proper Orthogonal Decomposition

Proper orthogonal decomposition (POD) is a method to derive reduced-order models for dynamical systems. In this section we introduce the POD method for nonlinear dynamical systems and propose the numerical realization of POD. Let us consider the following semilinear initial value problem:

$$\begin{aligned} \dot{x}(t) + \mathcal{A}x(t) &= f(t, x(t)) \quad \text{for } t \in (0, T), \\ x(0) &= x_\circ, \end{aligned} \tag{3.1}$$

where $-\mathcal{A}$ is the infinitesimal generator of a C_0-semigroup $S(t)$, $t > 0$, on a Hilbert space Ξ, $x_\circ \in \Xi$, and $f : [0, T] \times \Xi \to \Xi$ is continuous in t and uniformly Lipschitz-continuous on Ξ for every t. Problem (3.1) has a unique mild solution $x \in C([0, T]; \Xi)$ given implicitly by the integral representation

$$x(t) = S(t)x_\circ + \int_0^t S(t-s)f(s, x(s))\,\mathrm{d}s \quad \text{for } t \in (0, T); \tag{3.2}$$

see, for instance, [24, p. 184]. If, in addition, f is continuously differentiable, then the mild solution (3.2) with $x_\circ \in D(\mathcal{A}) = \{\varphi \in \Xi : \mathcal{A}\varphi \in \Xi\}$ is also a classical solution; i.e., $x \in C^1([0, T]; \Xi)$ holds and x satisfies (3.1) for all $t \in [0, T]$; see, e.g., [24, p. 187].

Let $\mathcal{U}$ and $\mathcal{W}$ be real separable Hilbert spaces and suppose that $\mathcal{U}$ is dense in $\mathcal{W}$ with compact embedding. Throughout we assume that Ξ denotes either the space $\mathcal{U}$ or $\mathcal{W}$ and that x denotes the unique solution to (3.1) with $x \in C^1([0, T]; \Xi)$. For given $n \in \mathbb{N}$ let

$$0 \le t_1 < t_2 < \cdots < t_n \le T$$

denote a grid in the interval $[0, T]$ and define

$$w_j = \begin{cases} x(t_j) & \text{for } j = 1, \dots, n, \\ \dot{x}(t_{j-n}) & \text{for } j = n+1, \dots, 2n. \end{cases} \tag{3.3}$$

Setting $n_1 = n$ and $n_2 = 2n$ we introduce two different linear spaces by

$$\mathcal{V}_1 = \operatorname{span}\{w_1, \dots, w_{n_1}\} \quad \text{and} \quad \mathcal{V}_2 = \operatorname{span}\{w_1, \dots, w_{n_2}\}.$$

We refer to $\mathcal{V}_1$ as the ensemble consisting of the so-called snapshots $\{x(t_j)\}_{j=1}^n$, cf. [25], whereas $\mathcal{V}_2 \supseteq \mathcal{V}_1$ also includes the time derivatives $\dot{x}(t)$ at $t = t_j$ for $1 \le j \le n$. By (3.1) and (3.3), we have

$$w_{n+j} = \dot{x}(t_j) = \mathcal{G}(t, x_j), \quad j = 1, \dots, n,$$

with $\mathcal{G}(t, x) = -\mathcal{A}x + f(t, x)$ for $(t, x) \in [0, T] \times \Xi$. Since we are interested in building up a reduced-order model for (3.1), we want to utilize snapshots that are best suited for reconstruction of the nonlinear dynamics $\mathcal{G}(t, x)$. Therefore, the set $\mathcal{V}_2$ contains, in addition to the set $\mathcal{V}_1$, information about the spatial derivatives and nonlinearities in the dynamical system; cf., e.g., [1, 15]. It follows that $\mathcal{V}_k \subset \Xi$ by construction, $k = 1$ or 2.

Let $\{\psi_i\}_{i=1}^{d_k}$ denote an orthonormal basis for $\mathcal{V}_k$ with $d_k = \dim \mathcal{V}_k$ for $k = 1$ or 2. Then each member of the ensemble can be expressed as

$$w_j = \sum_{k=1}^{d_k} \langle w_j, \psi_i \rangle_\Xi \psi_i \quad \text{for } j = 1, \dots, n_i. \tag{3.4}$$

For $k = 1$ or 2 the method of POD consists in choosing an orthonormal basis such that for every $\ell \in \{1, \dots, d_k\}$ the mean square error between the elements w_j, $1 \le j \le n_k$, and the corresponding ℓth partial sum of (3.4) is minimized on average:

$$\begin{gathered} \min \frac{1}{n_k} \sum_{j=1}^{n_k} \Big\| w_j - \sum_{i=1}^{\ell} \langle w_j, \psi_i \rangle_\Xi \psi_i \Big\|_\Xi^2 \\ \text{subject to } \langle \psi_i, \psi_j \rangle_\Xi = \delta_{ij} \quad \text{for } 1 \le i \le \ell,\ 1 \le j \le i. \end{gathered} \tag{3.5}$$

A solution $\{\psi_i\}_{i=1}^{\ell}$ to (3.5) is called *POD basis of rank* ℓ. The subspace spanned by the first ℓ POD basis functions is denoted by $\mathcal{V}^\ell$, i.e.,

$$\mathcal{V}^\ell = \text{span}\,\{\psi_1, \ldots, \psi_\ell\}. \tag{3.6}$$

Utilizing a Lagrangian framework the solution of (3.5) is characterized by the necessary optimality condition, which can be written as an eigenvalue problem (see, e.g., [11, pp. 88–91] and [27, Section 2]). For that purpose we introduce the positive semidefinite and symmetric matrix $\mathcal{K}_k \in \mathbb{R}^{n_k \times n_k}$, $k = 1$ or 2, with elements

$$\left(\mathcal{K}_k\right)_{ij} = \frac{1}{n_k}\,\langle w_j, w_i\rangle_\Xi.$$

The matrix $\mathcal{K}_k$ is often called a *correlation matrix*. Let $\lambda_1 \geq \lambda_2 \geq \cdots \geq \lambda_{d_k} > 0$ and $v_1, \ldots, v_{d_k} \in \mathbb{R}^{n_k}$ denote the d_k positive eigenvalues of $\mathcal{K}_k$ and the corresponding eigenvectors, respectively; i.e., we have

$$\mathcal{K}_k v_i = \lambda_i v_i, \qquad i = 1, \ldots, d_k. \tag{3.7}$$

Then the POD basis functions are given by

$$\psi_i = \frac{1}{n_k\sqrt{\lambda_i}} \sum_{j=1}^{n} v_i^j w_j \quad \text{for } i = 1, \ldots, \ell, \tag{3.8}$$

where v_i^j stands for the jth component of the eigenvector v_i.

Remark 3.1. POD is closely related to the singular value decomposition. This fact is very useful for implementation issues, in particular, for the computation of the POD basis functions ψ_i as well as of the corresponding eigenvalues λ_i, $1 \leq i \leq \ell$. The finite-dimensional case was studied in [14], whereas the infinite-dimensional case was analyzed in [22, 28].

If the POD basis $\{\psi_i\}_{i=1}^{\ell}$ is determined, e.g., by solving (3.7) and using (3.8), we obtain the following value for the cost functional in (3.5):

$$\frac{1}{n_k}\sum_{j=1}^{n_k}\left\| w_j - \sum_{i=1}^{\ell}\langle w_j, \psi_i\rangle_\Xi \psi_i\right\|_\Xi^2 = \sum_{\ell+1}^{d_k}\lambda_i;$$

see [27, Section 2], for instance. Thus, if the decay of the eigenvalues λ_i is very rapid, the term $\sum_{\ell+1}^{d_k}\lambda_i$ is small, even for a small value of ℓ. In this case only a few POD basis functions represent the ensemble $\mathcal{V}_1$ or $\mathcal{V}_2$ in a sufficient manner. One heuristic to choose the number ℓ of POD basis functions is

$$\mathcal{E}_k(\ell) = \frac{\sum_{i=1}^{\ell}\lambda_i}{\sum_{i=1}^{d_k}\lambda_i}\cdot 100\% \approx 99\%, \qquad k \in \{1, 2\}.$$

Obviously, we have $\mathcal{E}_k(d_k) = 100\%$.

In Section 3.4 we compare the performance of POD model reduction by utilizing either $\mathcal{V}_1$ or $\mathcal{V}_2$ and by taking either $\Xi = \mathcal{U}$ or $\Xi = \mathcal{W}$ with the choices $\mathcal{W} = L^2(\Omega)$ and $\mathcal{U} = H^1(\Omega)$.

3.3 Numerical Design of SOF Control Laws

We present a numerical strategy for the computation of a linear SOF control law for discretized PDE control systems. In example, a finite difference or finite element discretization of a PDE control problem yields a finite-dimensional control system of the following general form:

$$\begin{aligned} E\dot{x}(t) &= (A + \delta A)x(t) + G(x(t)) + B_1\tilde{w}(t) + Bu(t), \quad x(0) = x_\circ, \\ z(t) &= C_1x(t) + D_1u(t), \quad y(t) = Cx(t), \end{aligned} \tag{3.9}$$

where $x \in \mathbb{R}^{n_x}$ is the approximation of the state, $u \in \mathbb{R}^{n_u}$ is the control input, $y \in \mathbb{R}^{n_y}$ denotes the measurements, $\tilde{w} \in \mathbb{R}^{n_w}$ is a disturbance input, $z \in \mathbb{R}^{n_z}$ is the regulated output, $E \in \mathbb{R}^{n_x \times n_x}$ is a regular diagonal matrix, and $A \in \mathbb{R}^{n_x \times n_x}$, $B \in \mathbb{R}^{n_x \times n_u}$, $B_1 \in \mathbb{R}^{n_x \times n_w}$, $C \in \mathbb{R}^{n_y \times n_x}$, $C_1 = \sqrt{0.5c_1}\begin{bmatrix} I_{n_x} & 0_{n_u \times n_x}\end{bmatrix}^T$, $D_1 = \sqrt{0.5d_1}\begin{bmatrix} 0_{n_x \times n_u} & I_{n_u}\end{bmatrix}^T$ with $c_1, d_1 \in \mathbb{R}$ are given positive scalars. If $\delta A \equiv 0$, the system matrix A is not affected by a perturbation, and, if $G(x(t)) \equiv 0$, the system is linear. Depending on the corresponding PDE model, we get linear or nonlinear control systems which we want to control by a linear SOF control law of the form

$$u(t) = Fy(t), \quad F \in \mathbb{R}^{n_u \times n_y}. \tag{3.10}$$

If we neglect the nonlinear term $G(x(t))$ in (3.9) and substitute this linear SOF control into the state space plant (3.9), then we obtain the following linear closed-loop system:

$$\dot{x}(t) = A(F)x(t) + B(F)\tilde{w}(t), \quad z(t) = C(F)x(t), \tag{3.11}$$

where $A(F) := E^{-1}(A + \delta A + BFC)$, $B(F) := E^{-1}B_1$, $C(F) := C_1 + D_1FC$ are the closed-loop operators, respectively.

SOF Design and NSDP Formulation

One of the most basic static output feedback design problems is the SOF-$\mathcal{H}_2$ problem (see [17, 26]): *Find a SOF gain F such that the closed-loop matrix $A(F)$ is Hurwitz and the $\mathcal{H}_2$-norm of* (3.11) *is minimal.* It is well known that this problem can be rewritten as the following $\mathcal{H}_2$-NSDP; see, e.g., [19, 21, 22]:

$$\min \ \mathrm{Tr}(LB_1B_1^T) \ \text{ subject to } \ \begin{aligned} &A(F)^TV + VA(F) + I = 0, \ V \succ 0, \\ &A(F)^TL + LA(F) + C(F)^TC(F) = 0, \end{aligned} \tag{3.12}$$

where $L, V \in \mathcal{S}^{n_x}$. Note that (3.12) is bilinear in L, V, F and quadratic in F, and hence nonconvex in the free variables. Therefore different local minima might occur, and any suitable NSDP solver usually determines a local solution of the matrix optimization problem. A more attractive and realistic model-based control design tool is the mixed $\mathcal{H}_2/\mathcal{H}_\infty$ synthesis. It allows incorporation of model uncertainties in the control design. The SOF-$\mathcal{H}_2/\mathcal{H}_\infty$ problem can be formally stated in the following term; see, e.g., [3, 13, 18]: *For a given scalar $\gamma > 0$ find a SOF matrix F such that $A(F)$ is Hurwitz, the $\mathcal{H}_\infty$-norm of* (3.11) *is less than γ, and the $\mathcal{H}_2$-norm of* (3.11) *is minimal.* For computing the SOF-$\mathcal{H}_2/\mathcal{H}_\infty$ gain F, we consider the following known $\mathcal{H}_2/\mathcal{H}_\infty$-NSDP version; see, e.g., [19, 22]:

$$\begin{aligned} \min \ \mathrm{Tr}(LB_1B_1^T) \ \text{ subject to } \ &A(F)^TL + LA(F) + C(F)^TC(F) + \tfrac{1}{\gamma^2}LB_1B_1^TL = 0, \\ (A(F) + \tfrac{1}{\gamma^2}B_1B_1^TL)^TV + V(A(F) &+ \tfrac{1}{\gamma^2}B_1B_1^TL) + I = 0, \quad V \succ 0. \end{aligned} \tag{3.13}$$

Due to the bilinearity of the free matrix variables it is a nonconvex NSDP, too.

The NSDP Algorithm—Sketch of IPCTR

Our goal is to solve the NSDPs defined in the previous paragraph by IPCTR originally developed in [20] and augmented in [22] for solving NSDPs of the form

$$\min \quad J(X) \quad \text{subject to} \quad H(X) = 0, \quad G(X) = 0, \quad Y(X) \succeq 0, \tag{3.14}$$

where $X := (F, L, V) \in \mathcal{X} := \mathbb{R}^{n_u \times n_y} \times \mathcal{S}^{n_x} \times \mathcal{S}^{n_x}$. We assume that $J : \mathcal{X} \to \mathbb{R}$, $H, G, Y : \mathcal{X} \to \mathcal{S}^{n_x}$ are twice continuously (Frechét-) differentiable matrix functions and the mapping $H(\cdot)$ is only a function in the variables (F, L), i.e., $H_V(X) \equiv 0$. Moreover, for given $X \in \mathcal{F}_s$ we suppose that the linear operators $H_L(X)$ and $G_V(X)$ are invertible, where $\mathcal{F}_s := \{X \in \mathcal{X} \mid Y(X) \succ 0\}$. Obviously, the NSPDs (3.12) and (3.13) are in the form of (3.14). IPCTR is based on the approximate solution of a sequence of matrix equality constraint barrier problems

$$\min \quad \Phi^{\mu}(X) = J(X) - \mu \log\det(Y(X)) \quad \text{subject to} \quad H(X) = 0, \quad G(X) = 0, \tag{3.15}$$

where $\mu > 0$ and $Y(X)$ is (implicitly) assumed to be positive definite. The Lagrangian function associated with (3.15) is defined by

$$\ell^{\mu}(X, K) = \Phi^{\mu}(X) + \langle K_h, H(X)\rangle + \langle K_g, G(X)\rangle$$

where $K := (K_g, K_h) \in \mathcal{S}^n \times \mathcal{S}^n$ are Lagrange multipliers for the equality constraints. The basic IPCTR algorithm combines ideas of (primal) interior point and trust-region methods with a modified conjugate gradient (CG) procedure. Note that a primal-dual method can be used instead of a pure primal method. The primal approach has the advantage that we do not need to compute the dual matrix variables for the nonlinear matrix inequality constraints. In the primal-dual case, the computational complexity of a Newton-type method increases rapidly with the number of the dual matrix variables. In particular, if n_x is large, e.g., $n_x = 4000$, then the dual matrix variable for $Y(X) \preceq 0$, $Y(X) \in \mathcal{S}^{n_x}$ has several million unknown entries, e.g., $\frac{1}{2} \cdot n_x \cdot (n_x + 1) = 8.002 \cdot 10^6$.

Algorithm 3.2 (IPCTR, see, e.g., [20, 22]). Let $X_0 = (F_0, L_0, V_0)$ with $Y(X_0) \succ 0$ and $\mu_0, \epsilon_0 > 0$ be given. For $j = 0, 1, \ldots$ do:

Find a solution $X_{j+1} = (F_{j+1}, L_{j+1}, V_{j+1}) \in \mathcal{F}_s$ of (3.15) satisfying

$$||\nabla \ell_F^{\mu_j}(X_{j+1}, K_{j+1})|| + ||H(X_{j+1})|| + ||G(X_{j+1})|| \le \epsilon_j,$$

where the multipliers $K_{j+1} := (K_h, K_g)_{j+1}$ are the solutions of the adjoint (multiplier) equations $\nabla \ell_L^{\mu}(X, K_h, K_g) = 0, \nabla \ell_V^{\mu}(X, K_h, K_g) = 0$; e.g. ,

$$K_g = -(G_V^{-1}(\cdot))^* \nabla \Phi_V^{\mu}(\cdot), \quad K_h = -(H_L^{-1}(\cdot))^* (\nabla \Phi_L^{\mu}(\cdot) - G_L^*(\cdot) K_g).$$

Choose $\mu_{j+1} < \mu_j$ and $\epsilon_{j+1} < \epsilon_j$.

In an implementation of IPCTR, we have chosen the parameters μ_j and ϵ_j as stated in [22]. In particular, if the actual barrier parameter $\mu_j < 1$, we set $\mu_{j+1} = \Omega(\mu_j^{4m/(2m+1)})$, $m \in \mathbb{N}$. This is equivalent to $\mu_{j+1} \geq \mu_j^{4m/(2m+1)}$, where for related positive quantities α and β we write $\alpha = \mathcal{O}(\beta)$ if there is a constant $\kappa > 0$ such that $\alpha \geq \kappa\beta$ for all β sufficiently small and $\alpha = \Omega(\beta)$ if $\beta = \mathcal{O}(\alpha)$. Otherwise, we choose $\mu_{j+1} = a\mu_j$, $a \in (0,1)$. Using this rule, the rate at which the barrier parameter approaches zero can be made as close to quadratic as one desires. The updating rule for the inexact termination criterion is given by $\epsilon_{j+1} = \mathcal{O}(\mu_{j+1}^{1/m})$. In our practical implementation, we choose $m \in \{2, 3, 4, 5\}$. Making standard assumptions on problem (3.14), it can be proven that any cluster point of the sequence $\{X_j\}_{j\geq 0}$ generated by IPCTR is a KKT point of (3.14). For the proof we refer the reader to [20, Theorem 3.1]. The tool used in IPCTR for finding a solution of (3.15) is a tangent space trust-region method (see, e.g., [5, 6, 8, 20, 23]). For given X, in this variant the step $\Delta X = (\Delta F, \Delta L, \Delta V) = T(X)\Delta F + N(X)$ is decomposed into the tangential step $T(X)\Delta F$ and the normal step $N(X)$, respectively. Here we define $T(\cdot) = (\mathcal{I}, T_1(\cdot), T_2(\cdot)) \in \mathcal{L}(\mathbb{R}^{n_u \times n_y}, \mathcal{X})$, where $\mathcal{I} : \mathbb{R}^{n_u \times n_y} \to \mathbb{R}^{n_u \times n_y}$ is the identity mapping and T_1, T_2 are given by

$$T_1(\cdot) := -H_L^{-1}(\cdot)H_F(\cdot), \quad T_2(\cdot) := -G_V^{-1}(\cdot)\left(G_F(\cdot) - G_L(\cdot)H_L^{-1}(\cdot)H_F(\cdot)\right).$$

Moreover, $N(\cdot) = \left(0, -H_L^{-1}(\cdot)H(\cdot), -G_V^{-1}(\cdot)(G(\cdot) - G_L(\cdot)H_L^{-1}(\cdot)H(\cdot))\right) \in \mathcal{X}$, where 0 is the zero matrix. Using [22, Lemma 4.1] the normal step $N(\cdot)$ can be determined as follows: First, compute $(\Delta L^n, \Delta V^n)$ by solving the matrix equations

$$H_L(X)\Delta L^n + H(X) = 0, \quad G_V(X)\Delta V^n + G_L(X)\Delta L^n + G(X) = 0,$$

and second, control the size of $(\Delta L^n, \Delta V^n)$ such that they stay inside the current trust region. In this example, we compute the scalar $\beta \in (0, 1]$ by $\beta = 1$ if $||(\Delta L^n, \Delta V^n)|| \leq \omega\delta$, else $\beta = (\omega\delta)/||(\Delta L^n, \Delta V^n)||$ and set $(\Delta L^n, \Delta V^n) = \beta(\Delta L^n, \Delta V^n)$, where $\omega \in (0, 1)$ is a given scalar and $\delta > 0$ denotes the current trust-region radius. The tangential component depends on ΔF and, thus, if ΔF is known, we can compute $(\Delta L^t, \Delta V^t) = (T_1(\cdot)\Delta F, T_2(\cdot)\Delta F)$ as solutions of two linear matrix equations:

$$H_L(X)\Delta L^t + H_F(X)\Delta F = 0, \quad G_V(X)\Delta V^t + G_L(X)\Delta L^t + G_F(X)\Delta F = 0. \quad (3.16)$$

For computing ΔF we search a solution of the tangential trust-region subproblem

$$\begin{aligned} \min \quad & \Psi(\Delta F) = \langle \Delta F, T^*\nabla\Phi_X^\mu + T^*\nabla^2\ell_{XX}^\mu N\rangle + \tfrac{1}{2}\langle \Delta F, T^*\nabla^2\ell_{XX}^\mu T\Delta F\rangle, \\ & ||\Delta F|| \leq \delta, \quad Y(X + T(X)\Delta F + \Delta X^n) \succeq (1-\sigma)Y(X), \end{aligned} \quad (3.17)$$

where $\nabla\Phi_X^\mu = (\nabla\Phi_F^\mu(X), \nabla\Phi_L^\mu(X), \nabla\Phi_V^\mu(X)) \in \mathcal{X}$, $\ell^\mu = \ell^\mu(X, K_g, K_h)$, T and N are the tangential and the normal operator, respectively, $\sigma \in (0, 1)$ is given, $\Delta X^n = (0, \Delta L^n, \Delta V^n)$, and $X + T\Delta F + \Delta X^n = (F + \Delta F, L + T_1\Delta F + \Delta L^n, V + T_2\Delta F + \Delta V^n)$. We apply the modified CG algorithm as described in [20, Algorithm 2.1] for finding an approximate solution ΔF of (3.17). This CG approach has the following properties:

(i) It solves a reduced Newton-like equation in ΔF; i.e., on exit, the solution ΔF of (3.17) satisfies approximately the equation $\Psi_{\Delta F} = 0$, e.g.,

$$T^*(X)\nabla^2\ell_{XX}^\mu(\cdot)T(X)\Delta F = -T^*(X)(\nabla\Phi_X^\mu(X) + \nabla^2\ell_{XX}^\mu(\cdot)N(X)).$$

(ii) During each CG iteration, we compute a maximal scalar $\tau > 0$ such that the matrix inequality constraint in (3.17) is fulfilled. On exit, it is guaranteed that ΔF stays inside the trust region and $X + T(X)\Delta F + \Delta X^n$ satisfies the NSDP constraint in (3.17).

(iii) In every CG iteration the operator T has to be applied. In particular, for a given conjugate direction δF_{cg}, we solve the first linear matrix equation in (3.16) for ΔL^t (e.g., $\Delta L^t = T_1(\cdot)\delta F_{cg}$). Then we substitute this solution into the second equation of (3.16) and solve it for ΔV^t (e.g., $\Delta V^t = T_2(\cdot)\delta F_{cg}$).

(iv) There are different ways in which the CG method can terminate. (a) A direction of negative curvature is encountered in the CG iteration. In this case, we follow this direction until reaching the boundary of the intersection of the trust region and the NSDP constraints. Then the resulting step is returned as an approximate solution of (3.17). (b) The CG iterate has stepped outside of the intersection of the trust region and the NSDP constraints. In this case, we backtrack to this region and return the resulting step as a solution of (3.17). (c) The algorithm terminates with a prespecified inexact termination criterion.

The advantages of this strategy are the following: The CG loop works only in the space of the F variable, which is in general much smaller as the abstract state space $\mathcal{S}^n \times \mathcal{S}^n$, where the variables (L, V) live. There is no need for evaluating the Hessian of the Lagrangian explicitly. We need only to evaluate it applied to a direction. On exit, it is guaranteed that the matrix inequality is strictly satisfied. For given ΔF, the L- and V-part of the tangential component ΔX^t can be obtained by solving the linear (matrix) equations in (3.16). Note that the trust-region method in IPCTR uses a nonorthogonal decomposition of the step and the normal step is actually a so-called quasi-normal step. If the operators H_L and G_V are ill-conditioned, such steps can lead to poor performance of the method. In this case, the trust-region variant discussed in [8, Section 5.22] can be used alternatively. To keep the quasi-normal step within the trust region, the size factor β is introduced. If the operators are ill-conditioned, this may lead to a poor descent step compared to Byrd–Omojukun steps, too. Here a simple dogleg modification in the space of L and V could be used to overcome this problem. Finally, as shown in Wächter and Biegler [29], barrier methods using the step to the boundary rule (which is included in (3.17)) in combination with linearized equality constraints can fail by "crashing into bounds." Therefore, it seems that a breakdown may occur in IPCTR for some "degenerated" NSDPs. But, we have tested IPCTR extensively on the *COMPl$_e$ib* benchmark library [19] which contains actually 171 test examples. For all test runs we never observed a failure of IPCTR. In particular, for almost all *COMPl$_e$ib* test examples, IPCTR performs pretty well and very quickly. Even if the operators H_L and G_V are ill-conditioned, the performance of IPCTR was quite satisfactory. Moreover, we never observed the Wächter and Biegler "crashing into bounds" phenomenon during our test runs of IPCTR on *COMPl$_e$ib* . A whole convergence analysis of IPCTR is far beyond the scope of this chapter. For more algorithmic details and some convergence results, we refer the interested reader to [8, 20, 22].

3.4 Numerical Experiments

We conducted numerical experiments in computing linear SOF control laws for several discretized parabolic PDE control systems. To clarify our approach let us summarize the algorithmic steps:

1. solve the discrete dynamical system (3.9) of dimension n_x for a chosen/nominal control u, e.g., for $u = 0$ (uncontrolled dynamics) to get $x(t_j)$, $j = 1, \dots, n$ for a time grid in $[0, T]$;

2. define w_j, $j = 1, \dots, 2n$, according to (3.3) and choose $\mathcal{V}_1$ or $\mathcal{V}_2$ as well as $\Xi = \mathcal{U}$ or $\Xi = \mathcal{W}$ for (3.5);

3. compute the first ℓ POD basis functions by solving the eigenvalue problem (3.7) and using (3.8);

4. apply a Galerkin ansatz for the PDE utilizing the computed ℓ POD basis functions to derive a discrete dynamical system of the type (3.9) but with dimension $\ell \ll n_x$ (low-dimensional model);

5. neglect the nonlinear part $G(x(t))$ in (3.9), and solve SOF-$\mathcal{H}_2$ problem (3.12) to get a linear SOF control law of the form (3.10); see Section 3.3;

6. fit this SOF control into the high-dimensional discrete dynamical system (3.9) to obtain the closed-loop system

$$
\begin{aligned}
E\dot{x}(t) &= (A + BFC + \delta A)x(t) + G(x(t)) + B_1\tilde{w}(t), \quad x(0) = x_\circ, \\
z(t) &= (C_1 + D_1FC)x(t).
\end{aligned}
$$

In step 5 we neglect the nonlinear part in (3.9) and compute the SOF control law for the linear model. Alternatively, we linearize the nonlinear part instead of neglecting $G(x(t))$. In some cases, this could lead to a poor performance of the SOF controller, e.g., if $G(x(t))$ is highly nonlinear. For moderate nonlinear terms, the linear SOF control law is a good feedback control for the nonlinear system, too. For highly nonlinear models, it is likely that such a simple linear SOF controller cannot stabilize the unstable nonlinear system. In these cases, a nonlinear feedback should be used.

The computational performance of this method is pretty fast. The main work is the computation of the POD approximation of the large (nonlinear) system. Typically, we reduce the size from several thousand variables to 5–10 variables in the POD model. In this case, the corresponding NSDP (3.12) is a matrix optimization problem with approximately 25–100 unknowns and, typically, IPCTR computes a (local) solution of the small-sized NSDPs within some seconds on a Dell notebook.

3.4.1 Example (Linear Convection-Diffusion Model)

The first example is a two-dimensional model of a linear parabolic equation with n_u distributed control input functions in the domain $\Omega = [0, a] \times [0, b]$, where $a = 1, b = 1$.

Table 3.1. *Parameters for Example* 3.4.1.

n_x	n_u	n_y	ℓ	κ	c_1	d_1	ε_1	ε_2
3600	2	2	7	0.06119	10	100	0.29555	2.9555

The infinite-dimensional control problem of the convection-diffusion model is given by

$$\begin{aligned} v_t &= \kappa \Delta v - \varepsilon_1 (v_\xi + v_\eta) + \varepsilon_2 v + \sum_{i=1}^{n_u} u_i(t) b_i && \text{in } \Omega, t > 0, \\ v(\xi, \eta; t) &= 0 && \text{on } \partial\Omega, t > 0, \\ v(\xi, \eta, 0) &= v_0(\xi, \eta) && \text{in } \Omega, \end{aligned} \tag{3.18}$$

for the unknown function $v := v(\xi, \eta; t)$, $(\xi, \eta) \in \Omega$, $t > 0$, where Δ denotes the Laplace operator, $\varepsilon_1, \varepsilon_2 \geq 0$ are given constants, $\partial\Omega$ denotes the boundary of Ω, $\kappa > 0$ is the diffusion coefficient, $b_i, i = 1, \ldots, n_u$, are given shape functions for the control inputs $u_1, \ldots, u_{n_u}$, and $v_0(\cdot)$ is the initial state in Ω at $t = 0$. After a spatial finite difference discretization of (3.18) we end up with a linear control system of the form (3.9) with $E = I$, $\delta A \equiv 0$, $G(x(t)) \equiv 0$, and $n_x = 3600$ states. We choose $n_u = 2$ and $b_i = \chi_{\Omega_i^u}, i = 1, 2$, where $\chi_{\Omega_i^u}$ denotes the characteristic function on the control input domain $\Omega_i^u \subset \Omega$ of u_i and

$$\Omega_1^u = [0.1, 0.4] \times [0.1, 0.4], \quad \Omega_2^u = [0.6, 0.9] \times [0.7, 0.9].$$

Moreover, we set $n_y = 2$ and measure the state on the observation domains $\Omega_i^y \subset \Omega, i = 1, 2$ of $y(t) = (y_1(t), y_2(t))^T$, where

$$\Omega_1^y = [0.1, 0.4] \times [0.5, 0.7], \quad \Omega_2^y = [0.6, 0.9] \times [0.1, 0.4].$$

Hence, the data matrices $B \in \mathbb{R}^{n_x \times n_u}$, $C \in \mathbb{R}^{n_y \times n_x}$ contain only zeros and ones with ones at grid points within the control input and observation domains, respectively.

Table 3.1 lists the parameters that we have used in our numerical experiments. Note that due to the choice of ε_2 (see Table 3.1), the uncontrolled system is unstable. In this example, the real part of the largest eigenvalue of A is positive. Figure 3.1 shows the state of the linear convection-diffusion model (3.18) if we use SOF control law computed by IPCTR in combination with POD. We have computed the first $\ell = 7$ POD basis functions utilizing the (discrete) L^2-norm (left) as well as the discrete H^1-norm (right). This figure illustrates that the simple SOF controller stabilizes the unstable control system very well. The corresponding controller functions can be found in Figure 3.2. The plots in Figure 3.3 show the dynamical behavior of the controlled model for the choice of a different snapshot ensemble. In particular, for the computation of the POD basis function we use only every 10th snapshot and include the difference quotients in the snapshot ensemble. The corresponding control input functions are illustrated in Figure 3.4. On the other hand, Figure 3.5 visualizes the instability of this model. Moreover, Figure 3.6 illustrates the location of the two sensor (red) and control (green) domains in Ω. It turns out that in case of $\Xi = L^2(\Omega)$ and $\mathcal{V}_2$ the SOF control has the best stability characteristics. For $\Xi = H^1(\Omega)$ we observe that there is no big difference whether the time derivatives are included in the snapshots or not.

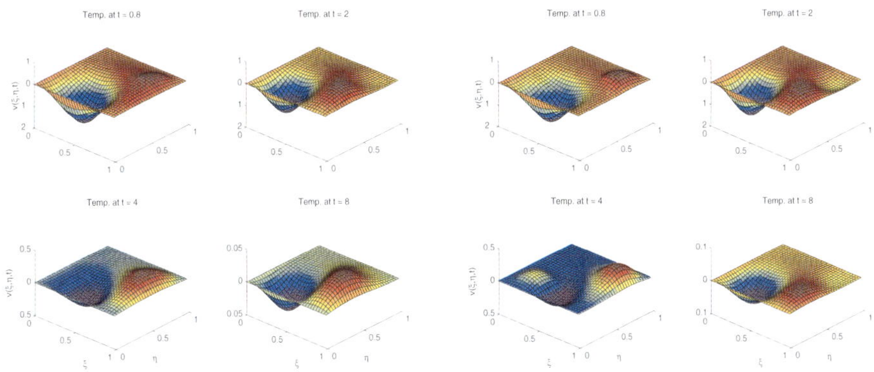

Figure 3.1. *Example* 3.4.1 *with SOF at* $t = 0.8, 2, 4, 8$: *POD with* $\mathcal{V}_1$, $\Xi = L^2(\Omega)$ *(left),* $\mathcal{V}_1$, $\Xi = H^1(\Omega)$ *(right).*

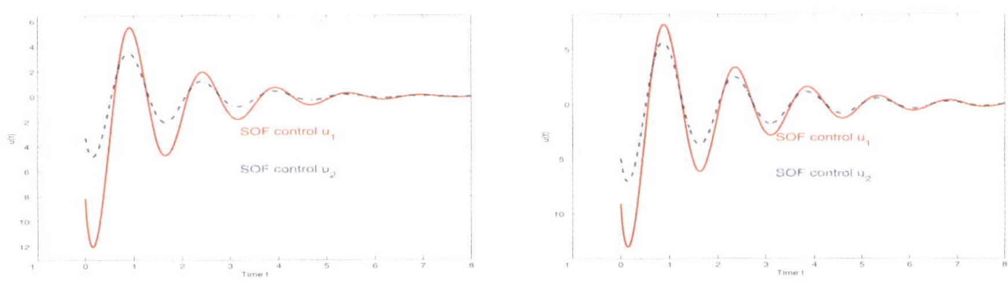

Figure 3.2. *SOF control* (3.10) *for Example* 3.4.1: *POD with* $\mathcal{V}_1$, $\Xi = L^2(\Omega)$ *(left),* $\mathcal{V}_1$, $\Xi = H^1(\Omega)$ *(right).*

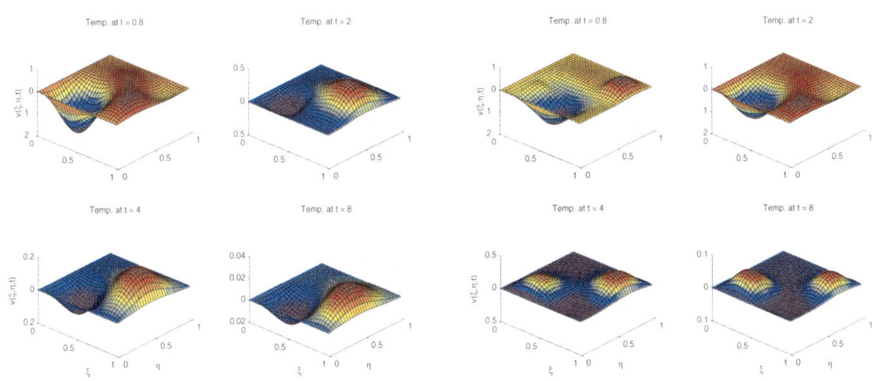

Figure 3.3. *Example* 3.4.1 *with SOF at* $t = 0.8, 2, 4, 8$: *POD with* $\mathcal{V}_2$, $\Xi = L^2(\Omega)$ *(left),* $\mathcal{V}_2$, $\Xi = H^1(\Omega)$ *(right).*

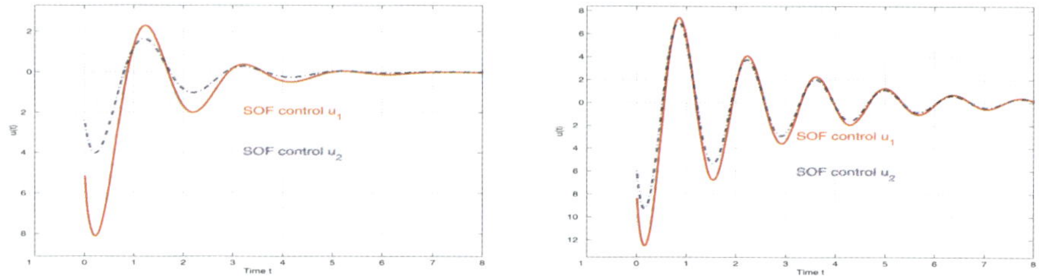

Figure 3.4. *SOF control* (3.10) *for Example* 3.4.1: *POD with* $\mathcal{V}_2$, $\Xi = L^2(\Omega)$ *(left)*, $\mathcal{V}_2$, $\Xi = H^1(\Omega)$ *(right).*

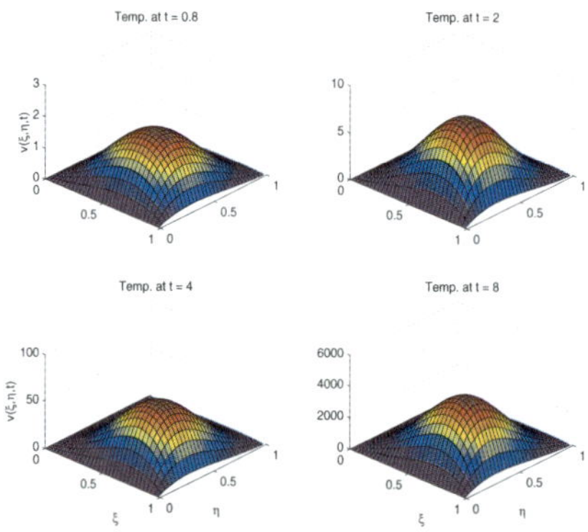

Figure 3.5. *Example* 3.4.1 *with no control at* $t = 0.8, 2, 4, 8$.

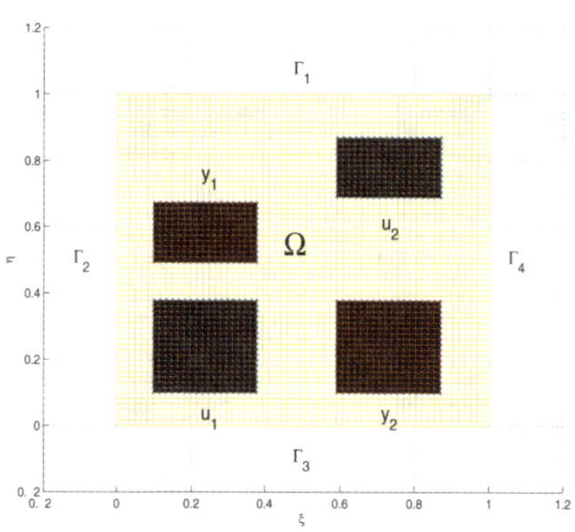

Figure 3.6. *Example* 3.4.1: *Sensor and control domains.*

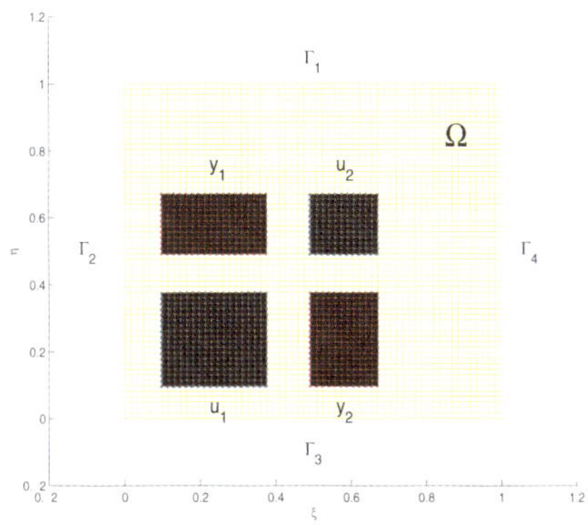

Figure 3.7. *Example* 3.4.2: *Sensor and control domains.*

3.4.2 Example (Nonlinear Unstable Heat Equation)

Our next example deals with a two-dimensional distributed control input model of a nonlinear unstable heat equation. In the domain $\Omega = [0, 1] \times [0, 1]$ we consider an initial boundary value problem of this nonlinear reaction-diffusion model for the unknown function $v(\xi, \eta; t)$, $(\xi, \eta) \in \Omega$, $t > 0$:

$$\begin{aligned} v_t &= \kappa \Delta v + \varepsilon_2 v - \varepsilon_1 v^3 + \sum_{i=1}^{n_u} u_i(t) b_i && \text{in } \Omega, t > 0, \\ v(\xi, \eta; t) &= 0 && \text{on } \partial\Omega, t > 0, \\ v(\xi, \eta, 0) &= v_0(\xi, \eta) && \text{in } \Omega, \end{aligned} \tag{3.19}$$

where $\varepsilon_1, \varepsilon_2 \geq 0$ are positive constants and the other quantities are defined as in the previous example. For $\varepsilon_1 = 0$ and $\varepsilon_2 = 0$, the open-loop system of (3.19) is the heat equation, which is asymptotically stable. However, it is unstable if $\varepsilon_2 > 0$ is large enough even if $\varepsilon_1 = 0$. We use a backstepping method for the finite difference semidiscretized approximation of (3.19) (with uniform mesh size $h = 0.01639$ to ensure at least a small error in the spatial approximation) which results in a nonlinear control system of the form (3.9) with $E = I$, $\delta A \equiv 0$, $G(x(t)) = -\varepsilon_1 x(t)^3$, and $n_x = 3600$ state variables. Moreover, we assume that we have two fixed sensor ($n_y = 2$) and two fixed control ($n_u = 2$) domains located in Ω. In particular, Figure 3.7 visualizes the location of the two control (green) and sensor (red) domains in Ω, respectively. The instability of the linear (e.g., $\varepsilon_1 = 0$) uncontrolled model is illustrated in Figure 3.8. In our numerical experiments for this nonlinear model we have used the parameters in Table 3.2. For the computation of the low-dimensional POD approximation of the nonlinear model, we use snapshots of the linearized unstable and uncontrolled system as shown in Figure 3.8. In this case study we have computed the first $\ell = 5$ POD basis functions utilizing the (discrete) L^2-norm as well as the discrete H^1-norm. Figure 3.9 shows the results of our experiments. Therein, we have plotted the temperature distribution of the nonlinear model (3.19) if we use the SOF controller computed by IPCTR in combination with POD. At $t = 6$, Figure 3.9 illustrates that the controlled dynamics are closer to zero with respect to the L^2-norm than the H^1-norm POD approximation. The

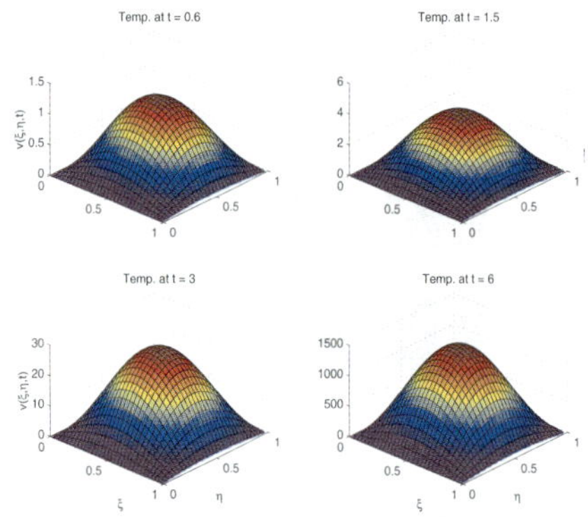

Figure 3.8. *Example* 3.4.2 *with no control at* $t = 0.8, 2, 4, 8$.

corresponding optimal output feedback control functions acting on the control domains in Ω can be found in Figure 3.10. The plots in Figure 3.11 contain the temperature distributions of the controlled nonlinear model for the choice of a different snapshot ensemble. In particular, for the computation of the POD basis function we use only every 10th snapshot and include the difference quotients in the snapshot ensemble. In this case, the temperature of the controlled system tends faster to the stable equilibrium state at zero with respect to the H^1-norm and included difference quotient than for all the other POD approximations. In this case, the output feedback control functions are given in Figure 3.12.

Table 3.2. *Parameters for Example* 3.4.2.

n_x	n_u	n_y	ℓ	κ	c_1	d_1	ε_1	ε_2
3600	2	2	5	0.06119	100	10	1.0	2.5

3.4.3 Example (Modified Burgers' Equation)

Now we turn to our third example, where we consider the viscous modified Burgers equation

$$\begin{aligned} v_t - \nu v_{\xi\xi} + v v_\xi - \varepsilon v &= 0 && \text{in } Q = (0,T) \times \Omega, \\ \nu v'(t,0) &= 0 && \forall t \in (0,T), \\ \nu v'(t,2\pi) &= u && \forall t \in (0,T), \\ v(0,\xi) &= \sin(\xi) && \forall \xi \in \Omega = (0,2\pi), \end{aligned} \tag{3.20}$$

where $\nu = 0.5$ denotes a viscosity parameter, $T = 2 > 0$ is the end time, $u \in L^2(0,T)$ is the control input, and $\varepsilon = 0.125$. In the context of feedback control for Burgers' equation with POD we refer the reader to [2, 16], for instance.

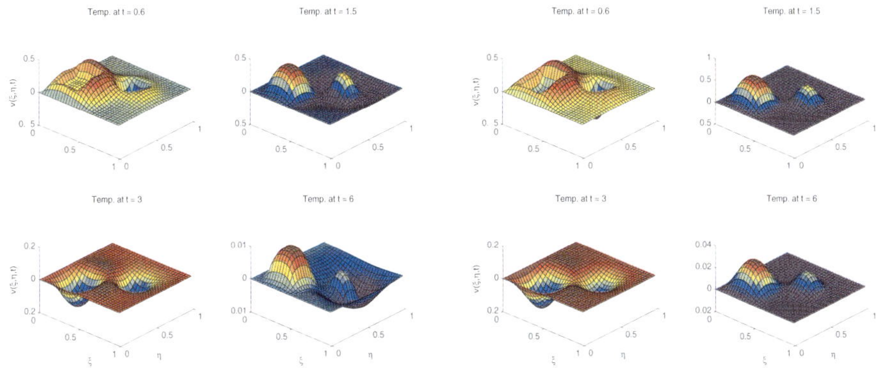

Figure 3.9. *Example* 3.4.2 *with SOF at* $t = 0.8, 2, 4, 8$: *POD with* $\mathcal{V}_1$, $\Xi = L^2(\Omega)$ *(left),* $\mathcal{V}_1$, $\Xi = H^1(\Omega)$ *(right).*

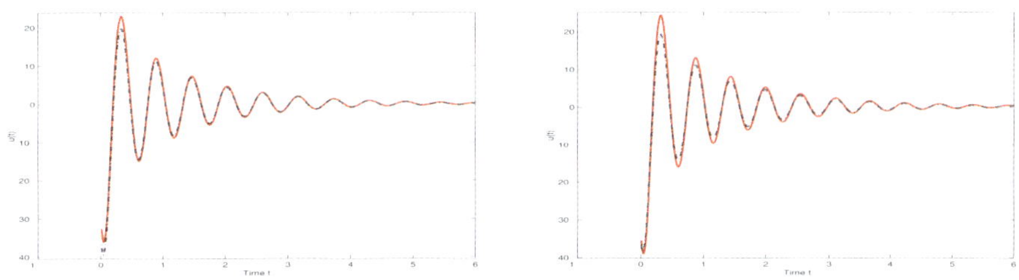

Figure 3.10. *SOF control* (3.10) *for Example* 3.4.2: *POD with* $\mathcal{V}_1$, $\Xi = L^2(\Omega)$ *(left),* $\mathcal{V}_1$, $\Xi = H^1(\Omega)$ *(right).*

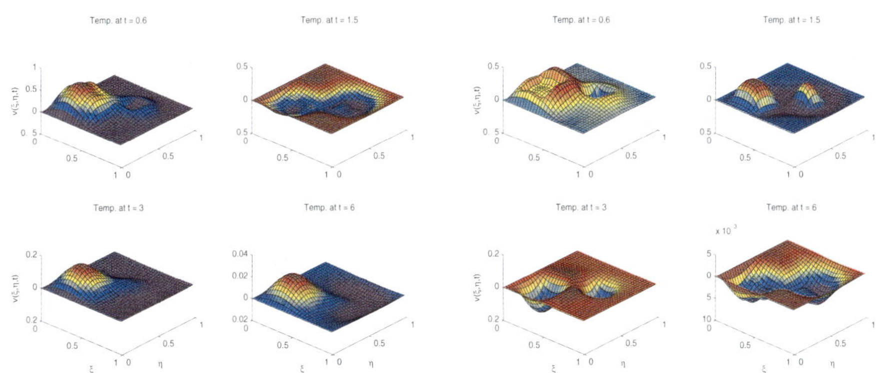

Figure 3.11. *Example* 3.4.2 *with SOF at* $t = 0.8, 2, 4, 8$: *POD with* $\mathcal{V}_2$, $\Xi = L^2(\Omega)$ *(left),* $\mathcal{V}_2$, $\Xi = H^1(\Omega)$ *(right).*

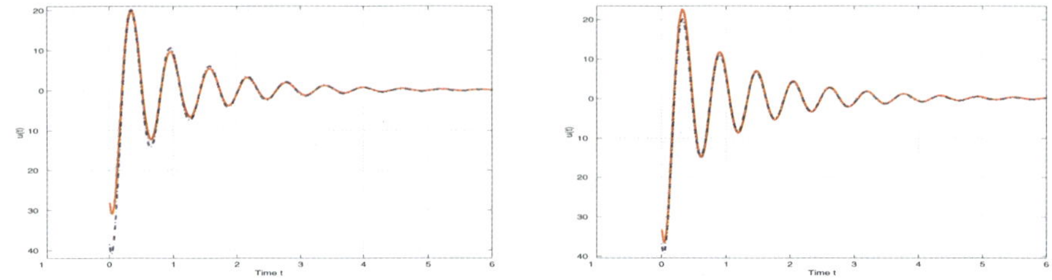

Figure 3.12. *SOF control* (3.10) *for Example* 3.4.2: *POD with* $\mathcal{V}_2$, $\Xi = L^2(\Omega)$ *(left),* $\mathcal{V}_2$, $\Xi = H^1(\Omega)$ *(right).*

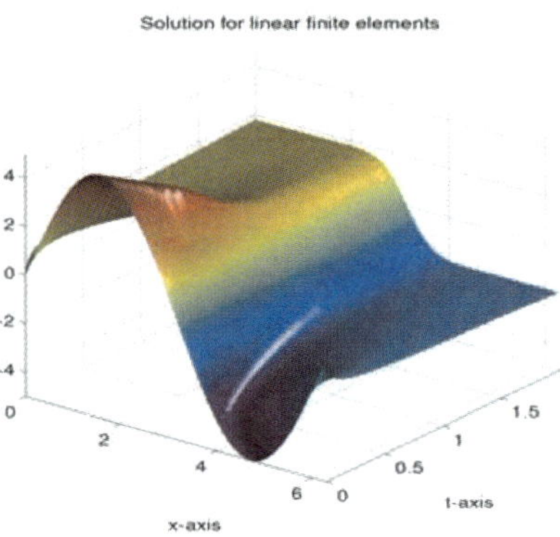

Figure 3.13. *Example* 3.4.3 *with no control.*

There is only one control input acting on the right end of the interval Ω. The only measured information available to this control is the state at time $t \in (0, T)$ at $\xi = 2\pi$, i.e.,

$$y(t) = v(t, 2\pi) \quad \forall t \in (0, T). \tag{3.21}$$

As in the two previous examples, we can express (3.20)–(3.21) by a dynamical system in an appropriate (infinite-dimensional) state space.

The goal of our optimal control problem is to compute a SOF control law to track the system to zero, i.e., to minimize the cost

$$J(v, u) = \frac{1}{2} \int_0^T \int_\Omega |v(t, \xi)|^2 + |v_\xi(t, \xi)|^2 \, d\xi \, dt + \frac{\beta}{2} \int_0^T |u(t)|^2 \, dt,$$

where $\beta = 10^{-6}$ is a fixed regularization parameter. Notice that for every $t \in (0, T)$ there is only one observation and one control point. Thus, the control influence is not too big and the SOF control law F is a real number.

For the finite element discretization we utilize the software Femlab, Version 2.2, where we took linear Lagrange elements with 1258 degrees of freedom. The uncontrolled dynamics are presented in Figure 3.13.

Utilizing $\ell = 3$ POD basis function we compute the SOF control law (3.10) and solve the closed-loop dynamics. Including more information into the snapshot set, e.g.,

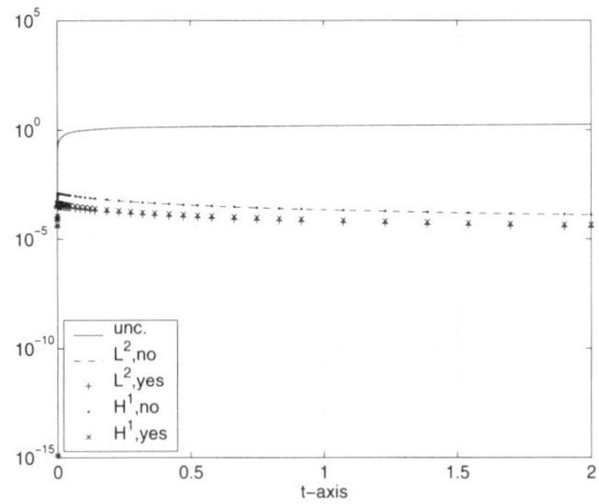

Figure 3.14. *Decay rates:* $t \to |v(t, 2\pi)|$.

Table 3.3. *Parameters for Example* 3.4.2.

	$\mathcal{E}(3)$	F	$J(v, u)$	$\int_0^T |v(t, 2\pi)|\, dt$
$\Xi = L^2, \mathcal{V}_1$	99.95%	−2060.07	69.08	0.0006
$\Xi = H^1, \mathcal{V}_2$	92.37%	−7536.10	69.12	0.0000
$\Xi = L^2, \mathcal{V}_1$	99.38%	−2017.68	69.09	0.0006
$\Xi = H^1, \mathcal{V}_2$	79.39%	−5657.58	69.06	0.0002
$u = 0$	—	—	73.34	3.0109

difference quotient or gradient norms, the value of $\mathcal{E}(\ell)$ decreases. In particular, for the choices $\Xi = H^1(\Omega)$ and $\mathcal{V}_2$ we have $\mathcal{E}(3) < 80\%$. However, the damping of $t \mapsto v(t, 2\pi)$ is better than for the choice $\mathcal{V}_1$, independent of Ξ; cf. Figure 3.14. Also the values of the SOF control law are quite different. The values for $\mathcal{V}_1$, i.e., without including the discrete-time derivatives, are of the same magnitude, whereas F is much greater if we take the ensemble $\mathcal{V}_2$. Due to the soft control input the cost is only reduced by about 6%. Results for different choices of Ξ are summarized in Table 3.3.

3.5 Conclusions

In this chapter we consider the SOF problem for PDE-constrained control problems in combination with the POD reduction method for deriving a low-dimensional control system and the interior point trust-region algorithm for solving the corresponding low-dimensional NSDP. The SOF control law can be constructed from the solution of the low-dimensional NSDP. In our numerical examples, we observe that this SOF can be used for controlling the large-dimensional PDE system. In particular, it turns out that including the difference quotients into the POD snapshot ensemble leads to better stability properties of the computed SOF control laws.

Bibliography

[1] A. Adrover and M. Giona, *Model reduction of PDE models by means of snapshot archetypes*, Phys. D, 182 (2003), pp. 23–45.

[2] J. A. Atwell, J. T. Borggaard, and B. B. King, *Reduced order controllers for Burgers' equation with a nonlinear observer*, Int. J. Appl. Math. Comput. Sci., 11 (2001), pp. 1311–1330.

[3] D. S. Bernstein and W. M. Haddad, *LQG control with an $\mathcal{H}_\infty$ performance bound: A Riccati equation approach*, IEEE Trans. Automat. Control, 34 (1989), pp. 293–305.

[4] S. Boyd, L. El Ghaoui, E. Feron, and V. Balakrishnan, *Linear Matrix Inequalities in System and Control Theory*, SIAM Stud. Appl. Math. 15, SIAM, Philadelphia, 1994.

[5] R. H. Byrd, *Robust trust region methods for nonlinearly constrained optimization*, in Proceedings of the SIAM Conference on Optimization, Houston, TX, 1987.

[6] A. R. Conn, N. I. M. Gould, and P. Toint, *Trust-Region Methods*, MPS/SIAM Ser. Optim. 1, SIAM, Philadelphia, 2000.

[7] R. F. Curtain and H. J. Zwart, *An Introduction to Infinite Dimensional Linear System Theory*, Texts Appl. Math. 21, Springer-Verlag, New York, 1995.

[8] J. E. Dennis, M. Heinkenschloss, and L. N. Vicente, *Trust-region interior-point SQP algorithms for a class of nonlinear programming problems*, SIAM J. Control Optim., 36 (1998), pp. 1750–1794.

[9] M. Hinze and S. Volkwein, *Proper orthogonal decomposition surrogate models for nonlinear dynamical systems: Error estimates and suboptimal control*, in Reduction of Large-Scale Systems, Lect. Notes Comput. Sci. Eng. 45, Springer-Verlag, Berlin, 2005, pp. 261–306.

[10] C. W. J. Hol, C. W. Scherer, E. G. van der Meché, and O. H. Bosgra, *A nonlinear SDP approach to fixed-order controller synthesis and comparison with two other methods applied to an active suspension system*, European J. Control, 9 (2003), pp. 13–28.

[11] P. Holmes, J. L. Lumley, and G. Berkooz, *Turbulence, Coherent Structures, Dynamical Systems and Symmetry*, Cambridge Monographs on Mechanics. Cambridge University Press, Cambridge, UK, 1996.

[12] T. Iwasaki, R. E. Skelton, and J. C. Geromel, *Linear quadratic suboptimal control with static output feedback*, Systems Control Lett., 23 (1994), pp. 421–430.

[13] P. P. Khargonekar and M. A. Rotea, *Mixed $\mathcal{H}_2/\mathcal{H}_\infty$ control: A convex optimization approach*, IEEE Trans. Automat. Control, 36 (1991), pp. 824–837.

[14] K. Kunisch and S. Volkwein, *Control of Burgers' equation by a reduced order approach using proper orthogonal decomposition*, J. Optim. Theory Appl., 102 (1999), pp. 345–371.

[15] K. Kunisch and S. Volkwein, *Galerkin proper orthogonal decomposition methods for a general equation in fluid dynamics*, SIAM J. Numer. Anal., 40 (2002), pp. 492–515.

[16] K. Kunisch, S. Volkwein, and L. Xie, *HJB-POD-based feedback design for the optimal control of evolution problems*, SIAM J. Appl. Dyn. Syst., 3 (2004), pp. 701–722.

[17] F. Leibfritz, *Static Output Feedback Design Problems*, Shaker-Verlag, Aachen, Germany, 1998.

[18] F. Leibfritz, *An LMI-based algorithm for designing suboptimal static $\mathcal{H}_2/\mathcal{H}_\infty$ output feedback controllers*, SIAM J. Control Optim., 39 (2001), pp. 1711–1735.

[19] F. Leibfritz, *COMPl$_e$ib: COnstrained Matrix-optimization Problem library – a collection of test examples for nonlinear semidefinite programs, control system design and related problems*, European J. Control, to appear.

[20] F. Leibfritz and E. M. E. Mostafa, *An interior point constrained trust region method for a special class of nonlinear semidefinite programming problems*, SIAM J. Optim., 12 (2002), pp. 1048–1074.

[21] F. Leibfritz and E. M. E. Mostafa, *Trust region methods for solving the optimal output feedback design problem*, Internat. J. Control, 76 (2003), pp. 501–519.

[22] F. Leibfritz and S. Volkwein, *Reduced order output feedback control design for PDE systems using proper orthogonal decomposition and nonlinear semidefinite programming*, Linear Algebra Appl., 415 (2006), pp. 542–575.

[23] E. O. Omojokun, *Trust Region Strategies for Optimization with Nonlinear Equality and Inequality Constraints*, Ph.D. thesis, University of Colorado, Boulder, CO, 1989.

[24] A. Pazy, *Semigoups of Linear Operators and Applications to Partial Differential Equations*, Springer-Verlag, New York, 1983.

[25] L. Sirovich, *Turbulence and the dynamics of coherent structures. Parts* I–III, Quart. Appl. Math., 45 (1987), pp. 561–590.

[26] V. L. Syrmos, C. T. Abdallah, P. Dorato, and K. Grigoriadis, *Static output feedback – A survey*, Automatica, 33 (1997), pp. 125–137.

[27] S. Volkwein, *Optimal control of a phase-field model using the proper orthogonal decomposition*, ZAMM Z. Angew. Math. Mech., 81 (2001), pp. 83–97.

[28] S. Volkwein, *Interpretation of proper orthogonal decomposition as singular value decomposition and HJB-based feedback design*, in Proceedings of the 16th International Symposium on Mathematical Theory of Networks and Systems (MTNS), Leuven, Belgien, 2004.

[29] A. WÄCHTER AND L. T. BIEGLER, *Failure of global convergence for a class of interior point methods for nonlinear programming*, Math. Program., 88 (2000), pp. 565–574.

[30] K. ZHOU, J. C. DOYLE, AND K. GLOVER, *Robust and Optimal Control*, Prentice–Hall, Upper Saddle River, NJ, 1996.

Chapter 4

A Least-Squares Finite Element Method for Optimization and Control Problems

Pavel B. Bochev and Max D. Gunzburger†*

4.1 Introduction

Optimization and control problems for systems governed by PDEs arise in many applications. Experimental studies of such problems go back 100 years [20]. Computational approaches have been applied since the advent of the computer age. Most of the efforts in the latter direction have employed elementary optimization strategies, but, more recently, there has been considerable practical and theoretical interest in the application of sophisticated local and global optimization strategies, e.g., Lagrange multiplier methods, sensitivity or adjoint-based gradient methods, quasi-Newton methods, evolutionary algorithms, etc.

The optimal control or optimization problems we consider consist of

- *state variables*, i.e., variables that describe the system being modeled;
- *control variables* or *design parameters*, i.e., variables at our disposal that can be used to affect the state variables;
- a *state system*, i.e., PDEs relating the state and control variables; and
- a *functional* of the state and control variables whose minimization is the goal.

*Computational Mathematics and Algorithms Department, Sandia National Laboratories, Albuquerque, NM 87185-1110.

†School of Computational Science and Information Technology, Florida State University, Tallahassee, FL 32306-4120.

Then, the problems we consider consist of finding state and control variables that minimize the given functional subject to the state system being satisfied. Here we restrict our attention to linear, elliptic state systems and to quadratic functionals.

The Lagrange multiplier rule is a standard approach for solving finite-dimensional optimization problems. It is not surprising then that several popular approaches to solving optimization and control problems constrained by PDEs are based on solving optimality systems deduced from the application of the Lagrange multiplier rule. In [8], least-squares finite element methods were used to develop methods for the approximate solution of the optimality systems; see also [18].

Penalty methods, other popular methods for solving optimization problems, have also been applied, albeit to a lesser extent, to problems such as those considered here. In particular, these algorithms enforce the PDEs constraints by using well-posed least-squares functionals as penalty terms that are added to the original cost functional. This type of penalty methods offers certain efficiency-related advantages compared to methods based on the solution of the Lagrange multiplier optimality system either by Galerkin or least-squares finite element methods. Such least-squares/penalty methods have been used for an optimal shape design problem [1], for controlling Stokes equations [7], for the Dirichlet control of the Navier–Stokes equations [2, 4], and for optimal control problems constrained by model first-order elliptic systems [19]. An extensive discussion of least-squares/penalty methods is given in [9].

In this chapter we discuss a new approach in which the cost functional is *constrained* by the least-squares functional. This approach is more effective than are least-squares/penalty methods. For the latter, one has methods that either require the satisfaction of discrete stability conditions or are prone to locking; see, e.g., [10]. Using the new approach, one can define a method that avoids both of these undesirable features. A more detailed discussion comparing these two approaches to incorporating least-square principles into PDE-constrained optimization problems as well as the approach of [8] is given in [9].

The chapter is organized as follows. In Section 4.2, we define an abstract quadratic optimization and control problem constrained by linear, elliptic PDEs. Then, in Section 4.3, we review results about least-squares finite element methods for the approximate solution of the constraint equations. In Section 4.4, we present and analyze the new approach that involves constraining the cost functional by the least-squares functional. Finally, in Section 4.5, we provide a concrete example of the abstract theory.

4.2 Quadratic Optimization and Control Problems in Hilbert Spaces with Linear Constraints

We begin with four given Hilbert spaces Θ, Φ, $\widehat{\Phi}$, and $\widetilde{\Phi}$ along with their dual spaces denoted by $(\cdot)^*$. We assume that $\Phi \subseteq \widehat{\Phi} \subseteq \widetilde{\Phi}$ with continuous embeddings and that $\widetilde{\Phi}$ acts as the pivot space for both the pair $\{\Phi^*, \Phi\}$ and the pair $\{\widehat{\Phi}^*, \widehat{\Phi}\}$ so that we not only have that $\Phi \subseteq \widehat{\Phi} \subseteq \widetilde{\Phi} \subseteq \widehat{\Phi}^* \subseteq \Phi^*$ but also

$$\langle \psi, \varphi \rangle_{\Phi^*,\Phi} = \langle \psi, \varphi \rangle_{\widehat{\Phi}^*,\widehat{\Phi}} = (\psi, \varphi)_{\widetilde{\Phi}} \quad \forall \psi \in \widehat{\Phi}^* \subseteq \Phi^* \quad \text{and} \quad \forall \varphi \in \Phi \subseteq \widehat{\Phi}, \tag{4.1}$$

where $(\cdot,\cdot)_{\widetilde{\Phi}}$ denotes the inner product on $\widetilde{\Phi}$. Next, we define the *functional*

$$\mathcal{J}(\varphi,\theta) = \frac{1}{2}a_1(\varphi - \widehat{\varphi}, \varphi - \widehat{\varphi}) + \frac{1}{2}a_2(\theta,\theta) \qquad \forall\, \varphi \in \Phi,\ \theta \in \Theta, \tag{4.2}$$

where $a_1(\cdot,\cdot)$ and $a_2(\cdot,\cdot)$ are symmetric bilinear forms on $\widehat{\Phi} \times \widehat{\Phi}$ and $\Theta \times \Theta$, respectively, and $\widehat{\varphi} \in \widehat{\Phi}$ is a given function. In the language of control theory, Φ is called the *state space,* φ the *state variable,* Θ the *control space,* and θ the *control variable.* In many applications, the control space is finite dimensional in which case θ is often referred to as the vector of *design variables.* We note that often Θ is chosen to be a bounded set in a Hilbert space but, for our purposes, we can consider the less general situation of Θ itself being a Hilbert space. The second term in the functional (4.2) can be interpreted as a penalty term[4] which limits the size of the control θ.

We make the following assumptions about the bilinear forms $a_1(\cdot,\cdot)$ and $a_2(\cdot,\cdot)$:

$$\left\{\begin{array}{ll} a_1(\varphi,\mu) \le C_1 \|\varphi\|_{\widehat{\Phi}} \|\mu\|_{\widehat{\Phi}} & \forall\, \varphi, \mu \in \widehat{\Phi}, \\ a_2(\theta,\nu) \le C_2 \|\theta\|_{\Theta} \|\nu\|_{\Theta} & \forall\, \theta, \nu \in \Theta, \\ a_1(\varphi,\varphi) \ge 0 & \forall\, \varphi \in \widehat{\Phi}, \\ a_2(\theta,\theta) \ge K_2 \|\theta\|^2_{\Theta} & \forall\, \theta \in \Theta, \end{array}\right. \tag{4.3}$$

where C_1, C_2, and K_2 are all positive constants.

Given another Hilbert space Λ, the additional bilinear forms $b_1(\cdot,\cdot)$ on $\Phi \times \Lambda$ and $b_2(\cdot,\cdot)$ on $\Theta \times \Lambda$, and the function $g \in \Lambda^*$, we define the *constraint equation*

$$b_1(\varphi,\psi) + b_2(\theta,\psi) = \langle g, \psi \rangle_{\Lambda^*,\Lambda} \quad \forall\, \psi \in \Lambda\,. \tag{4.4}$$

We make the following assumptions about the bilinear forms $b_1(\cdot,\cdot)$ and $b_2(\cdot,\cdot)$:

$$\left\{\begin{array}{ll} b_1(\varphi,\psi) \le c_1 \|\varphi\|_{\Phi} \|\psi\|_{\Lambda} & \forall\, \varphi \in \Phi,\ \ \psi \in \Lambda, \\ b_2(\theta,\psi) \le c_2 \|\theta\|_{\Theta} \|\psi\|_{\Lambda} & \forall\, \theta \in \Theta,\ \ \psi \in \Lambda, \\ \displaystyle\sup_{\psi \in \Lambda, \psi \neq 0} \frac{b_1(\varphi,\psi)}{\|\psi\|_{\Lambda}} \ge k_1 \|\varphi\|_{\Phi} & \forall\, \varphi \in \Phi, \\ \displaystyle\sup_{\varphi \in \Phi, \varphi \neq 0} \frac{b_1(\varphi,\psi)}{\|\varphi\|_{\Phi}} > 0 & \forall\, \psi \in \Lambda\,, \end{array}\right. \tag{4.5}$$

where c_1, c_2, and k_1 are all positive constants.

We consider the *optimal control problem*

$$\min_{(\varphi,\theta) \in \Phi \times \Theta} \mathcal{J}(\varphi,\theta) \quad \text{subject to} \quad b_1(\varphi,\psi) + b_2(\theta,\psi) = \langle g, \psi \rangle_{\Lambda^*,\Lambda} \quad \forall\, \psi \in \Lambda\,. \tag{4.6}$$

The following result is proved in, e.g., [8].

Theorem 4.1. *Let assumptions* (4.3) *and* (4.5) *hold. Then, the optimal control problem* (4.6) *has a unique solution* $(\varphi,\theta) \in \Phi \times \Theta$.

[4]The usage of the terminology "penalty term" in conjunction with the second term in (4.2) should not be confused with the usage of that terminology in Section 4.1. In particular, the second term in (4.2) has no connection with the terminology "least-squares/penalty" previously used.

It is instructive to rewrite the functional (4.2), the constraint (4.4), and the optimal control problem (4.6) in operator notation. To this end, we note that the bilinear forms serve to define operators

$$
\begin{aligned}
&A_1 : \widehat{\Phi} \to \widehat{\Phi}^*, \qquad && A_2 : \Theta \to \Theta^*, \qquad && B_1 : \Phi \to \Lambda^*,\\
&B_1^* : \Lambda \to \Phi^*, && B_2 : \Theta \to \Lambda^*, && B_2^* : \Lambda \to \Theta^*
\end{aligned}
$$

through the following relations:

$$
\begin{aligned}
&a_1(\varphi, \mu) = \langle A_1\varphi, \mu\rangle_{\widehat{\Phi}^*,\widehat{\Phi}} && \forall\, \varphi, \mu \in \widehat{\Phi},\\
&a_2(\theta, \nu) = \langle A_2\theta, \nu\rangle_{\Theta^*,\Theta} && \forall\, \theta, \nu \in \Theta,\\
&b_1(\varphi, \psi) = \langle B_1\varphi, \psi\rangle_{\Lambda^*,\Lambda} = \langle B_1^*\psi, \varphi\rangle_{\Phi^*,\Phi} && \forall\, \varphi \in \Phi, \quad \psi \in \Lambda,\\
&b_2(\psi, \theta) = \langle B_2\theta, \psi\rangle_{\Lambda^*,\Lambda} = \langle B_2^*\psi, \theta\rangle_{\Theta^*,\Theta} && \forall\, \theta \in \Theta, \quad \psi \in \Lambda\,.
\end{aligned}
\tag{4.7}
$$

Then, the functional (4.2) and the constraint (4.4), respectively, take the forms

$$
\mathcal{J}(\varphi, \theta) = \frac{1}{2}\big\langle A_1(\varphi - \widehat{\varphi}), (\varphi - \widehat{\varphi})\big\rangle_{\widehat{\Phi}^*,\widehat{\Phi}} + \frac{1}{2}\langle A_2\theta, \theta\rangle_{\Theta^*,\Theta} \qquad \forall\, \varphi \in \Phi,\ \theta \in \Theta \tag{4.8}
$$

and

$$
B_1\varphi + B_2\theta = g \qquad \text{in } \Lambda^* \tag{4.9}
$$

and the optimal control problem (4.6) takes the form

$$
\min_{(\varphi,\theta)\in\Phi\times\Theta} \mathcal{J}(\varphi, \theta) \qquad \text{subject to} \qquad B_1\varphi + B_2\theta = g \quad \text{in } \Lambda^*. \tag{4.10}
$$

Assumptions (4.3) and (4.5) imply that A_1, A_2, B_1, B_2, B_1^*, and B_2^* are bounded with

$$
\begin{aligned}
&\|A_1\|_{\widehat{\Phi}\to\widehat{\Phi}^*} \le C_1, \qquad && \|A_2\|_{\Theta\to\Theta^*} \le C_2, \qquad && \|B_1\|_{\Phi\to\Lambda^*} \le c_1,\\
&\|B_1^*\|_{\Lambda\to\Phi^*} \le c_1, && \|B_2\|_{\Theta\to\Lambda^*} \le c_2, && \|B_2^*\|_{\Lambda\to\Theta^*} \le c_2
\end{aligned}
$$

and that the operator B_1 is invertible with $\|B_1^{-1}\|_{\Lambda^*\to\Phi} \le 1/k_1$. See [8] for details. Note that, given $\theta \in \Theta$ and $g \in \Lambda^*$, the assumptions in (4.5) about the bilinear form $b_1(\cdot,\cdot)$ imply that the constraint equation (4.9) may be solved for $\varphi \in \Phi$ to yield $\varphi = B^{-1}(g - B_2\theta)$.

4.3 Least-Squares Formulation of the Constraint Equations

The constraint equations are given in variational form in (4.4) and in equivalent operator form in (4.9). They may also be defined through a least-squares minimization problem. Let $D : \Lambda \to \Lambda^*$ be a self-adjoint, strongly coercive operator;[5] i.e., there exist constants $c_d > 0$

[5]In what follows, we will also use the induced bilinear form

$$
d(\lambda, \psi) = \langle D\lambda, \psi\rangle_{\Lambda^*,\Lambda} \quad \forall\, \lambda, \psi \in \Lambda\,. \tag{4.11}
$$

The following results are immediate.

Proposition 4.2. *Assume that the operator D is symmetric and that* (4.13) *holds. Then, the bilinear form $d(\cdot,\cdot)$ is symmetric and*

$$
d(\lambda, \psi) \le c_d\|\lambda\|_\Lambda\|\psi\|_\Lambda \quad \forall\, \lambda, \psi \in \Lambda \quad \text{and} \quad d(\lambda, \lambda) \ge k_d\|\lambda\|_\Lambda^2 \quad \forall\, \lambda \in \Lambda\,. \tag{4.12}
$$

and $k_d > 0$ such that

$$\langle D\lambda, \psi\rangle_{\Lambda^*,\Lambda} \le c_d\|\lambda\|_\Lambda\|\psi\|_\Lambda \quad \text{and} \quad \langle D\lambda, \lambda\rangle_{\Lambda^*,\Lambda} \ge k_d\|\lambda\|_\Lambda^2 \qquad \forall \lambda, \psi \in \Lambda\,. \tag{4.13}$$

Note that then $k_d \le \|D\|_{\Lambda\to\Lambda^*} \le c_d$ and $1/c_d \le \|D^{-1}\|_{\Lambda^*\to\Lambda} \le 1/k_d$. Let[6]

$$\begin{aligned}&\mathcal{K}(\varphi; \theta, g)\\ &\quad = \big\langle B_1\varphi + B_2\theta - g,\, D^{-1}(B_1\varphi + B_2\theta - g)\big\rangle_{\Lambda^*,\Lambda} \quad \forall \varphi \in \Phi,\ \theta \in \Theta,\ g \in \Lambda^*.\end{aligned} \tag{4.14}$$

Given $\theta \in \Theta$ and $g \in \Lambda^*$, consider the problem

$$\min_{\varphi\in\Phi} \mathcal{K}(\varphi; \theta, g)\,. \tag{4.15}$$

Clearly, this problem is equivalent to (4.4) and (4.9); i.e., solutions of (4.15) are solutions of (4.4) or (4.9) and conversely. The Euler–Lagrange equation corresponding to problem (4.15) is given, in variational form, by

$$\widetilde{b}_1(\varphi, \mu) = \langle \widetilde{g}_1, \mu\rangle_{\Phi^*,\Phi} - \widetilde{b}_2(\theta, \mu) \quad \forall \mu \in \Phi\,, \tag{4.16}$$

where

$$\widetilde{b}_1(\varphi, \mu) = \big\langle B_1\mu, D^{-1}B_1\varphi\big\rangle_{\Lambda^*,\Lambda} = \big\langle B_1^*D^{-1}B_1\varphi, \mu\big\rangle_{\Phi^*,\Phi} \qquad \forall \varphi, \mu \in \Phi, \tag{4.17}$$

$$\widetilde{b}_2(\theta, \mu) = \big\langle B_1\mu, D^{-1}B_2\theta\big\rangle_{\Lambda^*,\Lambda} = \big\langle B_1^*D^{-1}B_2\theta, \mu\big\rangle_{\Phi^*,\Phi} \qquad \forall \theta \in \Theta,\ \mu \in \Phi, \tag{4.18}$$

and

$$\widetilde{g}_1 = B_1^*D^{-1}g \in \Phi^*\,. \tag{4.19}$$

The following proposition shows that the bilinear forms $\widetilde{b}_1(\cdot,\cdot)$ and $\widetilde{b}_2(\cdot,\cdot)$ are continuous and the former is strongly coercive.

Proposition 4.3. *Assume that* (4.5) *and* (4.13) *hold. Then, the bilinear form $\widetilde{b}_1(\cdot,\cdot)$ is symmetric and there exist positive constants $\widetilde{c}_1$, $\widetilde{c}_2$, and $\widetilde{k}_1$ such that*

$$\begin{cases} \widetilde{b}_1(\varphi, \mu) \le \widetilde{c}_1\|\varphi\|_\Phi\|\mu\|_\Phi & \forall \varphi, \mu \in \Phi,\\ \widetilde{b}_2(\theta, \mu) \le \widetilde{c}_2\|\mu\|_\Phi\|\theta\|_\Theta & \forall \theta \in \Theta,\quad \mu \in \Phi,\\ \widetilde{b}_1(\varphi, \varphi) \ge \widetilde{k}_1\|\varphi\|_\Phi^2 & \forall \varphi \in \Phi\,. \end{cases} \tag{4.20}$$

Moreover, $\|\widetilde{g}_1\|_{\Phi^*} \le \frac{c_1}{k_d}\|g\|_{\Lambda^*}$ and the problem (4.16), *or equivalently* (4.15), *has a unique solution.*

[6]Let $R : \Lambda \to \Lambda^*$ denote the Reisz operator; i.e., we have that if $\upsilon = R\lambda$ and $\chi = R\psi$ for $\lambda, \psi \in \Lambda$ and $\upsilon, \chi \in \Lambda^*$, then $\|\lambda\|_\Lambda = \|\upsilon\|_{\Lambda^*}$, $\|\psi\|_\Lambda = \|\chi\|_{\Lambda^*}$, and

$$(\psi, \lambda)_\Lambda = \langle R\psi, \lambda\rangle_{\Lambda^*,\Lambda} = \langle \chi, R^{-1}\upsilon\rangle_{\Lambda^*,\Lambda} = (\upsilon, \chi)_{\Lambda^*}.$$

Then, if one chooses $D = R$, the functional (4.14) reduces to $\mathcal{K}(\varphi; \theta, g) = (B_1\varphi + B_2\theta - g, B_1\varphi + B_2\theta - g)_{\Lambda^*} = \|B_1\varphi + B_2\theta - g\|_{\Lambda^*}^2$. Note that, in general, (4.14) can also be written as an inner product, i.e., $\mathcal{K}(\varphi; \theta, g) = (B_1\varphi + B_2\theta - g, RD^{-1}(B_1\varphi + B_2\theta - g))_{\Lambda^*}$.

Proof. The symmetry of the bilinear form $\widetilde{b}_1(\cdot,\cdot)$ follows immediately from the symmetry of the operator D. Since $b_1(\varphi,\psi) \le c_1\|\varphi\|_\Phi\|\psi\|_\Lambda$ for all $\varphi \in \Phi$ and $\psi \in \Lambda$, we have that

$$\langle B_1\varphi,\psi\rangle_{\Lambda^*,\Lambda} = b_1(\varphi,\psi) \le c_1\|\varphi\|_\Phi\|\psi\|_\Lambda \quad \forall\varphi\in\Phi,\ \ \psi\in\Lambda,$$

from which it easily follows that $\|B_1\varphi\|_{\Lambda^*} \le c_1\|\varphi\|_\Phi$ for all $\varphi\in\Phi$. We then have that

$$\widetilde{b}_1(\varphi,\mu) = \big\langle B_1\mu, D^{-1}B_1\varphi\big\rangle_{\Lambda^*,\Lambda} \le \|D^{-1}\|_{\Lambda\to\Lambda^*}\|B_1\varphi\|_{\Lambda^*}\|B_1\mu\|_{\Lambda^*} \le \frac{c_1^2}{k_d}\|\varphi\|_\Phi\|\mu\|_\Phi\,.$$

In a similar way, one shows that $\|B_2\theta\|_{\Lambda^*} \le c_2\|\theta\|_\Theta$ for all $\theta\in\Theta$ and

$$\widetilde{b}_2(\theta,\mu) = \big\langle B_1\mu, D^{-1}B_2\theta\big\rangle_{\Lambda^*,\Lambda} \le \frac{c_1c_2}{k_d}\|\theta\|_\Theta\|\mu\|_\Phi\,.$$

Also in a similar way, the bound for $\|\widetilde{g}_1\|_{\Phi^*}$ can be obtained.

Next, since $\sup_{\psi\in\Lambda,\psi\neq0}(b_1(\varphi,\psi)/\|\psi\|_\Lambda) \ge k_1\|\varphi\|_\Phi$ for all $\varphi\in\Phi$, we have that

$$\|B_1\varphi\|_{\Lambda^*} = \sup_{\psi\in\Lambda,\psi\neq0}\frac{\langle B_1\varphi,\psi\rangle_{\Lambda^*,\Lambda}}{\|\psi\|_\Lambda} = \sup_{\psi\in\Lambda,\psi\neq0}\frac{b_1(\varphi,\psi)}{\|\psi\|_\Lambda} \ge k_1\|\varphi\|_\Phi \qquad \forall\varphi\in\Phi\,.$$

We then have that, with $\lambda = D^{-1}B_1\varphi$ so that $\|B_1\varphi\|_{\Lambda^*} \le \|D\|_{\Lambda\to\Lambda^*}\|\lambda\|_\Lambda$,

$$\begin{aligned}\widetilde{b}_1(\varphi,\varphi) &= \big\langle B_1\varphi, D^{-1}B_1\varphi\big\rangle_{\Lambda^*,\Lambda} = \big\langle D\lambda,\lambda\big\rangle_{\Lambda^*,\Lambda} \ge k_d\|\lambda\|_\Lambda^2 \\ &\ge \frac{k_d}{\|D\|^2_{\Lambda\to\Lambda^*}}\|B_1\varphi\|^2_{\Lambda^*} \ge \frac{k_1^2k_d}{c_d^2}\|\varphi\|^2_\Lambda.\end{aligned}$$

Thus, (4.20) holds with $\widetilde{c}_1 = c_1^2/k_d$, $\widetilde{c}_2 = c_1c_2/k_d$, and $\widetilde{k}_1 = k_1^2k_d/c_d^2$.

The unique solvability of (4.16) then follows from the Lax–Milgram lemma. □

As an immediate consequence of Proposition 4.3, we have that the least-squares functional (4.14) is norm equivalent in the following sense.

Corollary 4.4. *Assume that* (4.13) *and the conditions on the bilinear form* $b_1(\cdot,\cdot)$ *in* (4.5) *hold. Then,*

$$\widetilde{k}_1\|\varphi\|^2_\Phi \le \mathcal{K}(\varphi;0,0) = \widetilde{b}_1(\varphi,\varphi) = \big\langle B_1\varphi, D^{-1}B_1\varphi\big\rangle_{\Lambda^*,\Lambda} \le \widetilde{c}_1\|\varphi\|^2_\Phi \quad \forall\varphi\in\Phi\,. \tag{4.21}$$

For all $\mu\in\Phi$, we can rewrite (4.16) as $\big\langle B_1\mu, D^{-1}(B_1\varphi+B_2\theta-g)\big\rangle_{\Lambda^*,\Lambda} = 0$ or $\big\langle B_1^*D^{-1}(B_1\varphi+B_2\theta-g),\mu\big\rangle_{\Phi^*,\Phi} = 0$ so that, in operator form, we have that (4.16) is equivalent to

$$\widetilde{B}_1\varphi + \widetilde{B}_2\theta = \widetilde{g}_1 \quad \text{in } \Phi^*, \tag{4.22}$$

where

$$\widetilde{B}_1 = B_1^*D^{-1}B_1 : \Phi\to\Phi^* \qquad\text{and}\qquad \widetilde{B}_2 = B_1^*D^{-1}B_2 : \Theta\to\Phi^*. \tag{4.23}$$

Note that (4.21) implies that the operator $\widetilde{B}_1 = B_1^*D^{-1}B_1$ in (4.22) is symmetric and coercive even when the operator B_1 in (4.9) is indefinite and/or nonsymmetric; these observations, of course, follow from the fact that the bilinear form $b_1(\cdot,\cdot)$ is weakly coercive (see (4.5)) while the bilinear form $\widetilde{b}_1(\cdot,\cdot)$ is strongly coercive (see (4.20)). It is also easy to see that (4.22) has the same solutions as (4.9).

Discretization of (4.16), or equivalently of (4.22), is accomplished in the standard manner. One chooses a subspace $\Phi^h \subset \Phi$ and then, given $\theta \in \Theta$ and $\widetilde{g} \in \Phi^*$, one solves the problem

$$\widetilde{b}_1(\varphi^h,\mu^h) = \langle \widetilde{g}_1, \mu^h\rangle_{\Phi^*,\Phi} - \widetilde{b}_2(\theta,\mu^h) \quad \forall \mu^h \in \Phi^h . \tag{4.24}$$

Then, (4.21) and the Lax–Milgram and Cea lemmas immediately imply the following results.

Proposition 4.5. *Assume that* (4.5) *and* (4.13) *hold. Then, problem* (4.24) *has a unique solution and, if φ denotes the solution of problem* (4.16), *or equivalently, of* (4.22), *there exists a constant $C > 0$ whose value is independent of h, φ, and φ^h such that*

$$\|\varphi - \varphi^h\|_\Phi \le C \inf_{\widetilde{\varphi}^h \in \Phi} \|\varphi - \widetilde{\varphi}^h\|_\Phi .$$

If $\{\varphi_j\}_{j=1}^J$ denotes a basis for Φ^h, then problem (4.24) is equivalent to the matrix problem

$$\widetilde{\mathbb{B}}_1 \vec{\boldsymbol{\varphi}} = \widetilde{\vec{\mathbf{g}}}_0 , \tag{4.25}$$

where $\vec{\boldsymbol{\varphi}}$ is the vector of coefficients for φ^h, $(\widetilde{\mathbb{B}}_1)_{ij} = \widetilde{b}_1(\varphi_i,\varphi_j) = \langle \widetilde{B}_1\varphi_i, \varphi_j\rangle_{\Phi^*,\Phi}$, and $\left(\widetilde{\vec{\mathbf{g}}}_0\right)_i = \langle \widetilde{g}_1, \varphi_i\rangle_{\Phi^*,\Phi} - \widetilde{b}_2(\theta,\varphi_i) = \langle \widetilde{g}_1 - \widetilde{B}_2\theta, \varphi_i\rangle_{\Phi^*,\Phi} = \langle B_1^* D^{-1} g - B_1^* D^{-1} B_2\theta\, , \, \varphi_i\rangle_{\Phi^*,\Phi}$.

The following result follows easily from Proposition 4.3 and Corollary 4.4.

Corollary 4.6. *Assume that* (4.13) *and the conditions on the bilinear form $b_1(\cdot,\cdot)$ in* (4.5) *hold. Then, the matrix $\widetilde{\mathbb{B}}_1$ is symmetric positive definite and spectrally equivalent to the Gramm matrix $\mathbb{G}$, $(\mathbb{G})_{i,j} = (\varphi_i,\varphi_j)_\Phi$.*

The main advantages of using a least-squares finite element method to solve the constraint equation (4.9) are that the matrix $\widetilde{\mathbb{B}}_1$ in (4.25) is symmetric and positive definite even when the operator B_1 in (4.9) is indefinite and/or nonsymmetric, and that the conforming finite element subspace $\Phi^h \subset \Phi$ is not subject to any additional discrete stability conditions.[7] In incorporating the least-squares formalism into the optimization setting of Section 4.2, we want to preserve these advantages.

In what follows, we will also use the bilinear form

$$c(\theta,\nu) = \left\langle B_2\nu, D^{-1}B_2\theta\right\rangle_{\Lambda^*,\Lambda} = \left\langle B_2^* D^{-1} B_2\theta, \nu\right\rangle_{\Theta^*,\Theta} \qquad \forall \theta,\nu \in \Theta \tag{4.27}$$

and the function

$$\widetilde{g}_2 = B_2^* D^{-1} g \in \Theta^* . \tag{4.28}$$

[7]A direct, conforming Galerkin finite element discretization of (4.9) requires that the discrete stability conditions

$$\begin{cases} \displaystyle\sup_{\psi^h\in\Lambda,\psi^h\ne 0} \frac{b_1(\varphi^h,\psi^h)}{\|\psi^h\|_\Lambda} \ge k_1^h \|\varphi^h\|_\Phi & \forall \varphi^h \in \Phi^h, \\ \displaystyle\sup_{\varphi^h\in\Phi^h,\varphi^h\ne 0} \frac{b_1(\varphi^h,\psi^h)}{\|\varphi^h\|_\Phi} > 0 & \forall \psi^h \in \Lambda^h \end{cases} \tag{4.26}$$

be satisfied.

The following results are immediate.

Proposition 4.7. *Assume that the operator D is symmetric and that* (4.13) *and the condition on the bilinear form $b_2(\cdot,\cdot)$ in* (4.5) *hold. Then, the bilinear form $c(\cdot,\cdot)$ is symmetric and, for some constant $C_c > 0$,*

$$c(\theta, \nu) \leq C_c \|\theta\|_\Theta \|\nu\|_\Theta \quad \forall \theta, \nu \in \Theta \quad \textit{and} \quad c(\theta,\theta) \geq 0 \quad \forall \theta \in \Theta\,. \tag{4.29}$$

Moreover, $\|\widetilde{g}_2\|_{\Theta^*} \leq \frac{c_2}{k_d}\|g\|_{\Lambda^*}$.

Associated with the bilinear form $c(\cdot,\cdot)$ we have the operator $C = B_2^* D^{-1} B_2 : \Theta \to \Theta^*$, i.e., $c(\theta,\nu) = \langle C\theta, \nu\rangle_{\Theta^*,\Theta}$ for all $\theta, \nu \in \Theta$.

4.4 Methods Based on Constraining by the Least-Squares Functional

A means for incorporating least-squares notions into a solution method for the constrained optimization problem of Section 4.2 is to solve, instead of (4.6) or its equivalent form (4.10), the bilevel minimization problem

$$\min_{(\varphi,\theta)\in\Phi\times\Theta} \mathcal{J}(\varphi,\theta) \qquad \text{subject to} \qquad \min_{\varphi\in\Phi} \mathcal{K}(\varphi;\theta,g)\,. \tag{4.30}$$

From (4.22), one sees that this is equivalent to the problem

$$\min_{(\varphi,\theta)\in\Phi\times\Theta} \mathcal{J}(\varphi,\theta) \qquad \text{subject to} \qquad \widetilde{B}_1\varphi + \widetilde{B}_2\theta = \widetilde{g} \quad \text{in } \Phi^*. \tag{4.31}$$

Using a Lagrange multiplier $\mu \in \Phi$ to enforce the constraint in (4.31) we obtain the Euler–Lagrange equations

$$\left\{\begin{array}{ccccccl} A_1\varphi & & & + & \widetilde{B}_1\mu & = A_1\widehat{\varphi} & \text{in } \Phi^*, \\ & & A_2\theta & + & \widetilde{B}_2^*\mu & = 0 & \text{in } \Theta^*, \\ \widetilde{B}_1\varphi & + & \widetilde{B}_2\theta & & & = \widetilde{g}_1 & \text{in } \Phi^* \end{array}\right. \tag{4.32}$$

for the saddle point $\{(\varphi,\theta),\mu\}$.

Problem (4.31) should be contrasted with problem (4.10). Both (4.10) and (4.31) involve the same functional $\mathcal{J}(\cdot,\cdot)$ but are constrained differently. As a result, the former leads to the optimality system (see [8])

$$\left\{\begin{array}{ccccccl} A_1\varphi & & & + & B_1\mu & = A_1\widehat{\varphi} & \text{in } \Phi^*, \\ & & A_2\theta & + & B_2^*\mu & = 0 & \text{in } \Theta^*, \\ B_1\varphi & + & B_2\theta & & & = g & \text{in } \Lambda^*, \end{array}\right. \tag{4.33}$$

while the latter leads to the optimality system (4.32). Although both optimality systems are of saddle point type, their internal structures are significantly different. For example, the operator B_1 that plays a central role in (4.33) may be nonsymmetric and indefinite; on the other hand, the operator $\widetilde{B}_1 = B_1^* D^{-1} B_1$ that plays the analogous role in (4.32) is always symmetric and positive definite whenever the assumptions (4.5) and (4.13) hold.

Penalization can be used to facilitate the solution of the system (4.32). To this end, we let $\widetilde{D} : \Phi \to \Phi^*$ be a self-adjoint, strongly coercive operator; i.e., there exist constants $\widetilde{c}_d > 0$ and $\widetilde{k}_d > 0$ such that

$$\langle \widetilde{D}\mu, \varphi \rangle_{\Phi^*,\Phi} \le \widetilde{c}_d \|\mu\|_\Phi \|\varphi\|_\Phi \qquad \text{and} \qquad \langle \widetilde{D}\mu, \mu \rangle_{\Phi^*,\Phi} \ge \widetilde{k}_d \|\mu\|_\Phi^2 \tag{4.34}$$

for all $\varphi, \mu \in \Phi$. Corresponding to the operator $\widetilde{D}$, we have the symmetric, coercive bilinear form

$$\widetilde{d}(\varphi, \mu) = \langle \widetilde{D}\mu, \varphi \rangle_{\Phi^*,\Phi} \qquad \forall \varphi, \mu \in \Phi .$$

We then consider the penalized functional

$$\widetilde{\mathcal{J}}_\epsilon(\varphi, \theta) = \mathcal{J}(\varphi, \theta) + \frac{1}{\epsilon}\big\langle \widetilde{B}_1\varphi + \widetilde{B}_2\theta - \widetilde{g}_1, \widetilde{D}^{-1}(\widetilde{B}_1\varphi + \widetilde{B}_2\theta - \widetilde{g}_1)\big\rangle_{\Phi^*,\Phi}$$

and the unconstrained optimization problem

$$\min_{\varphi\in\Phi,\, \theta\in\Theta} \widetilde{\mathcal{J}}_\epsilon(\varphi, \theta) . \tag{4.35}$$

The Euler–Lagrange equations corresponding to this problem are given by

$$\begin{cases} \left(A_1 + \dfrac{1}{\epsilon}\widetilde{B}_1\widetilde{D}^{-1}\widetilde{B}_1\right)\varphi_\epsilon + \dfrac{1}{\epsilon}\widetilde{B}_1\widetilde{D}^{-1}\widetilde{B}_2\theta_\epsilon = A_1\widehat{\varphi} + \dfrac{1}{\epsilon}\widetilde{B}_1\widetilde{D}^{-1}\widetilde{g}_1 & \text{in } \Phi^*, \\ \left(A_2 + \dfrac{1}{\epsilon}\widetilde{B}_2^*\widetilde{D}^{-1}\widetilde{B}_2\right)\theta_\epsilon + \dfrac{1}{\epsilon}\widetilde{B}_2^*\widetilde{D}^{-1}\widetilde{B}_1\varphi_\epsilon = \dfrac{1}{\epsilon}\widetilde{B}_2^*\widetilde{D}^{-1}\widetilde{g}_1 & \text{in } \Theta^* \end{cases} \tag{4.36}$$

or

$$\begin{cases} \left(A_1 + \dfrac{1}{\epsilon}B_1^*D^{-1}B_1\widetilde{D}^{-1}B_1^*D^{-1}B_1\right)\varphi_\epsilon & \\ +\dfrac{1}{\epsilon}B_1^*D^{-1}B_1\widetilde{D}^{-1}B_1^*D^{-1}B_2\theta_\epsilon = A_1\widehat{\varphi} + \dfrac{1}{\epsilon}B_1^*D^{-1}B_1\widetilde{D}^{-1}B_1^*D^{-1}g & \text{in } \Phi^*, \\ \left(A_2 + \dfrac{1}{\epsilon}B_2^*D^{-1}B_1\widetilde{D}^{-1}B_1^*D^{-1}B_2\right)\theta_\epsilon & \\ +\dfrac{1}{\epsilon}B_2^*D^{-1}B_1\widetilde{D}^{-1}B_1^*D^{-1}B_1\varphi_\epsilon = \dfrac{1}{\epsilon}B_2^*D^{-1}B_1\widetilde{D}^{-1}B_1^*D^{-1}g & \text{in } \Theta^*. \end{cases} \tag{4.37}$$

Letting $\mu_\epsilon = \widetilde{D}^{-1}(\widetilde{B}_1\varphi_\epsilon + \widetilde{B}_2\theta_\epsilon - \widetilde{g}_1)$, it is easy to see that (4.36) is equivalent to the following regular perturbation of (4.32):

$$\begin{cases} A_1\varphi_\epsilon & & + & \widetilde{B}_1\mu_\epsilon & = & A_1\widehat{\varphi} & \text{in } \Phi^*, \\ & A_2\theta_\epsilon & + & \widetilde{B}_2^T\mu_\epsilon & = & 0 & \text{in } \Theta^*, \\ \widetilde{B}_1\varphi_\epsilon & + \ \widetilde{B}_2\theta_\epsilon & - & \epsilon\widetilde{D}\mu_\epsilon & = & \widetilde{g}_1 & \text{in } \Phi^*. \end{cases} \tag{4.38}$$

The systems (4.36) and (4.38) are equivalent, but their discretizations are not, even if we use the same subspaces $\Phi^h \subset \Phi$ and $\Theta^h \subset \Theta$ to discretize both systems. However, unlike the situation that occurs when one directly penalizes the cost functional with a least-squares functional (see [9]), the discretization of either (4.36) or (4.38) will result in matrix systems (after elimination in the second case) that are uniformly (with respect to h) positive definite without regard to (4.26).

4.4.1 Discretize-Then-Eliminate

Let $\{\varphi_j\}_{j=1}^J$ and $\{\theta_k\}_{k-1}^K$, where $J = \dim(\Phi^h)$ and $K = \dim(\Theta^h)$, denote the chosen basis sets for Φ^h and Θ^h, respectively. In addition to the matrix $\widetilde{\mathbb{B}}_1$ defined previously in (4.25), we define the matrices

$$\begin{cases} (\mathbb{A}_1)_{ij} = a_1(\varphi_i, \varphi_j) & \text{for } i, j = 1, \dots, J, \\ (\mathbb{A}_2)_{k\ell} = a_2(\theta_k, \theta_\ell) & \text{for } k, \ell = 1, \dots, K, \\ (\widetilde{\mathbb{B}}_2)_{jk} = \widetilde{b}_2(\theta_k, \varphi_j) = \left\langle B_2\theta_k, D^{-1}B_1\varphi_j \right\rangle_{\Lambda^*,\Lambda} & \text{for } k = 1, \dots, K,\ j = 1, \dots, J, \\ (\widetilde{\mathbb{D}})_{ij} = \widetilde{d}(\varphi_i, \varphi_j) = \langle \widetilde{D}\varphi_i, \varphi_j \rangle_{\Phi^*,\Phi} & \text{for } i, j = 1, \dots, J \end{cases}$$

and the vectors

$$\begin{cases} (\vec{\mathbf{f}})_j = a_1(\widehat{\varphi}, \varphi_j) & \text{for } j = 1, \dots, J, \\ (\vec{\mathbf{g}}_1)_i = \langle \widetilde{g}_1, \varphi_i \rangle_{\Phi^*,\Phi} = \left\langle B_1\varphi_i, D^{-1}g \right\rangle_{\Lambda^*,\Lambda} & \text{for } k = 1, \dots, K. \end{cases}$$

Then, discretizing the equivalent weak formulation corresponding to (4.38) results in the matrix problem[8]

$$\begin{pmatrix} \mathbb{A}_1 & 0 & \widetilde{\mathbb{B}}_1 \\ 0 & \mathbb{A}_2 & \widetilde{\mathbb{B}}_2^T \\ \widetilde{\mathbb{B}}_1 & \widetilde{\mathbb{B}}_2 & -\epsilon\widetilde{\mathbb{D}} \end{pmatrix} \begin{pmatrix} \vec{\boldsymbol{\varphi}}_\epsilon \\ \vec{\boldsymbol{\theta}}_\epsilon \\ \vec{\boldsymbol{\mu}}_\epsilon \end{pmatrix} = \begin{pmatrix} \vec{\mathbf{f}} \\ \vec{\mathbf{0}} \\ \vec{\mathbf{g}}_1 \end{pmatrix}. \tag{4.40}$$

The system (4.40) is symmetric and indefinite, but it is uniformly (with respect to h) invertible without regard to (4.26). Indeed, we have that the matrices $\widetilde{\mathbb{B}}_1$ and $\mathbb{A}_2$ are symmetric and positive definite whenever (4.3), (4.5), and (4.13) hold.

The vector of coefficients $\vec{\boldsymbol{\mu}}_\epsilon$ may be eliminated from (4.40) to yield

$$\begin{cases} \left(\mathbb{A}_1 + \dfrac{1}{\epsilon}\widetilde{\mathbb{B}}_1\widetilde{\mathbb{D}}^{-1}\widetilde{\mathbb{B}}_1\right)\vec{\boldsymbol{\varphi}}_\epsilon + \dfrac{1}{\epsilon}\widetilde{\mathbb{B}}_1\widetilde{\mathbb{D}}^{-1}\widetilde{\mathbb{B}}_2\vec{\boldsymbol{\theta}}_\epsilon = \vec{\mathbf{f}} + \dfrac{1}{\epsilon}\widetilde{\mathbb{B}}_1\widetilde{\mathbb{D}}^{-1}\vec{\mathbf{g}}_1, \\ \left(\mathbb{A}_2 + \dfrac{1}{\epsilon}\widetilde{\mathbb{B}}_2^T\widetilde{\mathbb{D}}^{-1}\widetilde{\mathbb{B}}_2\right)\vec{\boldsymbol{\theta}}_\epsilon + \dfrac{1}{\epsilon}\widetilde{\mathbb{B}}_2^T\widetilde{\mathbb{D}}^{-1}\widetilde{\mathbb{B}}_1\vec{\boldsymbol{\varphi}}_\epsilon = \dfrac{1}{\epsilon}\widetilde{\mathbb{B}}_2^T\widetilde{\mathbb{D}}^{-1}\vec{\mathbf{g}}_1. \end{cases} \tag{4.41}$$

Theorem 4.8. *Let* (4.3), (4.5), *and* (4.13) *hold. Then,* (4.40) *has a unique solution* $\varphi_\epsilon^h \in \Phi^h$, $\theta_\epsilon^h \in \Theta^h$, *and* $\mu_\epsilon^h \in \Phi^h$. *Moreover, if* $\varphi \in \Phi$, $\theta \in \Theta$, *and* $\mu \in \Phi$ *denotes the unique solution of* (4.30), *or equivalently, of the optimization problem* (4.6), *then there exists a constant* $C > 0$ *whose value is independent of* ϵ *and h such that*

$$\begin{aligned} \|\varphi - \varphi_\epsilon^h\|_\Phi + \|\theta - \theta_\epsilon^h\|_\Theta + \|\mu - \mu_\epsilon^h\|_\Phi &\le C\epsilon\left(\|g\|_{\Lambda^*} + \|\widehat{\varphi}\|_{\widehat{\Phi}}\right) \\ &+ C\Big(\inf_{\widetilde{\varphi}^h\in\Phi^h}\|\varphi_\epsilon - \widetilde{\varphi}^h\|_\Phi + \inf_{\widetilde{\theta}^h\in\Theta^h}\|\theta_\epsilon - \widetilde{\theta}^h\|_\Theta + \inf_{\widetilde{\mu}^h\in\Phi^h}\|\mu_\epsilon - \mu^h\|_\Phi\Big). \end{aligned} \tag{4.42}$$

[8]Discretization of the unperturbed system (4.32) yields the related discrete system

$$\begin{pmatrix} \mathbb{A}_1 & 0 & \widetilde{\mathbb{B}}_1 \\ 0 & \mathbb{A}_2 & \widetilde{\mathbb{B}}_2^T \\ \widetilde{\mathbb{B}}_1 & \widetilde{\mathbb{B}}_2 & 0 \end{pmatrix} \begin{pmatrix} \vec{\boldsymbol{\varphi}} \\ \vec{\boldsymbol{\theta}} \\ \vec{\boldsymbol{\mu}} \end{pmatrix} = \begin{pmatrix} \vec{\mathbf{f}} \\ \vec{\mathbf{0}} \\ \vec{\mathbf{g}}_1 \end{pmatrix}. \tag{4.39}$$

It is usually the case that the approximation-theoretic terms on the right-hand side of (4.42) satisfy inequalities of the type

$$\inf_{\widetilde{\varphi}^h \in \Phi^h} \|\varphi_\epsilon - \widetilde{\varphi}^h\|_\Phi \le C h^\alpha, \qquad \inf_{\widetilde{\mu}^h \in \Phi^h} \|\mu_\epsilon - \widetilde{\mu}^h\|_\Phi \le C h^\alpha,$$

and

$$\inf_{\widetilde{\theta}^h \in \Theta^h} \|\theta_\epsilon - \widetilde{\theta}^h\|_\Theta \le C h^\beta,$$

where $\alpha > 0$ and $\beta > 0$ depend on the degree of the polynomials used for the spaces Φ^h and Θ^h and the regularity of the solution φ_ϵ, θ_ϵ, and μ_ϵ of (4.30). Combining with (4.42) we have that

$$\|\varphi - \varphi_\epsilon^h\|_\Phi + \|\theta - \theta_\epsilon^h\|_\Theta + \|\mu - \mu_\epsilon^h\|_\Phi \le C(\epsilon + h^\alpha + h^\beta) \tag{4.43}$$

so that if $\beta \ge \alpha$ and one chooses $\epsilon = h^\alpha$, one obtains the optimal error estimate

$$\|\varphi - \varphi_\epsilon^h\|_\Phi + \|\theta - \theta_\epsilon^h\|_\Theta + \|\mu - \mu_\epsilon^h\|_\Phi \le C\epsilon = C h^\alpha . \tag{4.44}$$

It is again important to note that the result (4.42) does not require that (4.26) is satisfied and that locking cannot occur.

4.4.2 Eliminate-Then-Discretize

Alternately, one could discretize (4.36) to obtain

$$\begin{pmatrix} \mathbb{A}_1 + \frac{1}{\epsilon}\mathbb{K}_1 & \frac{1}{\epsilon}\mathbb{K}_2 \\ \frac{1}{\epsilon}\mathbb{K}_2^T & \mathbb{A}_2 + \frac{1}{\epsilon}\widetilde{\mathbb{C}} \end{pmatrix} \begin{pmatrix} \vec{\boldsymbol{\varphi}}_\epsilon \\ \vec{\boldsymbol{\theta}}_\epsilon \end{pmatrix} = \begin{pmatrix} \vec{\mathbf{f}} + \frac{1}{\epsilon}\widetilde{\vec{\mathbf{g}}}_1 \\ \frac{1}{\epsilon}\widetilde{\vec{\mathbf{g}}}_2 \end{pmatrix}. \tag{4.45}$$

The matrices $\mathbb{A}_1$ and $\mathbb{A}_2$ and the vector $\vec{\mathbf{f}}$ are defined as before; we also have, in terms of the basis vectors for Φ^h and Θ^h, that

$$\begin{cases} (\mathbb{K}_1)_{ij} = \langle \widetilde{B}_1\varphi_i, \widetilde{D}^{-1}\widetilde{B}_1\varphi_j\rangle_{\Phi^*,\Phi} = \langle B_1^* D^{-1} B_1\varphi_i, \widetilde{D}^{-1} B_1^* D^{-1} B_1 \varphi_j\rangle_{\Phi^*,\Phi}, \\ (\mathbb{K}_2)_{jk} = \langle \widetilde{B}_2\theta_k, \widetilde{D}^{-1}\widetilde{B}_1\varphi_j\rangle_{\Phi^*,\Phi} = \langle B_1^* D^{-1} B_2\theta_k, \widetilde{D}^{-1} B_1^* D^{-1} B_1 \varphi_j\rangle_{\Phi^*,\Phi}, \\ (\widetilde{\mathbb{C}})_{k\ell} = \langle \widetilde{B}_2\theta_k, \widetilde{D}^{-1}\widetilde{B}_2\theta_\ell\rangle_{\Phi^*,\Phi} = \langle B_1^* D^{-1} B_2\theta_k, \widetilde{D}^{-1} B_1^* D^{-1} B_2 \theta_\ell\rangle_{\Phi^*,\Phi}; \end{cases}$$

and

$$\begin{cases} (\widetilde{\vec{\mathbf{g}}}_1)_j = \langle \widetilde{B}_1\varphi_j, \widetilde{D}^{-1}\widetilde{g}_1\rangle_{\Phi^*,\Phi} = \langle B_1^* D^{-1} B_1\varphi_j, \widetilde{D}^{-1} B_1^* D^{-1} g\rangle_{\Phi^*,\Phi}, \\ (\widetilde{\vec{\mathbf{g}}}_2)_k = \langle \widetilde{B}_2\theta_k, \widetilde{D}^{-1}\widetilde{g}_2\rangle_{\Phi^*,\Phi} = \langle B_1^* D^{-1} B_2\theta_k, \widetilde{D}^{-1} B_1^* D^{-1} g\rangle_{\Phi^*,\Phi} . \end{cases}$$

Theorem 4.9. *Let* (4.3), (4.5), *and* (4.13) *hold. Then, for* $0 < \epsilon \le 1$, (4.45) *has a unique solution* $\varphi_\epsilon^h \in \Phi^h$ *and* $\theta_\epsilon^h \in \Theta^h$. *Moreover, if* $\varphi \in \Phi$ *and* $\theta \in \Theta$ *denotes the unique solution of the optimization problem* (4.6), *then there exists a constant* $C > 0$ *whose value is independent of* ϵ *and* h *such that*

$$\begin{aligned} \|\varphi - \varphi_\epsilon^h\|_\Phi + \|\theta - \theta_\epsilon^h\|_\Theta &\le C\epsilon \left(\|g\|_{\Lambda^*} + \|\widehat{\varphi}\|_{\widehat{\Phi}}\right) \\ &+ C\left(1 + \frac{1}{\epsilon}\right)\left(\inf_{\widetilde{\varphi}^h \in \Phi^h} \|\varphi_\epsilon - \widetilde{\varphi}^h\|_\Phi + \inf_{\widetilde{\theta}^h \in \Theta^h} \|\theta_\epsilon - \widetilde{\theta}^h\|_\Theta\right). \end{aligned} \tag{4.46}$$

Clearly, (4.41) and (4.45) are not the same. However, the coefficient matrices of both systems are symmetric and uniformly (with respect to h) positive definite without regards to (4.26).

4.5 Example: Optimization Problems for the Stokes System

We use the same concrete examples as in [8]. Consider the velocity-vorticity-pressure formulation of the Stokes system:[9]

$$\left.\begin{aligned}\nabla\times\boldsymbol{\omega}+\nabla p+\boldsymbol{\theta} &= \mathbf{g}\\ \nabla\cdot\mathbf{u} &= 0\\ \nabla\times\mathbf{u}-\boldsymbol{\omega} &= \mathbf{0}\end{aligned}\right\}\quad \text{in } \Omega,\qquad \mathbf{u}=\mathbf{0}\quad \text{on } \Gamma,\qquad \int_\Omega p\,d\Omega=0 \tag{4.47}$$

and the functionals

$$\text{Case I:}\qquad \mathcal{J}_1(\boldsymbol{\omega},\boldsymbol{\theta})=\frac{1}{2}\int_\Omega|\boldsymbol{\omega}|^2\,d\Omega+\frac{\delta}{2}\int_\Omega|\boldsymbol{\theta}|^2\,d\Omega, \tag{4.48}$$

$$\text{Case II:}\qquad \mathcal{J}_2(\mathbf{u},\boldsymbol{\theta};\widehat{\mathbf{u}})=\frac{1}{2}\int_\Omega|\mathbf{u}-\widehat{\mathbf{u}}|^2\,d\Omega+\frac{\delta}{2}\int_\Omega|\boldsymbol{\theta}|^2\,d\Omega, \tag{4.49}$$

where Ω denotes an open, bounded domain in $\mathcal{R}^s$, $s=2$ or 3, with boundary Γ, $\mathbf{u}$, $\boldsymbol{\omega}$, and p denoting the velocity, vorticity, and pressure fields, respectively, $\boldsymbol{\theta}$ denotes a distributed control, and $\mathbf{g}$ and $\widehat{\mathbf{u}}$ are given functions. The optimization problems we study are to find $(\mathbf{u},\boldsymbol{\omega},p,\boldsymbol{\theta})$ that minimizes either the functional in (4.48) or (4.49), subject to the Stokes system in the form (4.47) being satisfied.

4.5.1 Precise Statement of Optimization Problems

We recall the space $L^2(\Omega)$ of all square integrable functions with norm $\|\cdot\|_0$ and inner product $(\cdot,\cdot)$, the space $L^2_0(\Omega)\equiv\{q\in L^2(\Omega):\int_\Omega p d\Omega=0\}$, the space $H^1(\Omega)\equiv\{v\in L^2(\Omega):\nabla v\in[L^2(\Omega)]^s\}$, and the space $H^1_0(\Omega)\equiv\{v\in H^1(\Omega):v=0 \text{ on } \Gamma\}$. A norm for functions $v\in H^1(\Omega)$ is given by $\|v\|_1\equiv(\|\nabla v\|^2+\|v\|_0^2)^{1/2}$. The dual space of $H^1_0(\Omega)$ is denoted by $H^{-1}(\Omega)$. The inner product in $H^{-1}(\Omega)$ is denoted by $(\cdot,\cdot)_{-1}$. Note that we may define $(\cdot,\cdot)_{-1}=\langle\cdot,(-\Delta)^{-1}(\cdot)\rangle_{\mathbf{H}^{-1}(\Omega),\mathbf{H}^1(\Omega)}=(\cdot,(-\Delta)^{-1}(\cdot))$, where $\Delta:H^1_0(\Omega)\to H^{-1}(\Omega)$ denotes the Laplace operator with respect to Ω with zero Dirichlet boundary conditions along Γ.

The corresponding spaces of vector-valued functions are denoted in boldface; e.g., $\mathbf{H}^1(\Omega)=[H^1(\Omega)]^s$ is the space of vector-valued functions each of whose components belongs to $H^1(\Omega)$. We note the following equivalence of norms [15]:

$$\widetilde{C}_1\|\mathbf{v}\|_1^2\le\|\nabla\times\mathbf{v}\|_0^2+\|\nabla\cdot\mathbf{v}\|_0^2\le\widetilde{C}_2\|\mathbf{v}\|_1^2\qquad\forall\,\mathbf{v}\in\mathbf{H}^1_0(\Omega) \tag{4.50}$$

[9]The reasons for using the velocity-vorticity-pressure formulation of the Stokes equations instead of, say, the standard primitive variable formulation, are discussed in [8].

for some constants $\widetilde{C}_1 > 0$ and $\widetilde{C}_2 > 0$.

Let $\Phi = \Lambda = \mathbf{H}_0^1(\Omega) \times \mathbf{L}^2(\Omega) \times L_0^2(\Omega)$ and $\Theta = \mathbf{L}^2(\Omega)$ so that $\Phi^* = \Lambda^* = \mathbf{H}^{-1}(\Omega) \times \mathbf{L}^2(\Omega) \times L_0^2(\Omega)$ and $\Theta^* = \mathbf{L}^2(\Omega)$. Let $\widehat{\Phi} = \widetilde{\Phi} = \mathbf{L}^2(\Omega) \times \mathbf{L}^2(\Omega) \times L_0^2(\Omega)$. Then, $\Phi \subset \widehat{\Phi} = \widetilde{\Phi} = \widetilde{\Phi}^* \subset \Phi^*$. For $\varphi = \{\mathbf{u}, \boldsymbol{\omega}, p\} \in \Phi$, we define the norm

$$\|\varphi\|_\Phi = \left(\|\mathbf{u}\|_1^2 + \|\boldsymbol{\omega}\|_0^2 + \|p\|_0^2\right)^{1/2}$$

and likewise for the other product spaces.

We make the associations of

$$\begin{array}{llll} \text{trial functions:} & \varphi = \{\mathbf{u}, \boldsymbol{\omega}, p\} \in \Phi, & \theta = \{\boldsymbol{\theta}\} \in \Theta, & \lambda = \{\mathbf{v}, \boldsymbol{\sigma}, q\} \in \Lambda\,, \\ \text{test functions:} & \mu = \{\widetilde{\mathbf{u}}, \widetilde{\boldsymbol{\omega}}, \widetilde{p}\} \in \Phi, & \nu = \{\widetilde{\boldsymbol{\theta}}\} \in \Theta, & \psi = \{\widetilde{\mathbf{v}}, \widetilde{\boldsymbol{\sigma}}, \widetilde{r}\} \in \Lambda\,, \\ \text{data:} & g = \{\mathbf{g}, \mathbf{0}, 0\} \in \Lambda^*, & \widehat{\varphi} = \left\{ \begin{pmatrix} \mathbf{0} \\ \widehat{\mathbf{u}} \end{pmatrix}, \mathbf{0}, 0 \right\} \in \widehat{\Phi} & \begin{array}{l} \text{for Case I} \\ \text{for Case II}\,. \end{array} \end{array}$$

We next define the bilinear forms

$$a_1(\varphi, \mu) = \begin{cases} (\widetilde{\boldsymbol{\omega}}, \boldsymbol{\omega}) & \text{for Case I} \\ (\widetilde{\mathbf{u}}, \mathbf{u}) & \text{for Case II} \end{cases} \qquad \forall\, \varphi = \{\mathbf{u}, \boldsymbol{\omega}, p\} \in \widehat{\Phi}, \quad \mu = \{\widetilde{\mathbf{u}}, \widetilde{\boldsymbol{\omega}}, \widetilde{p}\} \in \widehat{\Phi}\,,$$

$$a_2(\theta, \nu) = \delta(\boldsymbol{\theta}, \widetilde{\boldsymbol{\theta}}) \qquad \forall\, \theta = \{\boldsymbol{\theta}\} \in \Theta, \quad \nu = \{\widetilde{\boldsymbol{\theta}}\} \in \Theta\,,$$

$$b_1(\varphi, \psi) = (\boldsymbol{\omega}, \nabla \times \widetilde{\mathbf{v}}) - (p, \nabla \cdot \widetilde{\mathbf{v}}) + (\nabla \times \mathbf{u} - \boldsymbol{\omega}, \widetilde{\boldsymbol{\sigma}}) - (\nabla \cdot \mathbf{u}, \widetilde{r})$$
$$\forall\, \varphi = \{\mathbf{u}, \boldsymbol{\omega}, p\} \in \Phi, \quad \psi = \{\widetilde{\mathbf{v}}, \widetilde{\boldsymbol{\sigma}}, \widetilde{r}\} \in \Lambda\,,$$

$$b_2(\theta, \psi) = (\boldsymbol{\theta}, \widetilde{\mathbf{v}}) \qquad \forall\, \theta = \{\boldsymbol{\theta}\} \in \Theta, \quad \psi = \{\widetilde{\mathbf{v}}, \widetilde{\boldsymbol{\sigma}}, \widetilde{r}\} \in \Lambda\,.$$

For $\mathbf{g} \in \mathbf{H}^{-1}(\Omega)$, we also define the linear functional

$$\langle g, \psi \rangle_{\Lambda^*, \Lambda} = \langle \mathbf{g}, \widetilde{\mathbf{v}} \rangle_{\mathbf{H}^{-1}(\Omega), \mathbf{H}_0^1(\Omega)} \qquad \forall\, \psi = \{\widetilde{\mathbf{v}}, \widetilde{\boldsymbol{\sigma}}, \widetilde{r}\} \in \Lambda\,.$$

The operators associated with the bilinear forms are then

$$\begin{gathered} A_1 = \begin{pmatrix} 0 & 0 & 0 \\ 0 & I & 0 \\ 0 & 0 & 0 \end{pmatrix} \quad \text{for Case I,} \qquad A_1 = \begin{pmatrix} I & 0 & 0 \\ 0 & 0 & 0 \\ 0 & 0 & 0 \end{pmatrix} \quad \text{for Case II,} \\ A_2 = \delta I, \qquad B_1 = \begin{pmatrix} 0 & \nabla\times & \nabla \\ \nabla\times & -I & 0 \\ -\nabla\cdot & 0 & 0 \end{pmatrix}, \qquad \text{and} \qquad B_2 = \begin{pmatrix} I \\ 0 \\ 0 \end{pmatrix}. \end{gathered} \tag{4.51}$$

Note that $B_1^* = B_1$.

It is now easily seen that the functionals $\mathcal{J}_1(\cdot, \cdot)$ and $\mathcal{J}_2(\cdot, \cdot; \cdot)$ defined in (4.48) and (4.49), respectively, can be written in the form (4.2). Likewise, the Stokes system (4.47) can be written in the form (4.4). Thus, the two optimization problems for the Stokes system can both be written in the form (4.6), with $\mathcal{J}(\cdot, \cdot)$ being either $\mathcal{J}_1(\cdot, \cdot)$ or $\mathcal{J}_2(\cdot, \cdot)$ as appropriate.

Additional Notation Relevant to Example Problems

In Sections 4.3 and 4.4, additional bilinear forms, linear functionals, operators, and functions are introduced. In the context of the optimization problems considered in this section, the bilinear forms (4.17), (4.18), (4.27), (4.11), and (4.34) are, respectively, given by

$$
\begin{aligned}
\widetilde{b}_1\big(\{\mathbf{u},\omega,p\},\{\widetilde{\mathbf{u}},\widetilde{\omega},\widetilde{p}\}\big) &= \Big(\nabla\times\omega+\nabla p\,,\ \nabla\times\widetilde{\omega}+\nabla\widetilde{p}\Big)_{-1} \\
&\quad+\Big(\nabla\times\mathbf{u}-\omega\,,\ \nabla\times\widetilde{\mathbf{u}}-\widetilde{\omega}\Big)+\Big(\nabla\cdot\mathbf{u}\,,\ \nabla\cdot\widetilde{\mathbf{u}}\Big) \\
&= \Big(\nabla\times\omega+\nabla p\,,\ (-\Delta)^{-1}(\nabla\times\widetilde{\omega}+\nabla\widetilde{p})\Big) \\
&\quad+\Big(\nabla\times\mathbf{u}-\omega\,,\ \nabla\times\widetilde{\mathbf{u}}-\widetilde{\omega}\Big)+\Big(\nabla\cdot\mathbf{u}\,,\ \nabla\cdot\widetilde{\mathbf{u}}\Big), \\
\widetilde{b}_2\big(\{\boldsymbol{\theta},\{\widetilde{\mathbf{u}},\widetilde{\omega},\widetilde{p}\}\big) &= \Big(\boldsymbol{\theta}\,,\ \nabla\times\widetilde{\omega}+\nabla\widetilde{p}\Big)_{-1} = \Big(\boldsymbol{\theta}\,,\ (-\Delta)^{-1}(\nabla\times\widetilde{\omega}+\nabla\widetilde{p})\Big),
\end{aligned}
$$

$$
c\big(\{\boldsymbol{\theta}\},\{\widetilde{\boldsymbol{\theta}}\}\big) = \big(\boldsymbol{\theta},\widetilde{\boldsymbol{\theta}}\big)_{-1} = \Big(\widetilde{\boldsymbol{\theta}},(-\Delta)^{-1}\boldsymbol{\theta}\Big),
$$

$$
\begin{aligned}
d\big(\{\mathbf{u},\omega,p\},\{\widetilde{\mathbf{u}},\widetilde{\omega},\widetilde{p}\}\big) &= (\mathbf{u},\widetilde{\mathbf{u}})_{-1}+(\omega,\widetilde{\omega})+(p,\widetilde{p}) \\
&= \big((-\Delta)^{-1}\widetilde{\mathbf{u}}\,,\ \mathbf{u}\big)+(\omega,\widetilde{\omega})+(p,\widetilde{p}),
\end{aligned}
$$

$$
\widetilde{d}\big(\{\mathbf{u},\omega,p\},\{\widetilde{\mathbf{u}},\widetilde{\omega},\widetilde{p}\}\big) = d\big(\{\mathbf{u},\omega,p\},\{\widetilde{\mathbf{u}},\widetilde{\omega},\widetilde{p}\}\big).
$$

We also have the linear functionals

$$
\begin{aligned}
\langle\widetilde{g}_1,\{\widetilde{\mathbf{u}},\widetilde{\omega},\widetilde{p}\}\rangle_{\Phi^*,\Phi} &= \Big(\mathbf{g}\,,\ \nabla\times\widetilde{\omega}+\nabla\widetilde{p}\Big)_{-1} = \Big(\mathbf{g}\,,\ (-\Delta)^{-1}(\nabla\times\widetilde{\omega}+\nabla\widetilde{p})\Big) \\
&= \Big(\nabla\times(-\Delta)^{-1}\mathbf{g},\widetilde{\omega}\Big)-\Big(\nabla\cdot(-\Delta)^{-1}\mathbf{g},\widetilde{p}\Big)
\end{aligned}
$$

$$
\langle\widetilde{g}_2,\{\widetilde{\boldsymbol{\theta}}\}\rangle_{\Theta^*,\Theta} = \big(\mathbf{g},\widetilde{\boldsymbol{\theta}}\big)_{-1} = \Big(\mathbf{g}\,,\ (-\Delta)^{-1}\widetilde{\boldsymbol{\theta}}\Big) = \Big((-\Delta)^{-1}\mathbf{g},\widetilde{\boldsymbol{\theta}}\Big)
$$

and the operators and functions

$$
\widetilde{B}_1 = \begin{pmatrix}
\nabla\times\nabla\times-\nabla\nabla\cdot & -\nabla\times & 0 \\
-\nabla\times & I+\nabla\times(-\Delta)^{-1}\nabla\times & \nabla\times(-\Delta)^{-1}\nabla \\
0 & -\nabla\cdot(-\Delta)^{-1}\nabla\times & -\nabla\cdot(-\Delta)^{-1}\nabla
\end{pmatrix},
$$

$$
\widetilde{B}_2 = \begin{pmatrix} 0 \\ \nabla\times(-\Delta)^{-1} \\ -\nabla\cdot(-\Delta)^{-1} \end{pmatrix}, \qquad D=\widetilde{D}=\begin{pmatrix} -\Delta & 0 & 0 \\ 0 & I & 0 \\ 0 & 0 & I \end{pmatrix},
$$

$$
\widetilde{g}_1 = \begin{pmatrix} \mathbf{0} \\ \nabla\times(-\Delta)^{-1}\mathbf{g} \\ -\nabla\cdot(-\Delta)^{-1}\mathbf{g} \end{pmatrix}, \qquad \text{and} \qquad \widetilde{g}_2 = (-\Delta)^{-1}\mathbf{g}\,.
$$

With these definitions, all of the methods discussed in the preceding sections can be defined for the optimization problems considered in this section.

Verification of Hypotheses

In [8], the following results are proved.

Proposition 4.10. *Let the spaces* Φ, $\widehat{\Phi}$, Θ, *and* Λ *and the bilinear forms* $a_1(\cdot,\cdot)$, $a_2(\cdot,\cdot)$, $b_1(\cdot,\cdot)$, *and* $b_2(\cdot,\cdot)$ *be defined as above. Then, the assumptions* (4.3) *and* (4.5) *are satisfied with* $C_1 = 1$, $C_2 = \delta$, $K_2 = \delta$, $c_1 = 3$, $c_2 = 1$, *and*

$$k_1 = \min\{1, \widetilde{C}_1\} \Big/ \left(2\sqrt{\max\left\{\frac{1}{\widetilde{C}_1}, \widetilde{C}_2\right\}} \right).$$

We also easily have the following result.

Proposition 4.11. *Let the spaces* Φ *and* Λ *and the bilinear forms* $d(\cdot,\cdot)$ *and* $\widetilde{d}(\cdot,\cdot)$ *be defined as above. Then,* (4.13) *and* (4.34) *are satisfied with* $c_d = \widetilde{c}_d = 1$ *and* $k_d = \widetilde{k}_d = \frac{1}{2}\min\{1,\sigma\}$, *where* σ *denotes the smallest eigenvalue of* $-\Delta$ *with respect to* Ω *with zero Dirichlet boundary conditions along* Γ.

Having verified the assumptions (4.3), (4.5), (4.13), and (4.34) in the context of the two optimization problems for the Stokes system, we have that all the results of Sections 4.2–4.4 will apply to those systems.

4.5.2 Least-Squares Formulation of the Constraint Equations

Using the associations of spaces and variables defined in Section 4.5.1 as well as the operators defined there, it is easy to see that the least-squares functional (4.14) is given by, for the example problems we are considering,

$$\begin{aligned}\mathcal{K}(\{\mathbf{u},\boldsymbol{\omega},p\};\boldsymbol{\theta},\mathbf{g}) &= \|\nabla\times\boldsymbol{\omega}+\nabla p+\boldsymbol{\theta}-\mathbf{g}\|_{-1}^2 + \|\nabla\times\mathbf{u}-\boldsymbol{\omega}\|_0^2 + \|\nabla\cdot\mathbf{u}\|_0^2 \\ &= \big(\nabla\times\boldsymbol{\omega}+\nabla p+\boldsymbol{\theta}-\mathbf{g}, (-\Delta)^{-1}(\nabla\times\boldsymbol{\omega}+\nabla p+\boldsymbol{\theta}-\mathbf{g})\big) \\ &\qquad + \|\nabla\times\mathbf{u}-\boldsymbol{\omega}\|_0^2 + \|\nabla\cdot\mathbf{u}\|_0^2 .\end{aligned} \tag{4.52}$$

The Euler–Lagrange equation corresponding to the minimization of the least-squares functional (4.52) is given by (4.16), which now takes the following form: for given $\boldsymbol{\theta}\in\mathbf{L}^2(\Omega)$ and $\mathbf{g}\in\mathbf{H}^{-1}(\Omega)$, find $\{\mathbf{u},\boldsymbol{\omega},p\}\in\mathbf{H}_0^1(\Omega)\times\mathbf{L}^2(\Omega)\times L_0^2(\Omega)$ such that

$$\begin{aligned}\widetilde{b}_1\big(\{\mathbf{u},\boldsymbol{\omega},p\}, \{\widetilde{\mathbf{u}},\widetilde{\boldsymbol{\omega}},\widetilde{p}\}\big) &= \langle \widetilde{g}_1, \{\widetilde{\mathbf{u}},\widetilde{\boldsymbol{\omega}},\widetilde{p}\}\rangle_{\Phi^*,\Phi} - \widetilde{b}_2\big(\{\boldsymbol{\theta}, \{\widetilde{\mathbf{u}},\widetilde{\boldsymbol{\omega}},\widetilde{p}\}\big) \\ &\forall\,\{\widetilde{\mathbf{u}},\widetilde{\boldsymbol{\omega}},\widetilde{p}\}\in\mathbf{H}_0^1(\Omega)\times\mathbf{L}^2(\Omega)\times L_0^2(\Omega).\end{aligned} \tag{4.53}$$

We then have the following result; see [5, 6].

Proposition 4.12. *Let the spaces* Φ, $\widehat{\Phi}$, Θ, *and* Λ *and the bilinear forms* $a_1(\cdot,\cdot)$, $a_2(\cdot,\cdot)$, $b_1(\cdot,\cdot)$, *and* $b_2(\cdot,\cdot)$ *be defined as in Section* 4.5.1. *Let* $\mathcal{K}(\cdot,\cdot,\cdot;\cdot,\cdot)$ *be defined by* (4.52). *Then, we have the norm equivalence result*

$$\gamma_1\big(\|\mathbf{u}\|_1^2 + \|\boldsymbol{\omega}\|_0^2 + \|p\|_0^2\big) \le \mathcal{K}(\mathbf{u},\boldsymbol{\omega},p;\mathbf{0},\mathbf{0}) \le \gamma_2\big(\|\mathbf{u}\|_1^2 + \|\boldsymbol{\omega}\|_0^2 + \|p\|_0^2\big) \qquad (4.54)$$

for some constants $\gamma_1, \gamma_2 > 0$. *Moreover, the bilinear form* $\widetilde{b}_1(\{\cdot,\cdot,\cdot\},\{\cdot,\cdot,\cdot\})$ *defined in Section* 4.5.1 *is symmetric, continuous, and coercive and problem* (4.53) *has a unique solution; that solution is the unique minimizer of the least-squares functional* (4.52).

Proof. The results follow immediately from Corollary 4.4 and Propositions 4.3 and 4.10. □

To define least-squares finite element approximations of the constraint equations, we first choose conforming finite element subspaces $\mathbf{V}^h \subset \mathbf{H}_0^1(\Omega)$, $\mathbf{W}^h \subset \mathbf{L}^2(\Omega)$, and $S^h \subset L_0^2(\Omega)$. We then minimize the functional in (4.52) over the subspaces, or equivalently, solve the following problem: for given $\boldsymbol{\theta} \in \mathbf{L}^2(\Omega)$ and $\mathbf{g} \in \mathbf{H}^{-1}(\Omega)$, find $\{\mathbf{u}^h, \boldsymbol{\omega}^h, p^h\} \in \mathbf{V}^h \times \mathbf{W}^h \times S^h$ such that

$$\begin{aligned} &\widetilde{b}_1\big(\{\mathbf{u}^h,\boldsymbol{\omega}^h,p^h\},\ \{\widetilde{\mathbf{u}}^h,\widetilde{\boldsymbol{\omega}}^h,\widetilde{p}\}^h\big) \\ &\qquad = \big\langle \widetilde{g}_1, \{\widetilde{\mathbf{u}}^h,\widetilde{\boldsymbol{\omega}}^h,\widetilde{p}^h\}\big\rangle_{\Phi^*,\Phi} - \widetilde{b}_2\big(\{\boldsymbol{\theta},\{\widetilde{\mathbf{u}}^h,\widetilde{\boldsymbol{\omega}}^h,\widetilde{p}^h\}\big) \qquad (4.55)\\ &\qquad\qquad \forall\, \{\widetilde{\mathbf{u}}^h,\widetilde{\boldsymbol{\omega}}^h,\widetilde{p}\}^h \in \mathbf{V}^h\times\mathbf{W}^h\times S^h\,. \end{aligned}$$

We then have the following result.

Proposition 4.13. *Let* $\mathbf{V}^h \subset \mathbf{H}_0^1(\Omega)$, $\mathbf{W}^h \subset \mathbf{L}^2(\Omega)$, *and* $S^h \subset L_0^2(\Omega)$. *Then, the discrete problem* (4.55) *has a unique solution* $\{\mathbf{u}^h,\boldsymbol{\omega}^h,p^h\} \in \mathbf{V}^h\times\mathbf{W}^h\times S^h$. *Let* $\{\mathbf{u},\boldsymbol{\omega},p\} \in \mathbf{H}_0^1(\Omega)\times\mathbf{L}^2(\Omega)\times L_0^2(\Omega)$ *denote the unique solution of* (4.53). *Then,*

$$\begin{aligned} &\|\mathbf{u}-\mathbf{u}^h\|_1 + \|\boldsymbol{\omega}-\boldsymbol{\omega}^h\|_0 + \|p-p^h\|_0 \\ &\qquad \le C\Big(\inf_{\widetilde{\mathbf{u}}^h\in\mathbf{V}^h}\|\mathbf{u}-\widetilde{\mathbf{u}}^h\|_1 + \inf_{\widetilde{\boldsymbol{\omega}}^h\in\mathbf{W}^h}\|\boldsymbol{\omega}-\widetilde{\boldsymbol{\omega}}^h\|_0 + \inf_{\widetilde{p}^h\in S^h}\|p-\widetilde{p}^h\|_0\Big). \end{aligned} \qquad (4.56)$$

Proof. The results follow in a straightforward manner from Proposition 4.12. □

4.5.3 Discrete Systems for the Stokes Control Problem

One can also, in a straightforward manner, use the notation introduced in Section 4.5.1 to define the concrete discrete systems that correspond to the methods introduced in Section 4.4. Here, we consider, in the context of the Stokes equations, the discrete system defined by (4.41).

Let $\{\mathbf{u}_j\}_{j=1}^{J_1}$, $\{\boldsymbol{\omega}_j\}_{j=1}^{J_2}$, $\{p_j\}_{j=1}^{J_3}$, and $\{\boldsymbol{\theta}_k\}_{k=1}^{K}$, respectively, denote basis sets for the finite element spaces $\mathbf{V}^h \subset \mathbf{H}_0^1(\Omega)$, $\mathbf{W}^h \subset \mathbf{L}^2(\Omega)$, $S^h \subset L_0^2(\Omega)$, and $\boldsymbol{\Theta}^h \subset \mathbf{L}^2(\Omega)$, where $\boldsymbol{\Theta}^h$ is the finite element space introduced to approximate the control $\boldsymbol{\theta}$. Then, using the definitions given in Section 4.5.1 for the associated bilinear forms and linear functionals, the matrices and vectors appearing in (4.41) are given by

$$\mathbb{A}_1 = \begin{pmatrix} \mathbb{A}_{1,11} & 0 & 0 \\ 0 & \mathbb{A}_{1,22} & 0 \\ 0 & 0 & 0 \end{pmatrix}, \quad \widetilde{\mathbb{B}}_1 = \begin{pmatrix} \widetilde{\mathbb{B}}_{1,11} & \widetilde{\mathbb{B}}_{1,12} & 0 \\ \widetilde{\mathbb{B}}^T_{1,12} & \widetilde{\mathbb{B}}_{1,22} & \widetilde{\mathbb{B}}_{1,23} \\ 0 & \widetilde{\mathbb{B}}^T_{1,23} & \widetilde{\mathbb{B}}_{1,33} \end{pmatrix}, \quad \widetilde{\mathbb{B}}_2 = \begin{pmatrix} 0 \\ \widetilde{\mathbb{B}}_{2,21} \\ \widetilde{\mathbb{B}}_{2,31} \end{pmatrix},$$

$$\widetilde{\mathbb{D}} = \begin{pmatrix} \widetilde{\mathbb{D}}_{,11} & 0 & 0 \\ 0 & \mathbb{I} & 0 \\ 0 & 0 & \mathbb{I} \end{pmatrix}, \quad \vec{\mathbf{f}} = \begin{pmatrix} \vec{\mathbf{f}}_{,1} \\ \vec{\mathbf{0}} \\ \vec{\mathbf{0}} \end{pmatrix}, \quad \text{and} \quad \vec{\mathbf{g}}_1 = \begin{pmatrix} \vec{\mathbf{0}} \\ \vec{\mathbf{g}}_{1,2} \\ \vec{\mathbf{g}}_{1,3} \end{pmatrix},$$

where

$$\left(\mathbb{A}_{1,11}\right)_{ij} = \begin{cases} 0 & \text{(Case I)} \\ \left(\mathbf{u}_j\,,\,\mathbf{u}_i\right) & \text{(Case II)} \end{cases} \qquad \text{for } i,j = 1,\ldots,J_1,$$

$$\left(\mathbb{A}_{1,22}\right)_{ij} = \begin{cases} \left(\boldsymbol{\omega}_j\,,\,\boldsymbol{\omega}_i\right) & \text{(Case I)} \\ 0 & \text{(Case II)} \end{cases} \qquad \text{for } i,j = 1,\ldots,J_2,$$

$$\left(\mathbb{A}_2\right)_{k\ell} = \delta\left(\boldsymbol{\theta}_\ell\,,\,\boldsymbol{\theta}_k\right) \qquad \text{for } k,\ell = 1,\ldots,K,$$

$$\left(\widetilde{\mathbb{B}}_{1,11}\right)_{ij} = \left(\nabla\times\mathbf{u}_j\,,\,\nabla\times\mathbf{u}_i\right) + \left(\nabla\cdot\mathbf{u}_j\,,\,\nabla\cdot\mathbf{u}_i\right) \quad \text{for } i,j = 1,\ldots,J_1,$$

$$\left(\widetilde{\mathbb{B}}_{1,12}\right)_{ij} = -\left(\boldsymbol{\omega}_j\,,\,\nabla\times\mathbf{u}_i\right) \qquad \text{for } i = 1,\ldots,J_1,\ j = 1,\ldots,J_2,$$

$$\left(\widetilde{\mathbb{B}}_{1,22}\right)_{ij} = \left(\nabla\times\boldsymbol{\omega}_j\,,\,\nabla\times\boldsymbol{\omega}_i\right)_{-1} + \left(\boldsymbol{\omega}_j\,,\,\boldsymbol{\omega}_i\right) \qquad \text{for } i,j = 1,\ldots,J_2,$$

$$\left(\widetilde{\mathbb{B}}_{1,23}\right)_{ij} = \left(\nabla p_j\,,\,\nabla\times\boldsymbol{\omega}_i\right) \qquad \text{for } i = 1,\ldots,J_2,\ j = 1,\ldots,J_3,$$

$$\left(\widetilde{\mathbb{B}}_{1,33}\right)_{ij} = \left(\nabla p_j\,,\,\nabla p_i\right)_{-1} \qquad \text{for } i,j = 1,\ldots,J_3,$$

$$\left(\widetilde{\mathbb{B}}_{2,21}\right)_{jk} = \left(\boldsymbol{\theta}_k\,,\,\nabla\times\boldsymbol{\omega}_j\right) \qquad \text{for } j = 1,\ldots,J_2,\ k = 1,\ldots,K,$$

$$\left(\widetilde{\mathbb{B}}_{2,31}\right)_{jk} = \left(\boldsymbol{\theta}_k\,,\,\nabla p_j\right) \qquad \text{for } j = 1,\ldots,J_3,\ k = 1,\ldots,K,$$

$$\left(\widetilde{\mathbb{D}}_{,11}\right)_{ij} = \left(\nabla\mathbf{u}_j\,,\,\nabla\mathbf{u}_i\right) \qquad \text{for } i,j = 1,\ldots,J_1,$$

$$\left(\vec{\mathbf{f}}_{,1}\right)_i = \begin{cases} \vec{\mathbf{0}} & \text{(Case I)} \\ \left(\widehat{\mathbf{u}}\,,\,\mathbf{u}_i\right) & \text{(Case II)} \end{cases} \qquad \text{for } i = 1,\ldots,J_1,$$

$$\left(\vec{\mathbf{g}}_{1,2}\right)_i = \left(\mathbf{g}\,,\,\nabla\times\boldsymbol{\omega}_i\right)_{-1} \qquad \text{for } i = 1,\ldots,J_2,$$

$$\left(\vec{\mathbf{g}}_{1,3}\right)_i = \left(\mathbf{g}\,,\,\nabla p_i\right)_{-1} \qquad \text{for } i = 1,\ldots,J_3.$$

In this way, the matrix system (4.41) is completely defined.

Theorem 4.8 implies that if one uses continuous finite element spaces of degree r for all variables and if one chooses $\epsilon = h^r$, then

$$\|\mathbf{u} - \mathbf{u}_\epsilon^h\|_1 + \|\boldsymbol{\omega} - \boldsymbol{\omega}_\epsilon^h\|_0 + \|p - p_\epsilon^h\|_0 + \|\boldsymbol{\theta} - \boldsymbol{\theta}_\epsilon^h\|_0 = O(h^r)$$

provided that the solution of the optimization problems we are considering in this section are sufficiently smooth. One could use finite element spaces of one degree lower for the approximations of the vorticity $\boldsymbol{\omega}$, the pressure p, and the control $\boldsymbol{\theta}$ than that used for the velocity $\mathbf{u}$ and still obtain the same error estimate. However, one of the strengths of using least-squares finite element methods is that one can use any conforming finite element spaces and, in particular, one can use the same degree finite element spaces for all variables; we see that the methods introduced here inherit this strength, since they do not require the satisfaction of the discrete stability conditions in (4.26).

4.5.4 Some Practical Issues Arising in Implementations

One difficulty that arises in the implementation discussed in Section 4.5.3 is caused by the appearance of the $\mathbf{H}^{-1}(\Omega)$-norm in the least-squares functional (4.52). For example, this leads to the appearance of the $\mathbf{H}^{-1}(\Omega)$ inner product in the definition of the matrices and vectors that form the discrete system; see Section 4.5.3. The equivalence relation $(\cdot,\cdot)_{-1} = (\cdot, (-\Delta)^{-1}\cdot)$ is not of much help, since, in general, one cannot exactly invert the Laplace operator, even in the case of zero Dirichlet boundary conditions. Fortunately, there are several approaches available for ameliorating this difficulty; these are discussed in [8]; see also [3, 11, 12]. All the approaches discussed in [8] can be applied to the methods introduced in this chapter, with similar comparative effectiveness; thus, here, we do not consider this issue any further.

A second issue that needs to be discussed is the conditioning of the discrete systems. Actually, there are two issues here, i.e., the conditioning with respect to either h as $h \to 0$ or with respect to ϵ as $\epsilon \to 0$. First, let us discuss the $h \to 0$ issue. Least-squares finite element methods typically result in a "squaring" of operators, e.g., the normal equations in the linear algebra context. This is clearly indicated in (4.22) and (4.23) where one sees that the operator $\widetilde{B}_1$ that results from applying the least-squares principle (4.15) to the constraint equations involves the product of the operators B_1^* and B_1. It is well known that "squaring" operators can result in the squaring of the condition number of the corresponding matrices one obtains after discretization. This is the principal reason for using first-order formulations of the constraint equations, as was done for the Stokes equations in Section 4.5.1. The idea here is that after "squaring" first-order operators, one obtains a second-order operator so that the h-condition number of the resulting squared system is hopefully similar to that for Galerkin formulations of second-order equations. However, penalty formulations of optimal control problems can result in a second "squaring" of operators. For example, look at (4.37); we see there operators such as $B_1^* D^{-1} B_1 \widetilde{D}^{-1} B_1^* D^{-1} B_1$ which involves four copies of the operator B_1. However, that is not the whole story; that operator also involves two copies of the operator D^{-1} and also the operator $\widetilde{D}^{-1}$. Given the nature of all these operators, it is not at all clear that the h-condition number of the discrete systems of Sections 4.4 and 4.5.3 are similar to those that result from a naive double "squaring" of first-order operators; indeed, norm equivalence relations such as (4.21) and (4.54) can sometimes be used to show that h-condition numbers for least-squares-based methods are no worse than those for Galerkin-based methods.

The situation regarding the conditioning of the discrete systems as $\epsilon \to 0$ is problematic for all penalty methods, even for those for which locking does not occur. Note that to obtain a result such as (4.44), one chooses $\epsilon = h^\alpha$; with such a choice, ϵ is likely to be small. This situation can be greatly ameliorated by introducing an *iterated* penalty method; see, e.g., [14] and also [13, 16, 17]. To this end, let $\{\vec{\boldsymbol{\varphi}}_\epsilon, \vec{\boldsymbol{\theta}}_\epsilon, \vec{\boldsymbol{\mu}}_\epsilon\}$ denote the solution of (4.40) and set $\vec{\boldsymbol{\varphi}}^{(0)} = \vec{\boldsymbol{\varphi}}_\epsilon$, $\vec{\boldsymbol{\theta}}^{(0)} = \vec{\boldsymbol{\theta}}_\epsilon$, and $\vec{\boldsymbol{\mu}}^{(0)} = \vec{\boldsymbol{\mu}}_\epsilon$. Then, for $n \geq 1$, we solve the sequence of problems

$$\begin{pmatrix} \mathbb{A}_1 & 0 & \widetilde{\mathbb{B}}_1 \\ 0 & \mathbb{A}_2 & \widetilde{\mathbb{B}}_2^T \\ \widetilde{\mathbb{B}}_1 & \widetilde{\mathbb{B}}_2 & -\epsilon\widetilde{\mathbb{D}} \end{pmatrix} \begin{pmatrix} \vec{\boldsymbol{\varphi}}^{(n)} \\ \vec{\boldsymbol{\theta}}^{(n)} \\ \vec{\boldsymbol{\mu}}^{(n)} \end{pmatrix} = \begin{pmatrix} \vec{\mathbf{0}} \\ \vec{\mathbf{0}} \\ -\epsilon\widetilde{\mathbb{D}}\vec{\boldsymbol{\mu}}^{(n-1)} \end{pmatrix}. \tag{4.57}$$

Then, for any $N > 0$, we let

$$\vec{\boldsymbol{\varphi}}_{\epsilon,N} = \sum_{n=0}^{N} \vec{\boldsymbol{\varphi}}^{(n)}, \qquad \vec{\boldsymbol{\theta}}_{\epsilon,N} = \sum_{n=0}^{N} \vec{\boldsymbol{\theta}}^{(n)}, \qquad \text{and} \qquad \vec{\boldsymbol{\mu}}_{\epsilon,N} = \sum_{n=0}^{N} \vec{\boldsymbol{\mu}}^{(n)} \tag{4.58}$$

and we let $\varphi^h_{\epsilon,N} \in \Phi^h$, $\theta^h_{\epsilon,N} \in \Theta^h$, and $\mu^h_{\epsilon,N} \in \Phi^h$ be the finite element functions corresponding to the coefficients collected in the respective vectors in (4.58). Then, instead of the estimate (4.43), one obtains the estimate (see, e.g., [14] and also [13, 16])

$$\|\varphi - \varphi^h_{\epsilon,N}\|_\Phi + \|\theta - \theta^h_{\epsilon,N}\|_\Theta + \|\mu - \mu^h_{\epsilon,N}\|_\Phi \leq C(\epsilon^{N+1} + h^\alpha + h^\beta)$$

so that if $\beta \geq \alpha$ and one chooses $\epsilon = h^{\alpha/N+1}$, one obtains the optimal error estimate

$$\|\varphi - \varphi^h_{\epsilon,N}\|_\Phi + \|\theta - \theta^h_{\epsilon,N}\|_\Theta + \|\mu - \mu^h_{\epsilon,N}\|_\Phi \leq C\epsilon^{N+1} = Ch^\alpha$$

instead of (4.44). These estimates tell us that we make the error due to penalization as small as we want in two ways: we can choose either ϵ sufficiently small or N sufficiently large. Making the former choice, e.g., choosing $N = 0$ and $\epsilon = h^\alpha$, can lead to conditioning problems for the discrete systems, since $\epsilon \ll 1$. Making the latter choice allows us to obtain the same effect but with a much larger value for ϵ.

Note that $\vec{\boldsymbol{\mu}}^{(n)}$ may be eliminated from (4.57) to yield a reduced system with fewer unknowns. Thus, the iteration to compute the pairs $\{\vec{\boldsymbol{\varphi}}^{(n)}, \vec{\boldsymbol{\theta}}^{(n)}\}$ for $n = 0, 1, \ldots,$ using reduced systems proceeds as follows. Let $\vec{\boldsymbol{\varphi}}^{(0)} = \vec{\boldsymbol{\varphi}}_\epsilon$ and $\vec{\boldsymbol{\theta}}^{(0)} = \vec{\boldsymbol{\theta}}_\epsilon$, where $\vec{\boldsymbol{\varphi}}_\epsilon$ and $\vec{\boldsymbol{\theta}}_\epsilon$ denote the solution of (4.41), and then set

$$\vec{\mathbf{g}}^{(0)} = \widetilde{\mathbb{B}}_1\vec{\boldsymbol{\varphi}}^{(0)} + \widetilde{\mathbb{B}}_2\vec{\boldsymbol{\theta}}^{(0)} - \vec{\mathbf{g}}_1 .$$

Then, for $n = 1, 2, \ldots,$ solve the systems

$$\begin{cases} \left(\mathbb{A}_1 + \dfrac{1}{\epsilon}\widetilde{\mathbb{B}}_1\widetilde{\mathbb{D}}^{-1}\widetilde{\mathbb{B}}_1\right)\vec{\boldsymbol{\varphi}}^{(n)} + \dfrac{1}{\epsilon}\widetilde{\mathbb{B}}_1\widetilde{\mathbb{D}}^{-1}\widetilde{\mathbb{B}}_2\vec{\boldsymbol{\theta}}^{(n)} = \dfrac{1}{\epsilon}\widetilde{\mathbb{B}}_1\widetilde{\mathbb{D}}^{-1}\vec{\mathbf{g}}^{(n-1)}, \\ \left(\mathbb{A}_2 + \dfrac{1}{\epsilon}\widetilde{\mathbb{B}}_2^T\widetilde{\mathbb{D}}^{-1}\widetilde{\mathbb{B}}_2\right)\vec{\boldsymbol{\varphi}}^{(n)} + \dfrac{1}{\epsilon}\widetilde{\mathbb{B}}_2^T\widetilde{\mathbb{D}}^{-1}\widetilde{\mathbb{B}}_1\vec{\boldsymbol{\theta}}^{(n)} = \dfrac{1}{\epsilon}\widetilde{\mathbb{B}}_2^T\widetilde{\mathbb{D}}^{-1}\vec{\mathbf{g}}^{(n-1)}. \end{cases}$$

In order to define the next iterate, we set

$$\vec{\mathbf{g}}^{(n)} = \vec{\mathbf{g}}^{(n-1)} + \widetilde{\mathbb{B}}_1 \vec{\boldsymbol{\varphi}}^{(n)} + \widetilde{\mathbb{B}}_2 \vec{\boldsymbol{\theta}}^{(n)}.$$

4.6 Conclusions

In this chapter, a new least-squares method for optimization and control problems was formulated and analyzed. Instead of the more direct approach of penalizing the cost functional by least-squares terms, they are used to reformulate the constraint. This leads to a bilevel minimization problem with a number of attractive computational properties.

Most notably, the new method preserves the desirable features of least-squares methods such as being able to use discretization spaces for the state variables that are not subject to inf-sup stability conditions. The optimality system of the bilevel optimization problem can be solved by penalty methods without locking; this opens up possibilities for the design of more efficient solution methods for optimization and control problems constrained by PDEs.

Acknowledgment

Partial support by CSRI, Sandia National Laboratories, under contract 18407 is acknowledged.

Bibliography

[1] D. Bedivan and G. J. Fix, *Least-squares methods for optimal shape design problems*, Comput. Math. Appl., 30 (1995), pp. 17–25.

[2] P. Bochev, *Least-squares methods for optimal control*, Nonlinear Anal., 30 (1997), pp. 1875–1885.

[3] P. Bochev, *Negative norm least-squares methods for the velocity-vorticity-pressure Navier-Stokes equations*, Numer. Methods Partial Differential Equations, 15 (1999) pp. 237–256.

[4] P. Bochev and D. Bedivan, *Least-squares methods for Navier-Stokes boundary control problems*, Int. J. Comput. Fluid Dyn., 9 (1997), pp. 43–58.

[5] P. B. Bochev and M. D. Gunzburger, *Finite element methods of least-squares type*, SIAM Rev., 40 (1998), pp. 789–837.

[6] P. Bochev and M. Gunzburger, *Analysis of least-squares finite element methods for the Stokes equations*, Math. Comp., 63 (1994), pp. 479–506.

[7] P. Bochev and M. Gunzburger, *Least-squares finite-element methods for optimization and control problems for the Stokes equations*, Comput. Math. Appl., 48 (2004), pp. 1035–1057.

[8] P. Bochev and M. D. Gunzburger, *Least-squares finite element methods for optimality systems arising in optimization and control problems*, SIAM J. Numer. Anal., 43 (2006), pp. 2517–2543.

[9] P. Bochev and M. Gunzburger, *On least-squares variational principles for the discretization of optimization and control problems*, Methods Appl. Anal., 12 (2005), pp. 395–426.

[10] D. Braess, *Finite Elements*, Cambridge University Press, Cambridge, UK, 1997.

[11] J. Bramble, R. Lazarov, and J. Pasciak, *A Least Squares Approach Based on a Discrete Minus One Inner Product for First Order systems*, Technical report 94-32, Mathematical Science Institute, Cornell University, Ithaca, NY, 1994.

[12] J. Bramble and J. Pasciak, *Least-squares methods for Stokes equations based on a discrete minus one inner product*, J. Comput. Appl. Math., 74 (1996), pp. 155–173.

[13] F. Brezzi and M. Fortin, *Mixed and Hybrid Finite Element Methods*, Springer, New York, 1991.

[14] M. Fortin and A. Fortin, *A generalization of Uzawa's algorithm for the solution of the Navier-Stokes equations*, Appl. Numer. Meth., 1 (1985), pp. 205–208.

[15] V. Girault and P.-A. Raviart, *Finite Element Methods for Navier-Stokes Equations*, Springer-Verlag, Berlin, 1986.

[16] M. Gunzburger, *Iterative penalty methods for the Stokes and Navier-Stokes equations*, in Finite Element Analysis in Fluids, University of Alabama, Huntsville, AL, 1989, pp. 1040–1045.

[17] M. Gunzburger, *Finite Element Methods for Viscous Incompressible Flows*, Academic Press, Boston, 1989.

[18] M. Gunzburger and H. C. Lee, *Analysis and approximation of optimal control problems for first-order elliptic systems in three dimensions*, Appl. Math. Comp., 100 (1999), pp. 49–70.

[19] M. Gunzburger and H. C. Lee, *A penalty/least-squares method for optimal control problems for first-order elliptic systems*, Appl. Math. Comput., 107 (2000), pp. 57–75.

[20] H. Schlichting and K. Gersten, *Boundary Layer Theory*, Springer-Verlag, Berlin, 2000.

Part II

Fast PDE-Constrained Optimization Solvers

Chapter 5

Space-Time Multigrid Methods for Solving Unsteady Optimal Control Problems

*Alfio Borzì**

5.1 Introduction

In this chapter, recent developments [3, 4, 5, 6, 7, 9] on multigrid methods for time-dependent optimal control problems [18, 19, 21] and further results towards real-time optimal control of reaction-diffusion systems are presented. Real-time control means that the algorithm must guarantee a control response to an external event within a given time. To reach this objective, it seems mandatory to ask for optimal control algorithms with optimal computational complexity, such as the multigrid methods discussed here.

Control of reaction-diffusion systems has many applications in biology [5, 13], chemistry [6, 17], and physiology [1, 22]. To solve the related optimal control problems, the solution of the corresponding optimality systems which are governed by reaction-diffusion equations with opposite time orientation is considered. Of particular importance in applications and for benchmark purposes are singular optimal control problems [19] characterized by state equations with possible blowup at finite time. Especially in these cases, the coupling between state variables and controls must be realized in a robust way in order to guarantee convergence of the algorithms. Robust coupling is also necessary for designing algorithms that permit efficient solution to optimization problems with a computational performance that is independent of the value of the optimization parameters. The results presented in this chapter and in the references mentioned above show that the space-time multigrid methods considered here do meet these requirements. These objectives are achieved by using appropriate smoothing techniques and by solving optimality systems for distributed control

*Institut für Mathematik und Wissenschaftliches Rechnen, Karl-Franzens-Universität Graz, Heinrichstr. 36, A-8010 Graz, Austria.

or boundary control in one shot in the whole space-time cylinder.

Our approach represents an extension of space-time multigrid schemes for parabolic problems [12, 14, 24] to the case of reaction-diffusion systems with opposite time orientation. For this particular structure, two different smoothing schemes in combination with semicoarsening in space are considered. Other coarsening strategies are possible; see [3, 4]. In the case of tracking along trajectories a pointwise relaxation is presented which can be successfully applied [3, 7, 9] to solve singular parabolic optimal control problems. In the case of terminal observation, block relaxation is the most robust choice. Block smoothing is also advantageous in the case of reaction-diffusion problems with very small diffusion as it occurs in the chemical turbulence modeling; see [5, 6] and the references given there.

The resulting multigrid schemes show fast mesh independent convergence and robustness with respect to the weights of the cost functional. These features are confirmed by results of local Fourier analysis presented here and in a different form in [4, 5].

A disadvantage of any space-time approach is the requirement of storing the dependent variables for all time steps. This is certainly a limitation that arises when open-loop optimal control problems on a large time interval are considered. However, in the limiting case of very long time intervals, we overcome this difficulty by combining space-time multigrid schemes with receding horizon techniques [2, 15].

In the following section, reaction-diffusion optimal control problems and their approximation by finite differences and backward Euler schemes are introduced. To solve the resulting discrete optimality systems, space-time multigrid schemes formulated in the full approximation storage (FAS) framework [10] are defined in Section 5.3. This section is mostly dedicated to the derivation of two space-time smoothers and is completed with the formulation of a multigrid receding horizon algorithm. The smoothing property of the two iterative schemes is analyzed in Section 5.4 by means of local Fourier analysis. Section 5.5 is dedicated to numerical experiments for validating the numerical performance of the space-time multigrid algorithms. Typical multigrid convergence rates and robustness with respect to a large choice of optimization parameters are obtained. In the last section, the optimal control of cardiac arrhythmia is proposed. Also in this case, results of experiments suggest that the multigrid receding horizon method is a promising approach towards the development of real-time optimal control schemes.

5.2 Reaction-Diffusion Optimal Control Problems and Their Approximation

Depending on the application, reaction-diffusion processes can be controlled through source terms or through boundary terms. In the case of distributed control through source terms, the following optimal control problem is formulated:

$$\begin{cases} \min\limits_{u \in L^2(Q)} \; J(y,u), & & \\ -\partial_t y + G(y) + \sigma\,\Delta y &= u & \text{in } Q, \\ y &= y_0 & \text{in } \Omega \times \{t=0\}, \\ y &= 0 & \text{on } \Sigma, \end{cases} \tag{5.1}$$

where $Q = \Omega \times (0,T)$ and $\Sigma = \partial\Omega \times (0,T)$. To be specific we take $y_0 \in H_0^1(\Omega)$. In (5.1), the nonlinear term $G(y)$ models the reaction kinetics for the state y and u represents the control function. $\sigma > 0$ is the diffusion coefficient.

Alternatively, we consider the following optimal Neumann boundary control problem:

$$\left\{\begin{array}{rcll} \min_{u\in L^2(\Sigma)} & J(y,u), & & \\ -\partial_t y + G(y) + \sigma\,\Delta y & = & 0 & \text{in } Q, \\ y & = & y_0 & \text{in } \Omega\times\{t=0\}, \\ -\frac{\partial y}{\partial n} & = & u & \text{on } \Sigma. \end{array}\right. \tag{5.2}$$

Control may be required to track a desired trajectory given by $y_d(\mathbf{x},t) \in L^2(Q)$ or to reach a desired terminal state $y_T(\mathbf{x}) \in L^2(\Omega)$. For this purpose, the following cost functional is considered:

$$J(y,u) = \frac{\alpha}{2}||y - y_d||^2_{L^2(Q)} + \frac{\beta}{2}||y(\cdot,T) - y_T||^2_{L^2(\Omega)} + \frac{\nu}{2}||u||^2_{L^2(X)}, \tag{5.3}$$

where $X = Q$ or $X = \Sigma$ corresponds to distributed control or boundary control, respectively. Here $\nu > 0$ is the weight of the cost of the control and $\alpha \geq 0$, $\beta \geq 0$ are optimization parameters. For example, the case $\alpha = 1$, $\beta = 0$ corresponds to tracking without terminal observation.

Existence of solutions to the optimal control problems (5.1) and (5.2) can be established under suitable conditions for various forms of the nonlinearity; see, e.g., [7, 11, 16, 19, 21].

The solution to (5.1) is characterized by the following first-order optimality system:

$$\begin{array}{rl} -\partial_t y + G(y) + \sigma\Delta y = u & \text{in } Q, \\ \partial_t p + G'(y)p + \sigma\Delta p + \alpha(y - y_d) = 0 & \text{in } Q, \\ \nu u - p = 0 & \text{in } Q, \\ y = 0,\ p = 0 & \text{on } \Sigma. \end{array} \tag{5.4}$$

In the case of boundary control (5.2), the optimal solution satisfies the following:

$$\begin{array}{rl} -\partial_t y + G(y) + \sigma\Delta y = 0 & \text{in } Q, \\ \partial_t p + G'(y)p + \sigma\Delta p + \alpha(y - y_d) = 0 & \text{in } Q, \\ \nu u - p = 0 & \text{on } \Sigma, \\ -\dfrac{\partial y}{\partial n} = u, \quad -\dfrac{\partial p}{\partial n} = 0 & \text{on } \Sigma. \end{array} \tag{5.5}$$

In both cases we have the initial condition $y(\mathbf{x},0) = y_0(\mathbf{x})$ for the state variable (evolving forward in time). The terminal condition for the adjoint variable (evolving backward in time) is given by

$$p(\mathbf{x},T) = \beta(y(\mathbf{x},T) - y_T(\mathbf{x})). \tag{5.6}$$

5.2.1 Finite Difference Discretization

Consider the optimality systems (5.4) and (5.5) discretized by finite differences and the backward Euler scheme. Assume Ω is a square and $\overline{\Omega_h}$ is a uniform space mesh where h is the mesh size. Ω_h defines the set of interior mesh points, $(x_i, y_j) = ((i-1)h, (j-1)h)$, $2 \le i, j \le N_{\mathbf{x}}$. On this mesh $-\Delta_h$ denotes the negative Laplacian and is approximated by the common five-point stencil. For grid functions v_h and w_h defined on Ω_h, we have the discrete $L^2(\Omega)$ scalar product

$$(v_h, w_h)_{L_h^2(\Omega_h)} = h^2 \sum_{\mathbf{x}\in\Omega_h} v_h(\mathbf{x})\, w_h(\mathbf{x}),$$

with associated norm $|v_h|_0 = (v_h, v_h)_{L_h^2(\Omega_h)}^{1/2}$. Let $\delta t = T/N_t$ be the time stepsize. Define

$$Q_{h,\delta t} = \{(\mathbf{x}, t_m) : \mathbf{x} \in \Omega_h,\ t_m = (m-1)\delta t,\ 1 \le m \le N_t + 1\}.$$

On this grid, y_h^m and p_h^m denote grid functions at time level m. The action of the time difference operator on these functions is as follows:

$$\partial_t^+ y_h^m = \frac{y_h^m - y_h^{m-1}}{\delta t} \quad \text{and} \quad \partial_t^- p_h^m = \frac{p_h^{m+1} - p_h^m}{\delta t},$$

respectively; see [4] for details. For grid functions defined on $Q_{h,\delta t}$ we use the discrete $L^2(Q)$ scalar product with norm $||v_{h,\delta t}||_0 = (v_{h,\delta t}, v_{h,\delta t})_{L_{h,\delta t}^2(Q_{h,\delta t})}^{1/2}$. Later we use $\gamma = \delta t/h^2$.

For simplicity, we assume sufficient regularity of the data, y_d, y_T, such that these functions are properly approximated by their values at grid points. With this setting, the following discretization of (5.4) is obtained:

$$\begin{aligned} -\partial_t^+ y_h^m + G(y_h^m) + \sigma \Delta_h y_h^m &= u_h^m, \\ \partial_t^- p_h^m + G'(y_h^m) p_h^m + \sigma \Delta_h p_h^m + \alpha(y_h^m - y_{dh}^m) &= 0, \\ \nu u_h^m - p_h^m &= 0. \end{aligned} \tag{5.7}$$

The discretization of (5.5) is similar, requiring us, in addition, to specify the approximation of the Neumann boundary conditions. This is done by considering the optimality system on the boundary and discretizing the boundary derivative using second-order centered finite differences to eliminate the (ghost) variables outside of the domain.

While we do not discuss accuracy of solutions in this chapter, we refer the reader to [20] regarding accuracy estimates for (5.4) and (5.5) in a finite element context. For the present finite difference setting see [4] and the discussion in [8]. Thus, under suitable conditions the following estimates are obtained:

$$||y_h - y||_0 \le c\,h^2, \quad ||p_h - p||_0 \le c\,h^2, \quad \text{and } ||u_h - u||_0 \le c\,h^2,$$

assuming there exist positive constants $c_1 \le c_2$ such that $c_1 h^2 \le \delta t \le c_2 h^2$.

5.3 The FAS Multigrid Framework

The space-time multigrid algorithms discussed here are formulated based on the FAS multigrid framework [10]. The advantage of the FAS structure is its general applicability, since the full nonlinear problem and the corresponding solution are represented on all grids. The latter is essential also for linear problems where constraints on the control are present; see [3, 4].

To illustrate the FAS approach, consider L grid levels indexed by $k = 1, \ldots, L$, where $k = L$ refers to the finest grid. The mesh of level k is denoted by $Q_k = Q_{h_k,\delta t_k}$, where $h_k = h_1/2^{k-1}$ and $\delta t_k = \delta t$, and thus we have semicoarsening in space. Any operator and variable defined on Q_k is indexed by k.

The optimality system at level k with given initial, terminal, and boundary conditions is represented by the following nonlinear equation:

$$A_k(w_k) = f_k, \qquad w_k = (y_k, u_k, p_k). \tag{5.8}$$

In general, an initial approximation to the solution of this problem will differ from the exact solution because of errors involving high-frequency as well as low-frequency components. In order to solve for all frequency components of the error, the multigrid strategy combines two complementary schemes. The high-frequency components of the error are reduced by a smoothing iteration, denoted by S_k and defined in the following subsection, while the low-frequency error components are effectively reduced by a coarse-grid correction method as defined below.

The action of one FAS-cycle applied to (5.8) can be expressed in terms of a (nonlinear) multigrid iteration operator B_k. Starting with an initial approximation $w_k^{(0)}$ the result of one FAS-$V(\nu_1, \nu_2)$-cycle is then denoted by $w_k = B_k(w_k^{(0)})\, f_k$.

Algorithm 5.1 (Multigrid FAS-$V(\nu_1, \nu_2)$-Cycle).
Set $B_1(w_1^{(0)}) \approx A_1^{-1}$ (e.g., iterating with S_1 starting with $w_1^{(0)}$). For $k = 2, \ldots, L$ define B_k in terms of B_{k-1} as follows.

1. Set the starting approximation $w_k^{(0)}$.

2. Presmoothing. Define $w_k^{(l)}$ for $l = 1, \ldots, \nu_1$, by

$$w_k^{(l)} = S_k(w_k^{(l-1)}, f_k).$$

3. Coarse-grid correction. Set $w_k^{(\nu_1+1)} = w_k^{(\nu_1)} + I_{k-1}^k(w_{k-1} - \hat{I}_k^{k-1} w_k^{(\nu_1)})$, where

$$w_{k-1} = B_{k-1}(\hat{I}_k^{k-1} w_k^{(\nu_1)}) \left[I_k^{k-1}(f_k - A_k(w_k^{(\nu_1)})) + A_{k-1}(\hat{I}_k^{k-1} w_k^{(\nu_1)}) \right].$$

4. Postsmoothing. Define $w_k^{(l)}$ for $l = \nu_1 + 2, \ldots, \nu_1 + \nu_2 + 1$, by

$$w_k^{(l)} = S_k(w_k^{(l-1)}, f_k).$$

5. Set $B_k(w_k^{(0)})\, f_k = w_k^{(\nu_1+\nu_2+1)}$.

In our implementation, we choose I_k^{k-1} to be the full-weighted restriction operator [23] in space with no averaging in the time direction. The mirrored version of this operator applies also to the boundary points. The prolongation I_{k-1}^k is defined by bilinear interpolation in space. These two operators are given in stencil form by

$$I_k^{k-1} = \frac{1}{16}\begin{bmatrix} 1 & 2 & 1 \\ 2 & 4 & 2 \\ 1 & 2 & 1 \end{bmatrix} \quad \text{and} \quad I_{k-1}^k = \frac{1}{4}\begin{bmatrix} 1 & 2 & 1 \\ 2 & 4 & 2 \\ 1 & 2 & 1 \end{bmatrix}. \tag{5.9}$$

Typically, in order to transfer as much information as possible to the coarse grid, full-weighting is chosen to restrict residual. On the other hand, one chooses $\hat{I}_k^{k-1}$ to be straight injection consistently with the requirement that, at convergence, solutions on different levels be equal, $w_{k-1} - \hat{I}_k^{k-1} w_k^{(\nu_1)} = 0$. Therefore a zero coarse-grid correction results; see step 3 of Algorithm 5.1.

No interpolation in time is needed. Indeed, other choices of prolongation and restriction operators are possible; see [14].

5.3.1 Space-Time Smoothing Schemes

The key component in the development of multigrid solvers for optimal control problems is the design of robust collective smoothing schemes. In this section, we present two smoothing iterations and discuss their computational performance. The smoothing schemes discussed in this section are appropriate for a sequential computing environment. However, some comments are provided on possible extension to parallel computation.

For the formulation of these schemes, let us write (5.7) in expanded form. In case no constraints on the distributed control are present, we can eliminate the control variable by means of the optimality condition $\nu u_h^m - p_h^m = 0$. We have

$$\begin{aligned} &-\left[1 + 4\sigma\gamma\right] y_{i\,j\,m} + \sigma\gamma\left[y_{i+1\,j\,m} + y_{i-1\,j\,m} + y_{i\,j+1\,m} + y_{i\,j-1\,m}\right] + y_{i\,j\,m-1} \\ &+\delta t\, G(y_{i\,j\,m}) - \frac{\delta t}{\nu}\, p_{i\,j\,m} = 0, \quad 2 \le m \le N_t + 1, \end{aligned} \tag{5.10}$$

$$\begin{aligned} &-\left[1 + 4\sigma\gamma\right] p_{i\,j\,m} + \sigma\gamma\left[p_{i+1\,j\,m} + p_{i-1\,j\,m} + p_{i\,j+1\,m} + p_{i\,j-1\,m}\right] + p_{i\,j\,m+1} \\ &+\delta t\, G'(y_{i\,j\,m})\, p_{i\,j\,m} + \delta t\, \alpha\, (y_{i\,j\,m} - y_{d\,i\,j\,m}) = 0, \quad 1 \le m \le N_t. \end{aligned} \tag{5.11}$$

In case of terminal observation, at $t_m = T$ we have (5.6) in place of (5.11).

First, we define a pointwise smoothing scheme. Consider a collective Gauss–Seidel (–Newton) step which is applied at each space-time grid point to update $w_{i\,j\,m} = (y_{i\,j\,m}, p_{i\,j\,m})$. For this purpose consider (5.10) and (5.11) for the two variables $y_{i\,j\,m}$ and $p_{i\,j\,m}$ at the grid point $i\,j\,m$. We can refer to the left-hand sides of (5.10) and (5.11) as the negative of the residuals $r_y(w_{i\,j\,m})$ and $r_p(w_{i\,j\,m})$, respectively. A step of a collective smoothing iteration at this point consists of a local (Newton) update given by

$$\begin{pmatrix} y \\ p \end{pmatrix}_{i\,j\,m}^{(1)} = \begin{pmatrix} y \\ p \end{pmatrix}_{i\,j\,m}^{(0)} + \begin{bmatrix} -(1+4\sigma\gamma) + \delta t\, G' & -\delta t/\nu \\ \delta t(\alpha + G''\, p) & -(1+4\sigma\gamma) + \delta t\, G' \end{bmatrix}_{i\,j\,m}^{(0)\;-1} \begin{pmatrix} r_y \\ r_p \end{pmatrix}_{i\,j\,m}, \tag{5.12}$$

where r_y and r_p denote the residuals at $i\,j\,m$ prior to the update. While a sweep of this smoothing iteration can be performed in any ordering of i, j, the problem of how to proceed along the time direction arises.

To solve this problem two choices are possible. On the one hand, we could apply (5.12) in an iterative damped Jacobi fashion, requiring additional storage for the update procedure. This approach may be convenient in a parallel computing environment. On the other hand, in a sequential computing setting, a Gauss–Seidel-type update is the most efficient choice provided that the opposite time orientation of the state equation and of the adjoint equation is taken into account. For this purpose, to update the state variable we use the first vector component of (5.12) marching in the forward direction and the adjoint variable p is being updated using the second component of (5.12) marching backwards in time. In this way a robust iteration is obtained given by the following algorithm [3, 4, 7].

Algorithm 5.2 (Time-Splitted Collective Gauss–Seidel Iteration (TS-CGS)).

1. Set the starting approximation.
2. For $m = 2, \ldots, N_t$ do
3. For ij in, e.g., lexicographic order do

$$y^{(1)}_{i\,j\,m} = y^{(0)}_{i\,j\,m} + \frac{[-(1+4\sigma\gamma)+\delta t G']\, r_y(w) + \frac{\delta t}{\nu} r_p(w)}{[-(1+4\sigma\gamma)+\delta t G']^2 + \frac{\delta t^2}{\nu}(\alpha + G''p)}\Big|^{(0)}_{i\,j\,m},$$

$$p^{(1)}_{i\,j\,N_t-m+2} = p^{(0)}_{i\,j\,N_t-m+2} + \frac{[-(1+4\sigma\gamma)+\delta t G']\, r_p(w) - \delta t(\alpha + G''p)\, r_y(w)}{[-(1+4\sigma\gamma)+\delta t G']^2 + \frac{\delta t^2}{\nu}(\alpha + G''p)}\Big|^{(0)}_{i\,j\,N_t-m+2};$$

4. end.

The TS-CGS scheme applies with few modifications to the case of boundary control as follows. For $i, j \in \Omega_h$ the term $\delta t\, p/\nu$ in (5.10) is no longer present. On the boundary a similar term appears given by $(2\sigma/h)\,\delta t\, p/\nu$. To illustrate this fact consider (5.7) on a grid point i, j of the left-hand vertical side of $\overline{\Omega}$. At this point, we use the approximation $-\partial y/\partial n \approx (y_{i+1\,j} - y_{i-1\,j})/2h$ and obtain

$$\begin{aligned} -\left[1+4\sigma\gamma\right] y_{i\,j\,m} + \sigma\gamma\left[2\,y_{i+1\,j\,m} + y_{i\,j+1\,m} + y_{i\,j-1\,m}\right] + y_{i\,j\,m-1} \\ +\delta t\, G(y_{i\,j\,m}) - \left(\frac{2\sigma}{h}\right)\frac{\delta t}{\nu}\, p_{i\,j\,m} = 0, \quad 2 \le m \le N_t + 1, \end{aligned}$$

$$\begin{aligned} -\left[1+4\sigma\gamma\right] p_{i\,j\,m} + \sigma\gamma\left[2\,p_{i+1\,j\,m} + p_{i\,j+1\,m} + p_{i\,j-1\,m}\right] + p_{i\,j\,m+1} \\ +\delta t\, G'(y_{i\,j\,m})\, p_{i\,j\,m} + \delta t\, \alpha\,(y_{i\,j\,m} - y_{d\,i\,j\,m}) = 0, \quad 1 \le m \le N_t. \end{aligned}$$

Therefore to formulate the TS-CGS step on the boundary, the term $-\frac{\delta t}{\nu}$ in the Jacobian in (5.12) is replaced by $-\left(\frac{2\sigma}{h}\right)\frac{\delta t}{\nu}$. Notice that at corners we have $-\left(\frac{4\sigma}{h}\right)\frac{\delta t}{\nu}$ instead of $-\left(\frac{2\sigma}{h}\right)\frac{\delta t}{\nu}$.

Later in this chapter, results of local Fourier analysis show that the TS-CGS scheme has good smoothing properties, independently of the value of ν. This is also confirmed by results of numerical experiments. In the regime of small σ (or γ), however, the TS-CGS iteration cannot provide robust smoothing as we expect from our experience of multigrid methods for anisotropic problems. It is well known [23] that in order for an iteration scheme to be effective for smoothing, relaxation must be performed in the direction of strong coupling of the stencil of the operator. Now, in case σ is small the coupling in the space direction is weak, and therefore pointwise relaxation in space is not effective in reducing the high-frequency components of the error, as required. To overcome this problem, block relaxation of the variables that are strongly connected should be performed. In our case this means solving for the pairs of state and adjoint variables along the time direction for each space coordinate. This type of smoothing belongs to the class of block Gauss–Seidel relaxation [23].

To describe this Gauss–Seidel procedure, consider the discrete optimality system (5.7) at any i, j and for all time steps. For simplicity, we use the optimality condition to eliminate the control variable. Thus for each spatial grid point i, j a block-tridiagonal system is obtained, where each block is a 2×2 matrix corresponding to the pair (y, p). This block-tridiagonal system has the following form:

$$M = \begin{bmatrix} A_2 & C_2 & & & \\ B_3 & A_3 & C_3 & & \\ & B_4 & A_4 & C_4 & \\ & & & & \\ & & & & C_{N_t} \\ & & & B_{N_t+1} & A_{N_t+1} \end{bmatrix}. \tag{5.13}$$

Centered at t_m, the entries B_m, A_m, C_m refer to the variables (y, p) at t_{m-1}, t_m, and t_{m+1}, respectively. The block A_m, $m = 2, \dots, N_t$, is given by

$$A_m = \begin{bmatrix} -(1 + 4\sigma\gamma) + \delta t G' & -\frac{\delta t}{\nu} \\ \delta t(\alpha + G'' p) & -(1 + 4\sigma\gamma) + \delta t G' \end{bmatrix}, \tag{5.14}$$

where all functions within the brackets [ ] are evaluated at t_m. Correspondingly, the B_m and C_m blocks are given by

$$B_m = \begin{bmatrix} 1 & 0 \\ 0 & 0 \end{bmatrix} \quad \text{and} \quad C_m = \begin{bmatrix} 0 & 0 \\ 0 & 1 \end{bmatrix}. \tag{5.15}$$

Clearly, for each time step, the variables neighboring the point ij are taken as constant and contribute to the right-hand side of the system.

It remains to discuss the block A_{N_t+1} for $\beta \neq 0$. At $t_m = T$, we have the terminal condition (5.6) which we rewrite as

$$\beta\,(y_h^m - y_{Th}^m) - p_h^m = 0, \quad m = N_t + 1.$$

Thus, the block A_{N_t+1} is given by

$$A_{N_t+1} = \begin{bmatrix} -(1 + 4\sigma\gamma) + \delta t G' & -\frac{\delta t}{\nu} \\ \beta & -1 \end{bmatrix}. \tag{5.16}$$

For each i, j we have to solve a tridiagonal system $Mw = r$, where $w = (y_h^2, p_h^2, \dots, y_h^{N_t+1}, p_h^{N_t+1})$ and $r = (r_y(w^2), r_p(w^2), \dots, r_y(w^{N_t+1}), r_p(w^{N_t+1}))$. In particular we have $r_p(w^{N_t+1}) = p_h^{N_t+1} - \beta\,(y_h^{N_t+1} - y_{Th}^{N_t+1})$. Block-tridiagonal systems can be solved efficiently with $\mathcal{O}(N_t)$ effort. A block-tridiagonal solver is given in [5]. Summarizing our collective t-line relaxation is given by the following algorithm [5, 6].

Algorithm 5.3 (Time-Line Collective Gauss–Seidel Iteration (TL-CGS)).

1. Set the starting approximation.
2. For ij in, e.g., lexicographic order do
$$\begin{pmatrix} y \\ p \end{pmatrix}_{ij}^{(1)} = \begin{pmatrix} y \\ p \end{pmatrix}_{ij}^{(0)} + M^{-1} \begin{pmatrix} r_y \\ r_p \end{pmatrix}_{ij};$$
3. end.

Also in this case r_y and r_p denote the residuals at i, j and for all m prior to the update. Since the solution in time is exact, no time splitting is required. Notice that it is possible to formulate parallel TL-CGS schemes applying parallel algorithms for block tridiagonal systems.

5.3.2 Receding Horizon Approach

Results of numerical experiments and Fourier analysis estimates demonstrate the ability of the multigrid schemes presented here to solve tracking and terminal observation optimal control problems. This fact suggests combining these multigrid schemes with receding horizon techniques [2, 15] to develop an efficient optimal control algorithm for tracking a desired trajectory over very long time intervals. In the following, we sketch the implementation of the multigrid receding horizon scheme. For validation see the section on numerical experiments.

Consider the optimal control problem of tracking y_d for $t \geq 0$. Define time windows of size Δt. In each time window, an optimal control problem with tracking ($\alpha = 1$) and terminal observation ($\beta = 1$) is solved. The resulting optimal state at $n\Delta t$ defines the initial condition for the next optimal control problem defined in $(n\Delta t, (n+1)\Delta t)$ with desired terminal state given by $y_T(\mathbf{x}) = y_d(\mathbf{x}, (n+1)\Delta t)$. The following algorithm results.

Algorithm 5.4 (Multigrid Receding Horizon Scheme (MG-RH)).

1. Set $y(\mathbf{x}, 0) = y_0(\mathbf{x})$ and $n = 0$.
2. Set $y_T(\mathbf{x}) = y_d(\mathbf{x}, (n+1)\Delta t)$.
3. MG Solve (5.4) or (5.5) and (5.6) in $(n\Delta t, (n+1)\Delta t)$.
4. Update $n := n + 1$, set $y_0(\mathbf{x}) = y(\mathbf{x}, n\Delta t)$ and goto 2.

5.4 Fourier Smoothing Analysis

In this section, we perform local Fourier analysis [10, 14, 23] of the iterative schemes presented above. The resulting estimates are in agreement with the observed computational behavior of the multigrid algorithms discussed in this chapter, even when strong nonlinearities are present.

For the analysis that follows, we consider linear problems, infinite grids, and one space dimension. On the fine grid, consider the Fourier components $\varphi(\boldsymbol{j}, \boldsymbol{\theta}) = e^{i\boldsymbol{j}\cdot\boldsymbol{\theta}}$, where i is the imaginary unit, $\boldsymbol{j} = (j_x, j_t) \in \mathbf{Z} \times \mathbf{Z}$, $\boldsymbol{\theta} = (\theta_x, \theta_t) \in [-\pi, \pi)^2$, and $\boldsymbol{j} \cdot \boldsymbol{\theta} = j_x\theta_x + j_t\theta_t$.

In a semicoarsening setting, the frequency domain is spanned by the following two sets of frequencies:

$$\varphi \text{ low-frequency component } \Longleftrightarrow \qquad \theta_x \in \left[-\frac{\pi}{2}, \frac{\pi}{2}\right), \quad \theta_t \in [-\pi, \pi),$$

$$\varphi \text{ high-frequency component } \Longleftrightarrow \quad \theta_x \in [-\pi, \pi) \Big\backslash \left[-\frac{\pi}{2}, \frac{\pi}{2}\right), \quad \theta_t \in [-\pi, \pi).$$

Let $\tilde{w}(\boldsymbol{j}) = (\tilde{y}(\boldsymbol{j}), \tilde{p}(\boldsymbol{j})) = \sum_{\boldsymbol{\theta}} \tilde{W}_{\boldsymbol{\theta}}\, \varphi(\boldsymbol{j}, \boldsymbol{\theta})$ denote the errors on the space-time grid and $\tilde{W}_{\boldsymbol{\theta}} = (\tilde{Y}_{\boldsymbol{\theta}}, \tilde{P}_{\boldsymbol{\theta}})$ be the corresponding Fourier coefficients. The action of one smoothing step on $\tilde{w}$ can be expressed by $\tilde{W}_{\boldsymbol{\theta}}^{(1)} = \hat{S}(\boldsymbol{\theta})\, \tilde{W}_{\boldsymbol{\theta}}^{(0)}$, where $\hat{S}(\boldsymbol{\theta})$ is the Fourier symbol [23] of the smoothing operator.

One way to measure the ability of the smoothing scheme to reduce the high-frequency components of the error is given by the following definition of the smoothing factor. We have

$$\mu = \sup\{\, |\kappa(\hat{S}(\boldsymbol{\theta}))| \;:\; \boldsymbol{\theta} \text{ high frequency } \},$$

where κ denotes the spectral radius. Determining μ and assuming an ideal coarse-grid correction, which annihilates the low-frequency error components and leaves the high-frequency error components unchanged, one obtains the estimate $\rho \approx \mu^{\nu_1+\nu_2}$ of the multigrid convergence factor. An improved estimate can be obtained by the two-grid Fourier analysis presented in [4].

Notice that by local Fourier analysis the problem of computing the smoothing factor is reduced to that of determining the spectral radius of $\hat{S}(\boldsymbol{\theta})$, a 2×2 matrix. This task may be performed using any symbolic package (we use Mathematica).

Now consider applying the TS-CGS step for solving a distributed control problem with tracking. Substituting $\tilde{w}$ into (5.10)–(5.11) we obtain

$$\begin{pmatrix} -(1+2\sigma\gamma) + \sigma\gamma e^{-i\theta_x} & -\frac{\delta t}{\nu} \\ \alpha\delta t & -(1+2\sigma\gamma) + \sigma\gamma e^{-1\theta_x} \end{pmatrix} \begin{pmatrix} \tilde{Y}_{\boldsymbol{\theta}}^{(1)} \\ \tilde{P}_{\boldsymbol{\theta}}^{(1)} \end{pmatrix}$$
$$= \begin{pmatrix} -(e^{-i\theta_t} + \sigma\gamma e^{i\theta_x}) & 0 \\ 0 & -(e^{i\theta_t} + \sigma\gamma e^{i\theta_x}) \end{pmatrix} \begin{pmatrix} \tilde{Y}_{\boldsymbol{\theta}}^{(0)} \\ \tilde{P}_{\boldsymbol{\theta}}^{(0)} \end{pmatrix}.$$

Hence

$$\hat{S}(\boldsymbol{\theta}) = \begin{pmatrix} -(1+2\sigma\gamma) + \sigma\gamma e^{-i\theta_x} & -\frac{\delta t}{\nu} \\ \alpha\delta t & -(1+2\sigma\gamma) + \sigma\gamma e^{-i\theta_x} \end{pmatrix}^{-1} \tag{5.17}$$
$$\times \begin{pmatrix} -(e^{-i\theta_t} + \sigma\gamma e^{i\theta_x}) & 0 \\ 0 & -(e^{i\theta_t} + \sigma\gamma e^{i\theta_x}) \end{pmatrix}.$$

In Figure 5.1 (left) we depict the smoothing factor of the TS-CGS scheme as a function of ν and γ. It appears that μ is independent of the value of the weight ν and of the discretization parameter γ, as long as γ is sufficiently large. For $\gamma \to 0$ or, equivalently, $\sigma \to 0$ and moderate values of ν, worsening of the smoothing factor can be observed. Similar results are obtained with different choices of the time-step size and in case $\alpha = 0$, $\beta \neq 0$.

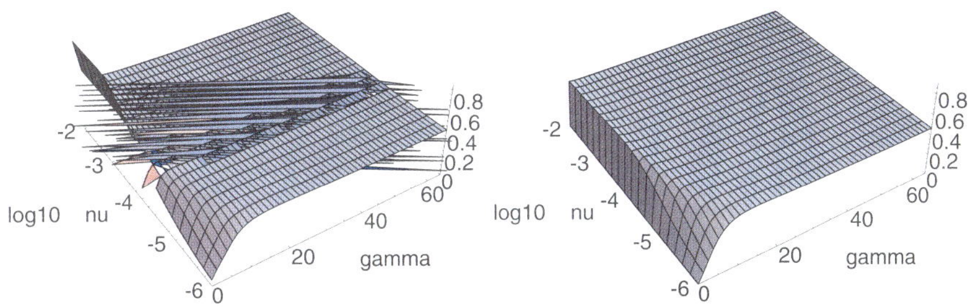

Figure 5.1. *Smoothing factors of TS-CGS (left) and TL-CGS (right) schemes as functions of* ν *and* γ; $\delta t = 1/64$, $\alpha = 1$, *and* $\sigma = 1$.

Next consider the case of TL-CGS relaxation. The Fourier symbol of the smoothing operator is given by the following 2×2 matrix:

$$\hat{S}(\boldsymbol{\theta}) = -(A + B\, e^{-i\theta_t} + C\, e^{i\theta_t} + \tilde{I}\, e^{-i\theta_x})^{-1}(\tilde{I}\, e^{i\theta_x}),$$

where

$$A = \begin{bmatrix} -(1+2\sigma\gamma) & -\frac{\delta t}{\nu} \\ \delta t \alpha & -(1+2\sigma\gamma) \end{bmatrix}, \quad B_m = \begin{bmatrix} 1 & 0 \\ 0 & 0 \end{bmatrix}, \text{ and } C_m = \begin{bmatrix} 0 & 0 \\ 0 & 1 \end{bmatrix},$$

and $\tilde{I} = \sigma\gamma\, I$, I is the 2×2 identity matrix.

In Figure 5.1 (right), the smoothing factor of the TL-CGS scheme as a function of ν and γ is shown. Notice that the smoothing factor of this scheme is independent of ν and γ. For $\sigma = 0$ no spatial coupling is present and the TL-CGS scheme becomes an exact solver; i.e., $\mu = 0$ results.

5.5 Numerical Experiments

In this section, a numerical investigation of the computational performance of TS-CGS and TL-CGS based space-time multigrid schemes is presented. The results reported here demonstrate optimal computational complexity of the proposed algorithm. We show that the convergence factors are independent of the discretization and optimization parameters, and therefore the work is linear in the problem size. (The FORTRAN code of the multigrid algorithm presented in this chapter is available at MGNet, http://www.mgnet.org/.)

We use two pre- and two postsmoothing steps ($\nu_1 = \nu_2 = 2$) and $h = 1/4$ is the coarsest space mesh size. We take $\Omega = (0, 1) \times (0, 1)$ and $T = 1$.

Table 5.1. *Results for tracking:* $\alpha = 1$, $\beta = 0$, *and* $\sigma = 1$.

$\nu = 10^{-3}$	TS-CGS			TL-CGS		
γ	ρ	$\|\|y - y_d\|\|_0$	CPU time(s)	ρ	$\|\|y - y_d\|\|_0$	CPU time(s)
32	0.06	$1.79\,10^{-2}$	0.6	0.06	$1.79\,10^{-2}$	0.8
64	0.06	$1.75\,10^{-2}$	5.7	0.06	$1.79\,10^{-2}$	7.3
128	0.06	$1.73\,10^{-2}$	53.8	0.06	$1.73\,10^{-2}$	58.1
$\nu = 10^{-6}$	TS-CGS			TL-CGS		
γ	ρ	$\|\|y - y_d\|\|_0$	CPU time(s)	ρ	$\|\|y - y_d\|\|_0$	CPU time(s)
32	0.08	$1.51\,10^{-4}$	0.8	0.08	$1.51\,10^{-4}$	1.1
64	0.06	$1.59\,10^{-4}$	6.4	0.06	$1.59\,10^{-4}$	8.8
128	0.06	$1.61\,10^{-4}$	53.5	0.06	$1.61\,10^{-4}$	71.3

Table 5.2. *Results for terminal observation with TL-CGS:* $\alpha = 0$ *and* $\beta = 1$.

	$\nu = 10^{-3}$			$\nu = 10^{-6}$		
$\sigma\,\gamma$	ρ	$\|y - y_T\|_0$	CPU time(s)	ρ	$\|y - y_T\|_0$	CPU time(s)
0.32	0.10	$8.80\,10^{-4}$	1.1	0.10	$8.85\,10^{-7}$	1.1
0.64	0.09	$9.31\,10^{-4}$	8.9	0.09	$9.40\,10^{-7}$	8.9
32	0.06	$6.97\,10^{-4}$	0.8	0.06	$7.28\,10^{-7}$	0.9
64	0.06	$7.56\,10^{-4}$	7.4	0.06	$7.94\,10^{-7}$	7.8

To describe the results of the experiments we report the multigrid convergence factor ρ, defined as the asymptotic value of the ratio of the norm of the dynamic residuals given by $||r_y||_0 + ||r_p||_0/\nu$ resulting from two successive multigrid cycles. The stopping criterion is $||r_y||_0 + ||r_p||_0/\nu < 10^{-10}$. The tracking ability of the space-time multigrid algorithms will be expressed in terms of the norms of the tracking error $||y - y_d||_0$ and of the terminal observation error $|y - y_T|_0$.

Three different grids $N_x \times N_y \times N_t$ are considered: $32 \times 32 \times 32$, $64 \times 64 \times 64$, and $128 \times 128 \times 128$, which result in $\gamma = 32$, $\gamma = 64$, and $\gamma = 128$, respectively. To investigate the dependence of the convergence factor on σ we report results with $\sigma = 1$ and with $\sigma = 0.01$.

As nonlinearity we choose $G(y) = e^y$ which is used to model explosive combustion phenomena; see [7, 9]. Depending on σ the resulting uncontrolled reaction-diffusion problem may have a stable solution or blowup at finite time. For example, with zero initial condition and $\sigma < 0.1$ we have blowup at $t < 1$.

The desired target trajectory is given by

$$y_d(\mathbf{x}, t) = (1 + t)\,(x_1 - x_1^2)\,(x_2 - x_2^2)\,\cos(4\pi\, t).$$

This is an oscillating function whose amplitude increases linearly with time. We take $y_T(\mathbf{x}) = y_d(\mathbf{x}, T)$.

The results reported in Tables 5.1 and 5.2 in the case of distributed control illustrate efficiency and robustness of the proposed multigrid solvers. In the case of tracking, results in Table 5.1 demonstrate usual multigrid convergence speeds which appear to be independent of the value of γ and ν. These results are in agreement with the discussion and results of the previous section. Notice that the CPU times scale approximately as a factor of $2^3 = 8$, that is, linearly with the number of grid points.

The choice of TL-CGS iteration results in a superior computational performance when only terminal observation is considered. Results for this case are reported in Table 5.2. Notice that because $\alpha = 0$ the coupling between state and control variables is much weaker in the space-time domain, resulting in a less efficient TS-CGS-based multigrid solver. Despite this weaker coupling, the use of TL-CGS smoothing provides a multigrid algorithm that performs equally well for tracking or terminal observation. In Table 5.2, typical multigrid convergence factors are reported for the TL-CGS choice that show robustness with respect to the values of γ, ν, and σ.

Results reported in Tables 5.1 and 5.2 demonstrate the tracking ability of the optimal control formulation. As the value of ν decreases, smaller values of $||y - y_d||_0$ and of $|y - y_T|_0$ are obtained as desired. Further numerical experiments demonstrate that the multigrid convergence behavior appears to be insensitive to the particular choice of the desired tracking function, which may not be attainable.

A similar series of numerical experiments has been performed for the case of boundary control. For $\sigma = 1$, convergence factors that are approximately twice those of the distributed control case are obtained, while the dependence of ρ on ν and γ remains qualitatively the same. For $\sigma = 10^{-2}$, no satisfactory control performance is obtained. This occurrence can be expected, since in case of small diffusivity, control on the boundary can hardly influence the behavior of the solution in the interior of the domain. For this reason boundary control seems less effective to control chemical turbulence [6].

To validate the multigrid receding horizon approach, consider the problem of tracking y_d given above in $(0, T)$ with $T = 6$ using 20 time windows of size $\Delta t = 0.3$. In each time window the optimal control problem is solved by 4 FAS-$V(2, 2)$-cycles on a $64 \times 64 \times 16$ grid. Results of tracking are depicted in Figure 5.2 for the choice $\nu = 10^{-6}$ and $\sigma = 1$. Further numerical experiments show that the quality of the tracking remains equally good even when choosing smaller Δt and correspondingly smaller N_t (e.g., $N_t = 3$). As shown in Figure 5.2, a smooth optimal control is obtained. These results correspond to the choice $\alpha = 1$ and $\beta = 1$. Choosing $\beta = 0$, an optimal control is obtained which is zero at the interfaces between time windows.

Notice that the purpose of the multigrid receding horizon approach is to determine a (sub-) optimal control in the case of very long or open-end time intervals. For the present case, $T = 6$, we can actually compare the receding horizon computation with the full horizon results. For a meaningful comparison we require similar accuracy and therefore a common time-step size. This means choosing $N_t = 16$ in the former setting with 20 time windows and $N_t = 320$ in the latter setting, respectively. With the receding horizon computation we obtain $\|y - y_d\|_0 = 5.53\ 10^{-4}$ and CPU time(s) = 41.6, while with the full horizon calculation we have $\|y - y_d\|_0 = 5.72\ 10^{-4}$ and CPU time(s) = 47.2. In fact, the two solutions are barely distinguishable.

In Figure 5.3 results of the multigrid receding horizon scheme with boundary control are depicted. For the setting $\alpha = 1$ and $\beta = 0$ the obtained multigrid convergence performance is similar to that obtained in the case of distributed control, as reported in Table 5.1. For the setting $\alpha = 0$ and $\beta = 1$ the proposed multigrid approach is not robust with respect to the choice of β and eventually diverges for sufficiently small ν. For convergence a small value of β must be chosen. For $\beta = 10^{-2}$ convergence factors $\rho \approx 0.2$ are obtained independently of ν.

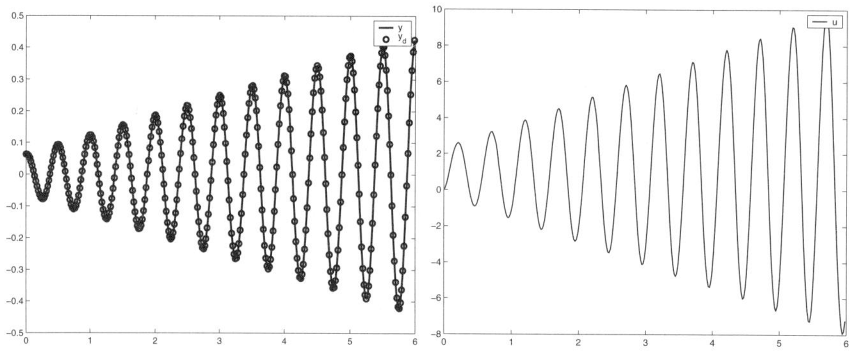

Figure 5.2. *Receding horizon solution for distributed control. Left: Time evolution of the state* y *(solid line) and the desired trajectory* y_d *(circles). Right: The optimal control* u *at* $(x_1, x_2) = (0.5, 0.5)$; $\alpha = 1$ *and* $\beta = 1$.

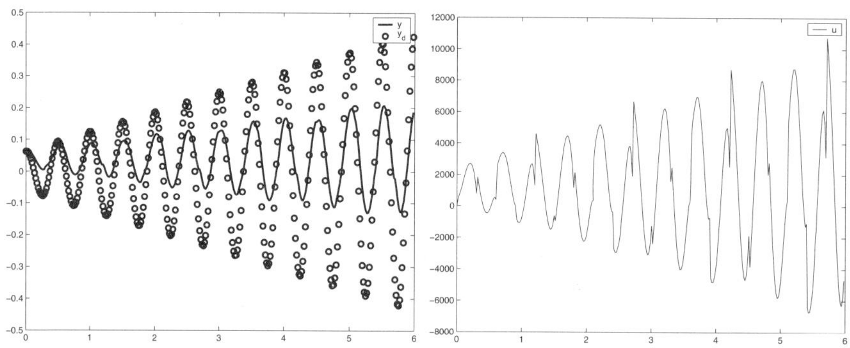

Figure 5.3. *Receding horizon solution for boundary control. Left: Time evolution of the state* y *(solid line) and the desired trajectory* y_d *(circles). Right: Evolution of the control* u *at* $(x_1, x_2) = (0.5, 0.5)$; $\alpha = 1$ *and* $\beta = 0.01$.

5.6 An Application in Physiology

We complete this chapter formulating a challenging optimal control problem related to cardiac arrhythmia. Arrhythmia is any deviation from normal heart rhythm due to disturbances of the normal sequence of impulse propagation through the heart. The resulting ventricular fibrillation appears as electrical turbulence showing spatial and temporal organization. An accurate model of cardiac cell tissue is the Aliev–Panfilov model [1, 22]. It is given by

$$\frac{\partial y_1}{\partial t} = -k y_1 (y_1 - a)(y_1 - 1) - y_1\, y_2 + \sigma\, \Delta y_1 + u, \tag{5.18}$$

$$\frac{\partial y_2}{\partial t} = \left[\epsilon_0 + \frac{\mu_1 y_2}{\mu_2 + y_1}\right] [-y_2 - k y_1 (y_1 - b - 1)], \tag{5.19}$$

where y_1 represents transmembrane potential and y_2 the membrane conductance. The parameters in this model can be adjusted to reproduce the characteristics of the given cardiac

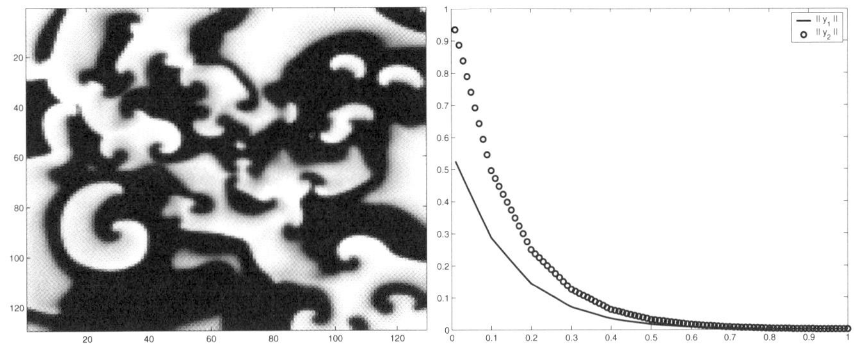

Figure 5.4. *Results of the Aliev–Panfilov model at $t = 1000$ (left) and time evolution of the L^2-norm of the controlled solution (right).*

tissue. Following [22] we take $k = 8$, $a = 0.1$, $b = 0.1$, $\epsilon_0 = 0.01$, $\mu_1 = 0.07$, $\mu_2 = 0.3$, and $\sigma = 2.5 \cdot 10^{-5}$. With this choice, the model evolves from an initial planar (half) wave as described in [1] to a turbulent electrical pattern as shown in Figure 5.4.

Our purpose is to determine a control response in the form of an electrical field, which can be realized by an implanted defibrillator, able to drive the system from a turbulent pattern to a uniform pattern as in the case of no stimulus. Notice that the control u appears only in (5.18) and no diffusion affects the conductance (other configurations are possible).

We compute the optimal control as follows. As initial state, the turbulent pattern as in Figure 5.4 is taken. We use the TL-CGS-based multigrid receding horizon scheme with no tracking and zero terminal state, $\alpha = 0$, $\beta = 1$. Ten time windows of size $\Delta t = 0.1$ are considered. On each window, the optimal control problem of minimizing (5.3) with $\nu = 0.1$ under the constraint given by (5.18)–(5.19) is solved efficiently on a $128 \times 128 \times 10$ grid by 3 FAS-$V(2,2)$-cycles. As a result, in Figure 5.4 (right) the desired fast decay of $|y_1|_0$ and $|y_2|_0$ in time can be observed.

Acknowledgments

I would like to thank Sasha Panfilov for helping me set up his cardiac excitation model. This work is supported in part by the Austrian Science Fund FWF project P18136-N13 "Quantum optimal control of semiconductor nanostructures."

Bibliography

[1] R. R. Aliev and A. V. Panfilov, *A simple two-variable model of cardiac excitation*, Chaos Solitons Fractals, 7 (1996), pp. 293–301.

[2] F. Allgöwer, T. A. Badgwell, J. S. Qin, J. B. Rawlings, and S. J. Wright, *Nonlinear predictive control and moving horizon estimation – An introductory overview*, in Advances in Control, Highlights of ECC'99, P. M. Frank, ed., Springer-Verlag, London, 1999, pp. 391–449.

[3] A. Borzì, *Fast multigrid methods for parabolic optimal control problems*, in Proceedings of the 18th Annual GAMM-Seminar, Leipzig, Germany, 2002, pp. 1–10.

[4] A. Borzì, *Multigrid methods for parabolic distributed optimal control problems*, J. Comput. Appl. Math., 157 (2003), pp. 365–382.

[5] A. Borzì and R. Griesse, *Distributed optimal control of lambda-omega systems*, J. Numer. Math., 14 (2006), pp. 17–40.

[6] A. Borzì and R. Griesse, *Experiences with a space-time multigrid method for the optimal control of a chemical turbulence model*, Internat. J. Numer. Methods Fluids, 47 (2005), pp. 879–885.

[7] A. Borzì and K. Kunisch, *A multigrid method for optimal control of time-dependent reaction diffusion processes*, in Fast Solution of Discretized Optimization Problems, Internat. Ser. Numer. Math. 138, K. H. Hoffmann, R. Hoppe, and V. Schulz, eds., Birkhäuser, Basel, 2001.

[8] A. Borzì and K. Kunisch, *The numerical solution of the steady state solid fuel ignition model and its optimal control*, SIAM J. Sci. Comput., 22 (2000), pp. 263–284.

[9] A. Borzì, K. Kunisch, and M. Vanmaele, *A multigrid approach to the optimal control of solid fuel ignition problems*, in Multigrid Methods, VI, Lect. Notes Computer Sci. Eng. 14, E. Dick, K. Riemslagh, and J. Vierendeels, eds., Springer-Verlag, Berlin, 2000, pp. 59–65.

[10] A. Brandt, *Multi-level adaptive solutions to boundary-value problems*, Math. Comp., 31 (1977), pp. 333–390.

[11] H. Goldberg and F. Tröltzsch, *Second-order sufficient optimality conditions for a class of nonlinear parabolic boundary control problems*, SIAM J. Control Optim., 31 (1993), pp. 1007–1025.

[12] W. Hackbusch, *Parabolic multigrid methods*, in Computing Methods in Applied Sciences and Engineering VI, R. Glowinski and J.-L. Lions, eds., North–Holland, Amsterdam, 1984, pp. 189–197.

[13] A. Hagberg, E. Meron, I. Rubinstein, and B. Zaltzman, *Controlling domain patterns far from equilibrium*, Phys. Rev. Lett., 76 (1996), pp. 427–430.

[14] G. Horton and S. Vandewalle, *A space-time multigrid method for parabolic partial differential equations*, SIAM J. Sci. Comput., 16 (1994), pp. 848–864.

[15] K. Ito and K. Kunisch, *Asymptotic properties of receding horizon optimal control problems*, SIAM J. Control Optim., 40 (2002), pp. 1585–1610.

[16] A. Kauffmann, *Optimal Control of the Solid Fuel Ignition Model*, Ph.D. thesis, Technical University of Berlin, Berlin, Germany, 1998.

[17] Y. Kuramoto, *Chemical Oscillations, Waves, and Turbulence*, Springer-Verlag, Berlin, 1984.

[18] J. L. Lions, *Optimal Control of Systems Governed by Partial Differential Equations*, Springer-Verlag, Berlin, 1971.

[19] J. L. Lions, *Control of Distributed Singular Systems*, Gauthier-Villars, Paris, 1985.

[20] K. Malanowski, *Convergence of approximations vs. regularity of solutions for convex, control-constrained optimal-control problems*, Appl. Math. Optim., 8 (1981), pp. 69–95.

[21] P. Neittaanmäki and D. Tiba, *Optimal Control of Nonlinear Parabolic Systems*, Marcel Dekker, New York, 1994.

[22] A. V. Panfilov, *Spiral breakup in an array of coupled cells: The role of intercellular conductance*, Phys. Rev. Lett., 88 (2002), pp. 1–4.

[23] U. Trottenberg, C. Oosterlee, and A. Schüller, *Multigrid*, Academic Press, London, 2001.

[24] S. Vandewalle, *Parallel Multigrid Waveform Relaxation for Parabolic Problems*, B. G. Teubner, Stuttgart, 1993.

Chapter 6

A Time-Parallel Implicit Methodology for the Near-Real-Time Solution of Systems of Linear Oscillators

Julien Cortial and Charbel Farhat†*

6.1 Introduction

Systems of ordinary differential equations (ODEs) are frequently encountered in computational sciences and engineering. Their solution can be often organized around two loops: (a) an inner loop which evaluates the solution and/or related quantities at a specific time instance, and (b) an outer one which advances the solution from a given time instance to the next one. The vocabulary used here assumes the derivative to be a temporal one. Similarly, the computational cost associated with the solution of systems of ODEs can be divided into two main components: (a) the CPU time incurred by the inner loop—that is, by the evaluation of the solution and/or related quantities at a given time instance, and (b) that incurred by the outer loop—that is, by advancing the computed solution in time. To a large extent, parallel algorithms for the solution of systems of ODEs have mainly addressed the reduction of the CPU time associated with the inner loop. When the focus system of ODEs arises from the semidiscretization of a partial differential-evolution equation (PDE), such parallel algorithms can be described as space-parallel algorithms.

By comparison, the parallelization of the outer loop of an ODE solver—or time parallelism—has received little attention in the literature. This is because, in general, time parallelism where the solution is computed simultaneously at different time instances is harder to achieve than space parallelism, due to the inherently sequential nature of the time-integration process. However, time parallelism can be of paramount importance to fast

*Ecole Nationale des Ponts et Chaussées, Champs-Sur-Marne, France. This work was performed while this author was performing an internship at the Center for Aerospace Structures at the University of Colorado at Boulder.

†Department of Aerospace Engineering Sciences and Center for Aerospace Structures, University of Colorado at Boulder, Boulder, CO 80309-0429.

computations, for example, when space parallelism is unfeasible or cannot exploit all the processors available on a given massively parallel system, or when computing the solution of a system of ODEs in near-real time requires exploiting both space and time parallelisms. Usually, space parallelism is unfeasible when the system of ODEs is too small to allow amortization of the interprocessor communication overhead. Robotic and protein folding applications exemplify this scenario as they typically involve only a few degrees of freedom (dofs); yet, these applications can be CPU intensive when time integration is performed over a relatively large time interval. Reduced-order models, which are often constructed to enable near-real-time computations, exemplify the same scenario. Predicting in near-real-time the time evolution of the variables governed by such models calls for time parallelism. Finally, when considering computations on massively parallel computers with thousands of processors such as the Department of Energy's ASC machines, space parallelism alone is not always the optimal strategy for accelerating the solution of reconstruction-evolution problems. Indeed, for many such problems—turbulence being a notorious exception—the mesh resolution dictated by accuracy considerations can be accommodated by a fraction of the total number of available processors. It is well known that in such cases, because of hardware-related issues such as memory bandwidth effects and interprocessor communication costs, the solution time typically increases when the number of processors N_p is increased beyond a certain critical value, N_p^{cr}, while the mesh size is maintained constant. In this sense, the smaller is N_p^{cr} compared to the maximum number of available processors, N_p^{max}, the less optimal is the space-parallel implementation strategy, and the more appealing becomes the approach that combines both space and time parallelisms in order to reduce as much as possible the solution time by exploiting all N_p^{max} processors [1].

Three approaches were proposed in the last decade for implementing time parallelism in an ODE solver. The first one is based on waveform relaxation [2, 3] and was introduced in [4, 5, 6, 7, 8]. In this approach and in the linear case, the matrix governing a system of linear ODEs expressed in first-order form is split into its lower, diagonal, and upper components, and the problem of solving a system of coupled ODEs is transformed into that of iteratively solving a sequence of uncoupled scalar ones. For ODEs arising from the semidiscretization of parabolic problems, multigrid methods were shown to accelerate the convergence rate of the iterative process. As such, waveform relaxation introduces essentially an opportunity for space parallelism, since each uncoupled scalar ODE can be solved concurrently with other ones. However, as described in [6, 8], further algebraic manipulations can be introduced in the waveform relaxation to enable a form of time parallelism. Unfortunately, it was shown in [9] that waveform relaxation does not perform well for hyperbolic problems. Moreover, even for parabolic problems, only the parallel efficiency of the time-parallel waveform relaxation appears to have been studied in the literature. Little if any information appears to be available regarding its intrinsic performance—that is, its serial performance compared to the performance of the serial version of the underlying ODE solver. Since waveform relaxation introduces time parallelism in an ODE solver at the expense of transforming it into a necessarily iterative process, its advantage in terms of performance remains undocumented.

In [10, 11], a second approach was proposed for time-parallelizing ODE solvers. This approach consists in transforming an initial value problem into a boundary value problem, and therefore benefits from all methodologies developed for space parallelism. However, a major drawback of this approach is that it results in a time-parallel ODE solver that is radically different from the underlying serial ODE solver.

The third approach proposed so far for introducing time parallelism in an ODE solver is based on the classical principles of domain decomposition [12] applied to the time domain. In this approach, the time domain of interest is first partitioned into time slices whose boundary points are treated as a coarse time grid. First, the solution is approximated on this coarse time grid to provide a seed—that is, an initial condition—to each time slice. Then, the ODE solver is applied independently and therefore concurrently in each time slice to advance the solution from the starting point of this time slice to its end point. Finally, an iterative process is invoked to improve the seeds and eliminate the jumps of the solution on the coarse time grid. Hence, like waveform relaxation, this approach is an iterative one. To the authors' best knowledge, this type of domain decomposition-based methodology for implementing time parallelism was first proposed in [13], albeit in a form specialized to quasi-Hamiltonian problems. It was then independently rediscovered in [14] in the less restrictive form of the parareal (for parallel real-time) algorithm. This algorithm featured, however, the concept of propagating the jumps of the solution on the coarse time grid in order to improve the accuracy of the seeds and accelerate convergence. The original form of the parareal algorithm [14] was not derived from any specific theory and assumed a single-stage ODE solver. Like waveform relaxation, it was applied to ODEs arising from the semidiscretization of parabolic problems. In [1], the parareal framework for time-parallel computations was formally shown to result from the application of Newton's method to the correction of each local solution associated with a time slice. Using the theory of distributions, this framework was then extended to the case of multistage ODE solvers and applied to the design of a new family of time-parallel implicit ODE solvers named PITA (for parallel implicit time integrator). Simultaneously, a different approach was followed in [15] to revise the original parareal algorithm in order to make it applicable to virtually any ODE solver. In this chapter, it will be shown that the PITA [1] and revised parareal [15] algorithm are identical in the linear case but different in the nonlinear one.

In [1], the feasibility of the PITA and its intrinsic performance were evaluated for three model problems that are representative of fluid, structure, and fluid-structure applications. These studies identified both the physical and simulation parameters that can make or break the objective of time-parallel ODE solvers. It was found that for parabolic and first-order hyperbolic problems, the PITA introduces an effective time parallelism in ODE solvers and offers a viable approach for reducing the total solution time of time-dependent PDE problems to below what space parallelism alone can achieve. However, it was also found that for ODEs resulting from the semidiscretization of second-order hyperbolic problems—such ODEs are usually referred to as "oscillators"—the PITA framework for time-parallel computations generates a parasitic beating phenomenon that severely restricts the stability time step and limits the potential of this time-parallel methodology to shorter-term simulations. Similarly, it was found in [16] then [17] that for linear oscillators, the original [14] and revised [15] parareal algorithms exhibit poor stability properties. Hence, the objective of this chapter is to address this specific issue and present a new framework for time-parallel computations that retains the desired stability properties of a preferred ODE solver, even for systems arising from the semidiscretization of second-order harmonic problems. Because of space limitation, focus is set on the linear case. The extension of this work to nonlinear oscillators will be published elsewhere.

To this effect, the remainder of this chapter is organized as follows. In Section 6.2, the parareal and original PITA frameworks for time-parallel computations are reviewed and contrasted. In Section 6.3, the origin of the spurious beating phenomenon associated with the solution by a parareal algorithm or a PITA of a system of linear oscillators is mathematically identified. A new PITA framework that addresses this issue and its source is presented in Section 6.4, and analyzed in Section 6.5. In particular, this new PITA framework is shown to have an interesting subspace convergence property that makes it attractive for second-order hyperbolic problems. In Section 6.6, the new PITA framework is applied to two different systems of linear oscillators. Superior performance over the parareal [15] and original PITA [1] frameworks and excellent potential for simple as well as complex linear hyperbolic problems are demonstrated.

6.2 Two Frameworks for Time-Parallel Algorithms

6.2.1 Notation and Definitions

Throughout this chapter, the system of ODEs of interest is written in first-order form as follows:

$$\begin{aligned} \frac{dy}{dt} &= F(y) + b(t), \\ y(t_0) &= y_0, \end{aligned} \tag{6.1}$$

where $y \in \mathbb{R}^{N_{dof}}$, N_{dof} denotes the number of dofs or the size of problem (6.1), $F(y)$ is a linear function and therefore can be written as

$$F(y) = Ay, \tag{6.2}$$

$A \in \mathbb{R}^{N_{dof} \times N_{dof}}$ and arises, for example, from the semidiscretization of a PDE, $b \in \mathbb{R}^{N_{dof}}$, t denotes time, and $y_0 \in \mathbb{R}^{N_{dof}}$ denotes an initial condition for y. A preferred *implicit* time integrator for solving (6.1) in the time domain $\Omega_t = [t^0, t^f]$—that is, a preferred implicit ODE solver—is denoted by ITA . The emphasis on implicit schemes is explained in Section 6.2.3. Given a constant time step Δt, Ω_t is partitioned into N_{ts} time slices $\{S^i = [T^i,\ T^{i+1}]\}_{i=0}^{N_{ts}-1}$ of constant size ΔT, where

$$\Delta T = J \Delta t, \ J \in \mathbb{N}, \ J > 1. \tag{6.3}$$

T^{i-} denotes the right boundary of the time slice $S^{i-1} = [T^{i-1},\ T^i]$ and T^{i+} denotes the left boundary of the time slice $S^i = [T^i,\ T^{i+1}]$.

The sampling of Ω_t by Δt is referred to as the *fine* time grid, and the sampling of this time domain by ΔT is referred to as the *coarse* time grid. The correspondence between these two time grids is given by

$$T^i = t^{Ji}, \quad 0 \leq i \leq N_{ts}. \tag{6.4}$$

Hence, Δt and ΔT will be referred to as the fine and coarse time steps, respectively.

The following additional notation is borrowed from [1].

- A quantity associated with a kth iteration is designated by the subscript k.

- A quantity evaluated on the fine time grid at the time instance $t^i = i\Delta t$ is denoted by a lower case and the superscript i.
- A quantity evaluated on the coarse time grid at the time instance $T^i = t^{Ji} = Ji\Delta t = i\Delta T$ is denoted by an upper case and the superscript i.
- A function defined in the time slice S^i is denoted by a lower case, the superscript i, and a pair of parentheses ().
- The result obtained by applying an ITA to the solution of problem (6.1) on the fine time grid is designated by the subscript ITA.
- The exact solution of problem (6.1) is designated by the subscript ex.

For example, the initial approximation of $Y^i = y(T^i)$ is denoted by Y_0^i.

The amplification matrices of a given ITA associated with the coarse and fine time grids are denoted by G_{ITA}^C and G_{ITA}^F, respectively. Using these amplification matrices, the numerical solutions of problem (6.1) on the coarse and fine time grids can be written as

$$Y_k^{i+1} = G_{ITA}^C Y_k^i + B^i \tag{6.5}$$

and

$$y_k^i(t^{Ji+j+1}) = G_{ITA}^F y_k^i(t^{Ji+j}) + b^{Ji+j}, \tag{6.6}$$

respectively, where $B^i \in \mathbb{R}^{N_{dof}}$ and $b^{Ji+j} \in \mathbb{R}^{N_{dof}}$ depend on b and the chosen ITA, and $B^i = b^{Ji+j} = 0$ if $b(t) = 0$. Since $t^{Ji} = T^i$ and $t^{J(i+1)} = T^{i+1}$, it follows that

$$\begin{aligned} y_k^i(T^{(i+1)}) &= (G_{ITA}^F)^J y_k^i(T^i) + \underbrace{\sum_{j=0}^{J-1} (G_{ITA}^F)^{J-j-1}\, b^{Ji+j}}_{\overline{B}^i} \\ &= (G_{ITA}^F)^J y_k^i(T^i) + \overline{B}^i, \end{aligned} \tag{6.7}$$

where i identifies the time slice, $\overline{B}^i \in \mathbb{R}^{N_{dof}}$ depends on b and the chosen ITA, and $\overline{B}^i = 0$ if $b(t) = 0$.

A common objective of the parareal algorithm and PITA is to "process" concurrently as many time slices S^i as there are processors that can be assigned to this task. A common mechanism for achieving this objective is to generate as fast as possible an increasingly accurate sequence of seeds $\{Y_k^i\}_{i=1}^{N_{ts}-1}$ that can be used to initiate embarrassingly parallel computations across the time slices. In the description of both algorithms, the symbol $\ominus$ will be used to designate the time-sequential steps—and these will be performed on the coarse time grid—and the symbol $\oplus$ to designate the time-parallel steps—and these will be performed on the decomposed fine time grid.

As stated in the introduction, the emphasis of this chapter is on linear oscillators, or ODEs that arise from the semidiscretization of second-order hyperbolic problems. In this case, y can be written as

$$y = \begin{pmatrix} \frac{dq}{dt} \\ q \end{pmatrix}, \tag{6.8}$$

where $q \in \mathbb{R}^{N'_{dof}}$, $N'_{dof} = \frac{N_{dof}}{2}$, and A and b can be written as

$$A = \begin{pmatrix} -M^{-1}D & -M^{-1}K \\ I & 0 \end{pmatrix}, \quad b = \begin{pmatrix} M^{-1}f \\ 0 \end{pmatrix}, \tag{6.9}$$

where I denotes the identity matrix, $M \in \mathbb{R}^{N'_{dof} \times N'_{dof}}$, $D \in \mathbb{R}^{N'_{dof} \times N'_{dof}}$, and $K \in \mathbb{R}^{N'_{dof} \times N'_{dof}}$ are the usual symmetric mass, damping, and stiffness matrices, and $f \in \mathbb{R}^{N'_{dof}}$ is a forcing vector.

6.2.2 The Parareal Framework

The parareal framework for time-parallel algorithms considered here is the recent revision proposed in [15] of the original version presented in [14]. Given a preferred ITA for the serial solution of a *linear* system of ODEs written in the form given in (6.1), the corresponding parareal algorithm can be summarized as follows.

- $\ominus$ Step 0: *provide initial seeds* Y_0^i, $0 \leq i \leq N_{ts} - 1$, *for example, by solving problem* (6.1) *on the coarse time grid.*

For $k = 0,\ 1,\ \ldots$

- $\oplus$ Step 1: *using the updated seeds as initial conditions, apply the ITA to the solution of problem* (6.1) *on the decomposed fine time grid*

$$\begin{aligned} \frac{dy_k^i()}{dt} &= Ay_k^i() + b && \text{in } S^i, \\ y_k^i(T^i) &= Y_k^i, && 0 \leq i \leq N_{ts} - 1. \end{aligned}$$

- $\oplus$ Step 2: *evaluate the jumps*

$$\Delta_k^i = y_k^{i-1}(T^i) - Y_k^i, \quad 1 \leq i \leq N_{ts} - 1, \tag{6.10}$$

  on the coarse time grid. Stop if all these jumps are sufficiently small.
- $\ominus$ Step 3: *improve the seeds by propagating the jumps computed above on the coarse time grid*

$$Y_{k+1}^i = (G_{ITA}^C Y_{k+1}^{i-1} + B^{i-1}) + \sum_{l=0}^{k} \Delta_l^i, \quad 0 \leq i \leq N_{ts} - 1. \tag{6.11}$$

Remark: As written in (6.11), Step 3 of the parareal algorithm described above exploits the linear assumption made in this chapter for problem (6.1). The reader is referred to [15] for a general expression of this step that is valid for systems of nonlinear ODEs.

6.2.3 The PITA Framework

The PITA framework for time-parallel computations considered in *this section* is that developed in [1]. Given a preferred ITA for the serial solution of a *linear* system of ODEs written in the form given in (6.1), this algorithm can be described as follows.

- $\ominus$ Step 0: *provide initial seeds* Y_0^i, $0 \le i \le N_{ts} - 1$, *for example, by solving problem* (6.1) *on the coarse time grid.*

For $k = 0, 1, \ldots$

- $\oplus$ Step 1: *using the updated seeds as initial conditions, apply the ITA to the solution of problem* (6.1) *on the decomposed fine time grid*

$$\begin{aligned} \frac{dy_k^i()}{dt} &= Ay_k^i() + b \qquad && \text{in } S^i, \\ y_k^i(T^i) &= Y_k^i, && 0 \le i \le N_{ts} - 1. \end{aligned}$$

- $\oplus$ Step 2: *evaluate the jumps*

$$\Delta_k^i = y_k^{i-1}(T^i) - Y_k^i, \quad 1 \le i \le N_{ts} - 1, \tag{6.12}$$

on the coarse time grid. Stop if all these jumps are sufficiently small.

- $\ominus$ Step 3: *apply the ITA to the solution on the coarse time grid of the following correction problem:*

$$\begin{aligned} &\frac{dc_k}{dt} = F_y(y_k)c_k = Ac_k, \\ &c_k(T^0) = 0, \\ &c_k(T^{i+}) = c_k(T^{i-}) + \Delta_k^i, \quad 1 \le i \le N_{ts} - 1, \end{aligned} \tag{6.13}$$

in order to compute the correction coefficients $C_k^i = c_k(T^{i-})$, $1 \le i \le N_{ts} - 1$.

- $\oplus$ Step 4: *update the seeds*

$$Y_{k+1}^i = y_k^{i-1}(T^i) + C_k^i = Y_k^i + \Delta_k^i + C_k^i, \quad 0 \le i \le N_{ts} - 1. \tag{6.14}$$

Remark: The emphasis on implicit schemes is due to Step 0 and Step 3 of the PITA framework outlined above. In these two steps, the serial ODE solver is applied on a coarse time grid and therefore with a relatively large time step. Because of their numerical stability limitation, explicit schemes cannot operate on time grids that are sufficiently coarse to maximize the parallel efficiency of a PITA, unless the spatial discretization is derefined to keep the CFL number within the given stability limit. This can be quite cumbersome in realistic applications. On the other hand, a carefully chosen implicit ODE solver can operate on most coarse time grids dictated by the parallel efficiency of the PITA framework [1].

6.2.4 Equivalence for Linear Problems

From (6.10) and (6.7), it follows that

$$\Delta_k^i = \left((G_{ITA}^F)^J Y_k^{i-1} + \overline{B}^{i-1}\right) - Y_k^i, \tag{6.15}$$

and from (6.11) it follows that

$$Y_k^i = G_{ITA}^C Y_k^{i-1} + B^{i-1} + \sum_{l=0}^{k-1} \Delta_l^i. \tag{6.16}$$

Substituting (6.15) into (6.11) gives

$$Y_{k+1}^{i} = G_{ITA}^{C} Y_{k+1}^{i-1} + B^{i-1} + \sum_{l=0}^{k-1} \Delta_l^i + \Delta_k^i \tag{6.17}$$

$$= G_{ITA}^{C} Y_{k+1}^{i-1} + B^{i-1} + \sum_{l=0}^{k-1} \Delta_l^i + (G_{ITA}^{F})^J Y_k^{i-1} + \overline{B}^{i-1} - Y_k^i, \tag{6.18}$$

which in view of (6.16) leads to

$$Y_{k+1}^{i} = G_{ITA}^{C} Y_{k+1}^{i-1} + (G_{ITA}^{F})^J Y_k^{i-1} + \overline{B}^{i-1} - G_{ITA}^{C} Y_k^{i-1}. \tag{6.19}$$

Hence, Step 3 of the parareal algorithm can be rewritten as

$$Y_{k+1}^{i} = G_{ITA}^{C} (Y_{k+1}^{i-1} - Y_k^{i-1}) + (G_{ITA}^{F})^J Y_k^{i-1} + \overline{B}^{i-1}. \tag{6.20}$$

Now, from (6.14), problem (6.13), and (6.7) it follows that Step 4 of the PITA framework for time-parallel algorithms can be rewritten as

$$Y_{k+1}^{i} = (G_{ITA}^{F})^J Y_k^{i-1} + \overline{B}^{i-1} + C_k^i. \tag{6.21}$$

Since the solution of problem (6.13) satisfies

$$C_k^i = G_{ITA}^{C} (C_k^{i-1} + \Delta_k^{i-1}) = G_{ITA}^{C} (Y_{k+1}^{i-1} - Y_k^{i-1}), \tag{6.22}$$

(6.21) can be rewritten as

$$Y_{k+1}^{i} = (G_{ITA}^{F})^J Y_k^{i-1} + \overline{B}^{i-1} + G_{ITA}^{C} (Y_{k+1}^{i-1} - Y_k^{i-1}), \tag{6.23}$$

which in view of (6.20) proves that for linear problems, the revised parareal framework presented in [15] and the original PITA framework proposed in [1] generate identical time-parallel algorithms. For nonlinear problems, the two frameworks differ by the manner in which they propagate the jumps on the coarse time grid. The original PITA [1] adopts a Newton method for this purpose—which explains the occurrence of the directional derivative F_y in problem (6.13)—whereas the parareal framework of [15] relies on the user for providing a preferred coarse propagator.

6.3 Unstable Behavior for Linear Oscillators

The accuracy properties of the PITA methodology and some aspects of its numerical stability were formally analyzed in [1]. Reference [1] also includes a report on the feasibility and performance of several PITAs for the solution of linear and nonlinear, fluid, structure, and fluid-structure interaction problems. It was found that for first-order hyperbolic problems exemplified by gas dynamics applications, PITAs offer a significant potential for speeding up solution time on massively parallel processors. However, it was also found that for second-order linear or linearized hyperbolic problems exemplified by linear or linearized structural dynamics applications, the PITA methodology generates a parasitic "beating" phenomenon that severely restricts its stability time step. More specifically, it was numerically shown in [1] that for a linear oscillator, the jumps (6.10) generated by a PITA—and for that matter, by a parareal algorithm too—exhibit a harmonic behavior characterized by a frequency that is very close to a natural frequency of the linear oscillator. Since these jumps drive the coarse correction system (6.13), and this system has the same mathematical structure as the original system (6.1), it follows that this system—and therefore Step 3 of a PITA—undergoes a beating phenomenon and exhibits large-amplitude oscillations when the coupled system of ODEs (6.1) represents a linear oscillator. Since the parareal algorithm and PITA are identical for linear problems, it also follows that the parareal algorithm suffers in the same manner from the consequences of this parasitic beating phenomenon. Indeed, it was shown in [16] that for linear oscillators, parareal algorithms exhibit numerical instabilities at any reasonably large time step. The same conclusion was reached in [17], where the author advocated using artificial numerical dissipation for restoring numerical stability when a parareal algorithm is applied to the solution of a second-order hyperbolic problem. However, artificial dissipation is detrimental to accuracy and unpopular among structural dynamicists. For this reason, the objective of this chapter is to present an alternative approach for restoring numerical stability when the parareal algorithm and PITA are applied to linear oscillators. But first, the parasitic beating phenomenon observed in [1] is proved next.

In the remainder of this section, it is assumed that G^C_{ITA} and G^F_{ITA} commute. This assumption is not restrictive as it holds, for example, for the entire family of Newmark methods. Furthermore, the case where $b(t) = 0$, and therefore $\overline{B}^i = B^i = 0$, is considered for simplicity but without any loss of generality. In this case, (6.15) simplifies to

$$\Delta^i_k = (G^F_{ITA})^J Y^{i-1}_k - Y^i_k. \tag{6.24}$$

First, note that the initial seeds obtained by solving problem (6.1) on the coarse time grid satisfy

$$Y^i_0 = (G^C_{ITA})^i y_0. \tag{6.25}$$

Next, recall that it was proved in [1] that a PITA is characterized by the following finite termination properties:

$$\begin{array}{rcll} \Delta^i_k & = & 0, & 0 \leq i \leq k, \\ C^i_k & = & 0, & 0 \leq i \leq k+1, \\ Y^i_{k+1} & = & (A^F_{ITA})^{Ji} y_0, & 0 \leq i \leq k+1. \end{array} \tag{6.26}$$

Finally, let

$$\widetilde{G}_{ITA} = (G^F_{ITA})^J (G^C_{ITA})^{-1}, \tag{6.27}$$

and consider the following convention:

$$\forall (n,k) \in \mathbb{N}^2, \quad n < k \Rightarrow \binom{n}{k} = 0. \tag{6.28}$$

Proposition 6.1. *Given a problem of the form given in* (6.1) *with* $b(t) = 0$ *and a serial ITA for which* G^C_{ITA} *and* G^F_{ITA} *commute, the corresponding PITA satisfies*

$$Y^i_k = \left[\sum_{l=0}^{k} \binom{i}{l} (\widetilde{G}_{ITA} - I)^l \right] Y^i_0, \tag{6.29a}$$

$$\Delta^i_k = \binom{i-1}{k} (\widetilde{G}_{ITA} - I)^{k+1} Y^i_0, \quad i > 0, \tag{6.29b}$$

$$C^i_k = \binom{i-1}{k+1} (\widetilde{G}_{ITA} - I)^{k+1} Y^i_0, \quad i > 0. \tag{6.29c}$$

Proof. The proof of the above proposition is given in Appendix A. □

From (6.29b), (6.25), (6.27), and the fact that G^C_{ITA} and G^F_{ITA} commute it follows that

$$\begin{aligned}
\Delta^i_k &= \binom{i-1}{k} (\widetilde{G}_{ITA} - I)^{k+1} (G^C_{ITA})^i y_0 \\
&= \binom{i-1}{k} (G^C_{ITA})^i (\widetilde{G}_{ITA} - I)^{k+1} y_0 \\
&= \binom{i-1}{k} (G^C_{ITA})^i (\widetilde{G}_{ITA} - I)^{k+1} (G^C_{ITA})^{-(k+1)} Y^{k+1}_0 \\
&= \binom{i-1}{k} (G^C_{ITA})^{i-(k+1)} \Delta^{k+1}_k.
\end{aligned} \tag{6.30}$$

For a system of linear oscillators, the exact solution $c_{ex}(t)$ of problem (6.13) with $b(t) = 0$ can be expressed as the superposition of harmonic functions. Each one of these harmonic functions is characterized by a natural frequency of the oscillating system and can be approximated by solving numerically a decoupled ODE obtained by diagonalizing problem (6.13). For each decoupled ODE, the amplification matrix on the coarse time grid of the chosen ITA can be written as $G^C_{ITA} = e^{I\varphi}$, where I is the imaginary number satisfying $I^2 = -1$, and φ is a constant that depends on the time step ΔT and the natural frequency characterizing that ODE. It follows that for a given k, $(G^C_{ITA})^{i-(k+1)} = e^{I(i-(k+1))\varphi}$, and therefore the jump Δ^i_k can also be expressed as the superposition of harmonic approximations characterized by the natural frequencies of the same oscillating system, and these are also natural frequencies of the correction problem (6.13). Hence, for a system of linear oscillators, Step 3 of the parareal algorithm (see Section 6.2.2) and Step 3 of the original PITA (see Section 6.2.3) encounter a resonance phenomenon which, because of numerical approximation and round-off effects, can behave as a beating phenomenon. This resonance or beating phenomenon was shown in [1] to severely restrict the stability time step of the parareal algorithm and PITA and limit their potential to shorter-term simulations.

6.4 A New Time-Parallel Framework for Second-Order Hyperbolic Problems

6.4.1 Propagation of the Jumps on Both Time Grids

For linear problems, propagating the jumps Δ_k^i on the fine time grid guarantees convergence in a single iteration of a PITA to the solution computed by the underlying serial ITA. However, such a strategy is unfeasible because it involves a serial computation of the same complexity as the original problem to be solved. For this reason, both the parareal algorithm and the original PITA improve the seeds Y_k^i by propagating the jumps of the solution on the coarse time grid, as highlighted in Step 3 of Section 6.2.2 and Step 3 of Section 6.2.3. In other words, convergence—and therefore accuracy and numerical stability—suggests propagating the jumps Δ_k^i on the fine time grid, whereas computational efficiency suggests propagating them on a coarse time grid. The idea introduced next attempts to extract the best of these two suggestions without inheriting their respective disadvantages.

From Step 3 described by (6.13), it follows that the original PITA propagates the jumps on the coarse time grid by computing there the sequence of corrections

$$C_k^{i+1} = G_{ITA}^C(C_k^i + \Delta_k^i) \tag{6.31}$$

with $C_k^0 = 0$. Here the main idea has two parts. Its first part consists of improving the convergence of the original PITA by propagating Δ_k^i on both the coarse and fine time grids. For this reason, it is proposed to compute the correction terms C_i^k as follows:

$$\begin{aligned} C_k^0 &= 0, \\ C_k^{i+1} &= \left((G_{ITA}^F)^J P_k + G_{ITA}^C(I - P_k)\right)(C_k^i + \Delta_k^i), \quad 0 \le i \le N_{ts} - 1, \end{aligned} \tag{6.32}$$

where P_k is an orthogonal projector onto a subspace $\mathcal{S}_k$. This implies that the information to be propagated, namely,

$$U_k^i = C_k^i + \Delta_k^i, \tag{6.33}$$

is divided into two components. The first component,

$$U_k^{i^{\mathcal{S}_k}} = P_k U_k^i, \tag{6.34}$$

belongs to $\mathcal{S}_k$ and is propagated on the fine time grid. The second component,

$$U_k^{i^\perp} = U_k^i - U_k^{i^{\mathcal{S}_k}} = (I - P_k)U_k^i, \tag{6.35}$$

belongs to the complementary of $\mathcal{S}_k$ in $\mathbb{R}^{N_{dof}}$ and is propagated on the coarse time grid.

The second part of the main idea consists in recognizing that for the purpose of computational efficiency, the subspace $\mathcal{S}_k$ must be chosen so that for all $s \in \mathcal{S}_k$, $(G_{ITA}^F)^J s$ can be evaluated without performing any additional computation on the fine time grid. The construction of such a subspace $\mathcal{S}_k$ is presented next.

6.4.2 Construction of the Subspace $\mathcal{S}_k$

The Homogeneous Case ($b(t) = 0$)

In the homogeneous case, $b(t) = 0$, $\overline{B}^i = B^i = 0$, $0 \le i \le N_{ts} - 1$, and therefore $y_k^i(T^{(i+1)}) = (G_{ITA}^F)^J Y_k^i$ (see (6.7) and the second of (6.12)).

Let $\mathcal{S}_k$ be the vector space defined by

$$\mathcal{S}_k = Vect(Y_l^j;\ 0 \leq l \leq k,\ 0 \leq j \leq N_{ts} - 1). \tag{6.36}$$

Hence,

$$\forall s \in \mathcal{S}_k,\ \ \exists\, (\alpha_l^j) \in R^{N_{ts}(k+1)}\ /\ s = \sum_{l,j} \alpha_l^j Y_l^j. \tag{6.37}$$

Since at the $(k+1)$th iteration the vectors $(G_{ITA}^F)^J Y_l^j = y_l^j(T^{(l+1)})$, $0 \leq l \leq k$, have already been computed on the fine time grid during the previous parallel steps, it follows that for all vectors s belonging to $\mathcal{S}_k$, the propagation

$$(G_{ITA}^F)^J s = \sum_{l,j} \alpha_j^l (G_{ITA}^F)^J Y_l^j = \sum_{l,j} \alpha_l^j y_l^j (T^{(j+1)}) \tag{6.38}$$

can be performed without any additional computation on the fine time grid. Hence, the choice of $\mathcal{S}_k$ specified in (6.36) meets the computational efficiency requirement outlined in Section 6.4.1.

The Nonhomogeneous Case ($b(t) \neq 0$)

If $b(t) \neq 0$, from (6.7) and the second of (6.12) it follows that $y_k^i(T^{(i+1)})$ can be written as

$$y_k^i(T^{(i+1)}) = (G_{ITA}^F)^J Y_k^i + \overline{B}^i. \tag{6.39}$$

Equation (6.39) above suggests that $\overline{B}_i$, which is independent of k, can be precomputed in a time-parallel fashion by initializing the seeds to $Y_{-1}^i = 0$ and applying the ITA to the solution of the time-decomposed problem (6.12). This gives

$$\overline{B}^i = y_{-1}^i(T^{(i+1)}),\ \ 0 \leq i \leq N_{ts} - 1, \tag{6.40}$$

where the notation $k = -1$ is used to designate a precomputation. Then, if the subspace $\mathcal{S}_k$ is chosen as in the homogeneous case—that is, as in (6.36)—from (6.37), (6.39), and (6.40) it follows that for all vectors s belonging to $\mathcal{S}_k$,

$$(G_{ITA}^F)^J s = \sum_{l,j} \alpha_j^l (G_{ITA}^F)^J Y_l^j = \sum_{l,j} \alpha_l^j \left(y_l^j(T^{(j+1)}) - y_{-1}^i(T^{(i+1)}) \right). \tag{6.41}$$

This shows that if $\overline{B}^i$ is precomputed as outlined in (6.40), at each iteration $k+1$ the desired propagation (6.41) can be performed without any additional computation on the fine time grid. It follows that the choice of $\mathcal{S}_k$ specified in (6.36) meets also in this case the computational efficiency requirement formulated in Section 6.4.1.

An Alternative Description of the Subspace $\mathcal{S}_k$

From (6.14) and definition (6.33) it follows that the choice made in (6.36) for the subspace $\mathcal{S}_k$ can also be written as

$$\mathcal{S}_k = Vect(Y_0^j, U_l^j;\ 0 \leq l \leq k-1,\ 0 \leq j \leq N_{ts} - 1), \tag{6.42}$$

which highlights the definition of $\mathcal{S}_k$ as the subspace generated by the initial values of the seeds, and the solution jumps and corrections on the coarse time grid.

6.4.3 Construction of the Projector P_k

If S_k denotes the matrix whose range spans the subspace $\mathcal{S}_k$, an orthogonal projector onto $\mathcal{S}_k$ can be written as

$$P_k = S_k(S_k^T Q S_k)^{-1} S_k^T Q, \tag{6.43}$$

where the superscript T denotes the transpose, and $Q \in \mathbb{R}^{N_{dof} \times N_{dof}}$ represents the inner product chosen for defining the orthogonality—or in other words, the metric. Different choices for Q lead to different projectors P_k and therefore to different PITAs, albeit within the same framework of time-parallel algorithms proposed in this chapter for addressing second-order hyperbolic problems.

As recalled in Section 6.3, it was found in [1] that for a system of linear oscillators, the jumps (6.10) generated by a parareal algorithm or the original PITA exhibit a harmonic behavior that is characterized by a natural frequency of the system and consequently drive the coarse correction problem (6.13) into a resonance or beating phenomenon that triggers numerical instability. Therefore, in order to restore numerical stability, it is important that whenever the U_k^i vectors (see (6.33)) contain natural or eigenmodes of the given system of linear oscillators, the projector P_k filters out these modes. Since the natural modes of a system of linear oscillators are the eigenmodes of the symmetric pencil (K, M), where K and M are the symmetric mass and stiffness matrices associated with this system (see Section 6.2.1), these modes are M- and K-orthogonal. Hence, they can be filtered out of U_k^i if the orthogonal projector P_k is based on the inner product implied by the choice

$$Q = \begin{pmatrix} M & 0 \\ 0 & K \end{pmatrix}. \tag{6.44}$$

When equipped with the choice of the Q-metric specified above, P_k minimizes the energy norm (potential and kinetic) of the $U_k^{i\perp}$ information (see (6.35)) that is propagated on the coarse time grid, which is also desirable.

Finally, it is noted that a few other choices for Q including $Q = I$ were considered; they failed either to restore numerical stability in the PITA methodology, or to perform as well as the choice (6.44), for many but the simplest structural dynamics applications.

6.4.4 Computational Remarks

The propagation (6.32) of U_k^i (6.33) involves the computation of $(G_{ITA}^F)^J P_k U_k^i$ and $G_{ITA}^C(U_k^i - P_k U_k^i)$. As emphasized by (6.38) and (6.41), the first of these two computations requires essentially the extraction of the coefficients α_l^j. From (6.37) and expression (6.43) of the projector P_k , it follows that these coefficients can be obtained in vector form by performing only the first three subcomputations of the product $P_k U_k^i$ as follows:

$$\alpha = \underbrace{\left(S_k^T Q S_k\right)^{-1} \underbrace{\left(S_k^T \underbrace{(Q U_k^i)}_{1}\right)}_{2}}_{3}. \tag{6.45}$$

On the other hand, the computation of $G_{ITA}^{C}(U_k^i - P_k U_k^i)$ requires evaluating the entire product

$$\left(\underbrace{S_k\,\underbrace{\left(S_k^T Q S_k\right)^{-1}\,\underbrace{\left(S_k^T\,\underbrace{(QU_k^i)}_{1}\right)}_{2}}_{3}}_{4}\right). \tag{6.46}$$

The parallelization of all subcomputations highlighted in (6.45) and (6.46) is trivial, except for the solution of the coarse problems of the form

$$(S_k^T Q S_k) V_k^i = W_k^i, \tag{6.47}$$

where V_k^i and W_k^i are two vectors. The parallelization of the above subcomputation depends on many factors including the chosen solution algorithm and the target number of processors. For examples of parallel coarse problem solvers, the reader is referred to [18, 19, 20].

6.4.5 PITA for Systems of Linear Oscillators

The new PITA framework proposed in this chapter for time-parallelizing a preferred ITA for the solution of a system of linear oscillators can be described as follows.

The Homogeneous Case ($b(t) = 0$)

- $\ominus$ Step 0: *provide initial seeds* Y_0^i, $0 \le i \le N_{ts} - 1$, *for example, by solving problem* (6.1) *on the coarse time grid.*

For $k = 0, 1, \ldots$

- $\oplus$ Step 1: *using the updated seeds as initial conditions, apply the ITA to the solution of problem* (6.1) *on the decomposed fine time grid.*
- $\oplus$ Step 2: *evaluate the jumps*

$$\Delta_k^i = y_k^{i-1}(T^i) - Y_k^i, \quad 1 \le i \le N_{ts} - 1, \tag{6.48}$$

on the coarse time grid. Stop if all these jumps are sufficiently small.

- $\oplus$ Step 3: *Construct the subspace* S_k *and the corresponding projector* P_k.
- $\ominus$ Step 4: *apply the ITA to the generation of the correction coefficients* $C_k^i = c_k(T^{i-})$, $1 \le i \le N_{ts} - 1$, *as summarized in* (6.32).
- $\oplus$ Step 5: *update the seeds*

$$Y_{k+1}^i = y_k^{i-1}(T^i) + C_k^i = Y_k^i + \Delta_k^i + C_k^i, \quad 0 \le i \le N_{ts} - 1. \tag{6.49}$$

The Nonhomogeneous Case ($b(t) \neq 0$)

- $\oplus$ Step -1: *set* $Y_{-1}^i = 0$, $0 \le i \le N_{ts} - 1$, *and apply the ITA to the solution of problem* (6.1) *on the decomposed fine time grid. This step generates the quantities* $\overline{B}^i$, $0 \le i \le N_{ts} - 1$.

- $\ominus$ Step 0: *provide initial seeds* Y_0^i, $0 \le i \le N_{ts} - 1$, *for example, by solving problem* (6.1) *on the coarse time grid.*

For $k = 0, 1, \ldots$

- $\oplus$ Step 1: *using the updated seeds as initial conditions, apply the ITA to the solution on the decomposed fine time grid of problem* (6.1).
- $\oplus$ Step 2: *evaluate the jumps*

$$\Delta_k^i = y_k^{i-1}(T^i) - Y_k^i, \quad 1 \le i \le N_{ts} - 1, \tag{6.50}$$

on the coarse time grid. Stop if all these jumps are sufficiently small.
- $\oplus$ Step 3: *Construct the subspace* S_k *and the corresponding projector* P_k.
- $\ominus$ Step 4: *apply the ITA to the generation of the correction coefficients* $C_k^i = c_k(T^{i-})$, $1 \le i \le N_{ts} - 1$, *as summarized in* (6.32).
- $\oplus$ Step 5: *update the seeds*

$$Y_{k+1}^i = y_k^{i-1}(T^i) + C_k^i = Y_k^i + \Delta_k^i + C_k^i, \quad 0 \le i \le N_{ts} - 1. \tag{6.51}$$

6.5 Mathematical Analysis

Here some of the mathematical properties of the new PITA framework are exposed. For simplicity, but without any loss of generality, it is assumed that $b(t) = 0$.

6.5.1 Preliminaries

The amplification matrix associated with computation (6.32) of the correction coefficients C_i^k is

$$\widehat{G}_{ITA,k}^C = (G_{ITA}^F)^J P_k + G_{ITA}^C (I - P_k). \tag{6.52}$$

From (6.49), (6.32), (6.52), and (6.33) it follows that

$$Y_{k+1}^i = y_k^{i-1}(T^i) + C_k^i = (G_{ITA}^F)^J Y_k^{i-1} + \widehat{G}_{ITA,k}^C U_k^{i-1}, \tag{6.53}$$

and from (6.49) and (6.33) it follows that

$$Y_{k+1}^{i-1} - Y_k^{i-1} = \Delta_k^{i-1} + C_k^{i-1} = U_k^{i-1}. \tag{6.54}$$

Substituting (6.53) into (6.24) written for $k + 1$, and (6.54) into the result, leads to the following expression of the jumps on the coarse time grid:

$$\Delta_{k+1}^i = \left((G_{ITA}^F)^J - \widehat{G}_{ITA,k}^C\right) U_k^{i-1}, \tag{6.55}$$

which in view of (6.52) can also be rewritten as

$$\Delta_{k+1}^i = \left((G_{ITA}^F)^J - G_{ITA}^C\right)(I - P_k) U_k^{i-1}. \tag{6.56}$$

6.5.2 Error Analysis of the Hybrid Coarse/Fine Propagation

The new PITA framework for time-parallel computations distinguishes itself from the original PITA framework [1], and from the parareal framework [15], by algorithm (6.32) for the computation of the correction coefficients. This algorithm propagates selective information on both the coarse and the fine time grids. It is characterized by the following result.

Proposition 6.2. *Let* $|\cdot|$ *denote the Euclidian norm for* $\mathbb{R}^{N_{dof}}$ *associated with the metric of the projector* P_k, *and let* G^C_{ex} *denote the exact amplification matrix on the coarse time grid of the solution of problem* (6.1). *Given a* pth-*order accurate ITA that is stable on both the fine and coarse time grids, there exists a constant* Θ_k *such that*

$$\forall y \in \mathbb{R}^{N_{dof}}, \quad \left|(\widehat{G}^C_{ITA,k} - G^C_{ex})y\right| \leq \Theta_k \Delta T^p \, |y| \, . \tag{6.57}$$

In other words, algorithm (6.32) *preserves the order of accuracy of the chosen ITA.*

Proof. The exact amplification matrix on the coarse time grid can be defined by

$$y_{ex}(T^{i+1}) = G^C_{ex} \; y_{ex}(T^i), \quad 0 \leq i \leq N_{ts}. \tag{6.58}$$

From (6.52), it follows that

$$\widehat{G}^C_{ITA,k} - G^C_{ex} = \left((G^F_{ITA})^J - G^C_{ex}\right)P_k + (G^C_{ITA} - G^C_{ex})(I - P_k). \tag{6.59}$$

If the chosen ITA is pth-order accurate and stable on both the fine and coarse time grids, there exist two constants $\theta_{1,k}$ and $\theta_{2,k}$ associated with its accuracy on the coarse and fine time grids, respectively, and a third constant $\theta_{3,k}$ associated with its stability on the fine time grid such that

$$\forall y \in \mathbb{R}^{N_{dof}}, \quad \left|\left((G^F_{ITA})^J - G^C_{ex}\right)P_k y\right| \leq J\theta_{2,k}\theta_{3,k}\Delta t^p \, |P_k y| \tag{6.60}$$

and

$$\forall y \in \mathbb{R}^{N_{dof}}, \quad \left|(G^C_{ITA} - G^C_{ex})(I - P_k)y\right| \leq \theta_{1,k}\Delta T^p \, |(I - P_k)y| \, . \tag{6.61}$$

(For further details, see [1].) Since P_k is an orthogonal projector with the same metric as $|\cdot|$, then

$$\forall y \in \mathbb{R}^{N_{dof}}, \quad |P_k y| \leq |y|, \text{ and } |(I - P_k)y| \leq |y|. \tag{6.62}$$

From (6.59)–(6.62) it follows that

$$\begin{aligned}
\forall y \in \mathbb{R}^{N_{dof}}, \quad \left|(\widehat{G}^C_{ITA,k} - G^C_{ex})y\right| &\leq \left|\left((G^F_{ITA})^J - G^C_{ex}\right)P_k y\right| + \left|(G^C_{ITA} - G^C_{ex})(I - P_k)y\right| \\
&\leq J\theta_{2,k}\theta_{3,k}\Delta t^p \, |y| + \theta_{1,k}\Delta T^p \, |y| \\
&\leq J^{1-p}\theta_{2,k}\theta_{3,k}\Delta T^p \, |y| + \theta_{1,k}\Delta T^p \, |y| \\
&\leq (J^{1-p}\theta_{2,k}\theta_{3,k} + \theta_{1,k})\Delta T^p \, |y| \\
&\leq \Theta_k \Delta T^p \, |y| \, ,
\end{aligned} \tag{6.63}$$

where $\Theta_k = J^{-p}\theta_{2,k}\theta_{3,k} + \theta_{1,k}$, which proves Proposition 6.2. ☐

6.5.3 Convergence in a Subspace

Let $\{\mathcal{E}_l\}_{l=1}^{N'_{dof}}$ denote the eigen subspaces of dimension equal to two that are associated with a given system of linear oscillators and orthogonal with respect to the inner product

$$\left\langle \begin{pmatrix} \dot{q}_1 \\ q_1 \end{pmatrix}, \begin{pmatrix} \dot{q}_2 \\ q_2 \end{pmatrix} \right\rangle_{MK} = \dot{q}_1^T M \dot{q}_2 + q_1^T K q_2. \tag{6.64}$$

In definition (6.64), a dot designates a time derivative, and M and K are the usual mass and stiffness matrices associated with the second-order form

$$\begin{aligned} M\ddot{q} + D\dot{q} + Kq &= f(t), \\ q(0) &= q_0 \in \mathbb{R}^{N'_{dof}}, \\ \dot{q}(0) &= \dot{q}_0 \in \mathbb{R}^{N'_{dof}} \end{aligned} \tag{6.65}$$

of a system of linear oscillators and were introduced in (6.9). These subspaces form a basis. Therefore,

$$\mathbb{R}^{N_{dof}} = \mathcal{E}_1 \oplus \mathcal{E}_2 \oplus \cdots \oplus \mathcal{E}_{N'_{dof}}. \tag{6.66}$$

Let

$$\mathbb{E}_r = \bigcup_{l=1}^{r} \mathcal{E}_l, \tag{6.67}$$

where the notation $\bigcup_{l=1}^{r}$ does not necessarily imply an ordering for l but denotes only the union of r subspaces. From (6.66), it follows that for any given initial condition $y_0 \in \mathbb{R}^{N_{dof}}$ of problem (6.1) (or any pair of initial conditions $q_0 \in \mathbb{R}^{N'_{dof}}$ and $\dot{q}_0 \in \mathbb{R}^{N'_{dof}}$ of the equivalent problem (6.65)), there exists an integer $r(y_0)$ such that $y_0 \in \mathbb{E}_{r(y_0)}$. The new PITA framework proposed in this chapter for time-parallelizing the solution of a system of linear ODEs associated with a second-order hyperbolic problem is characterized by the following result.

Proposition 6.3. *If the amplification matrix G_{ITA} of the chosen ITA satisfies*

$$\forall y_0 \in \mathbb{E}_{r(y_0)}, \quad \forall i \geq 0, \quad (G_{ITA})^i y_0 \in \mathbb{E}_{r(y_0)}, \tag{6.68}$$

the new PITA framework converges as soon as $\mathcal{S}_k = \mathbb{E}_{r(y_0)}$.

Proof. The proof of the above proposition is given in Appendix B. □

Note that assumption (6.68) is a weak one because it is satisfied by most implicit time integrators including the popular midpoint rule (which for linear problems is identical to the Newmark algorithm with $\beta = 1/4$ and $\gamma = 1/2$). Also note that in practice, $r(y_0)$ is small because, in most engineering applications, only a small fraction of the eigenmodes of a system of linear oscillators are excited. Hence, the result presented in Proposition 6.3 is significant as it implies that the new PITA framework can be expected to converge much faster for systems of linear oscillators than suggested by the finite termination properties (6.26).

6.6 Examples

The new PITA framework proposed in this chapter for time-parallelizing the solution of systems of ODEs arising from the semidiscretization of linear second-order hyperbolic problems is illustrated here with two sample applications. The first one focuses on the vibration of a simple three-dimensional space structure with 348 dofs only. Because of this small value of N_{dof}, this problem is representative of many candidate problems for near-real-time processing. It has also the advantage of being sufficiently simple to reproduce by the interested reader. The second application involves a detailed finite element structural model of an F-16 fighter aircraft with 168,799 dofs. It was mainly chosen to highlight the robustness of the new PITA framework for complex applications. Because of its size, it is representative of those moderate-scale problems whose space-based parallelization alone cannot exploit efficiently a large number of processors.

For both applications, the Newmark algorithm with $\beta = 1/4$ and $\gamma = 1/2$ is chosen as the ITA because of its second-order accuracy and unconditional stability. The fine time step, Δt, is set to meet the accuracy requirement of each application. The coarse time step, ΔT, is determined from the imposed ratio $J = \frac{\Delta T}{\Delta t}$. In [1], it was shown that the theoretical parallel speedup of a PITA, S_p, increases with J, as long as all the N_c target processors or clusters of processors are active (this implies that $N_c \geq N_{ts}$). It was also shown in [1] that S_p decreases with the number of iterations for convergence, N_{itr} (or $k^\star$), and that this number of iterations increases with J. Hence, J should be chosen sufficiently small to keep N_{itr} low, and sufficiently large to enable a good theoretical speedup, S_p. For many applications including the two examples considered here , $10 \leq J \leq 20$ is a reasonable choice.

As recalled in the introduction of this chapter, all attempts previously reported in the literature for time-parallelizing the solution of ODEs failed for applications related to second-order hyperbolic problems because of intrinsic performance issues, and not parallel implementation issues. For this reason, the remainder of this section focuses on the intrinsic performance of the new PITA framework which, after Δt and J are set as discussed above, is essentially governed by N_{itr}. Indeed, the lower is N_{itr}, the lower is the excess of computational work performed by the PITA in comparison with the computational work performed by the sequential ITA, and the lower is the computational overhead associated with constructing and operating the projector P_k. A complementary study of the parallel implementation strategies of a PITA and their CPU performances on various parallel processors will be published elsewhere.

6.6.1 Free Vibration of a Three-Dimensional Space Structure

The three-dimensional structure considered here is graphically depicted in Figure 6.1. It has 32 members represented by an undamped ($D = 0$) finite element structural model with 128 two-noded beam elements, 116 nodes, and 672 dofs. It is clamped at one end (on the left in Figure 6.1). This structure is excited by an initial displacement corresponding to a linear combination of the first ten mode shapes of its finite element model. The fine time step, Δt, is chosen such that the period of the highest excited frequency (tenth mode) is sampled by 30 points. This approach for choosing Δt is typical for implicit time integrators. The ratio of the coarse and fine time steps is set to $J = 10$, and the dynamic response of the structure set in free vibrations $\big(b(t) = 0\big)$ is computed during $N_{ts} = 24$ time slices.

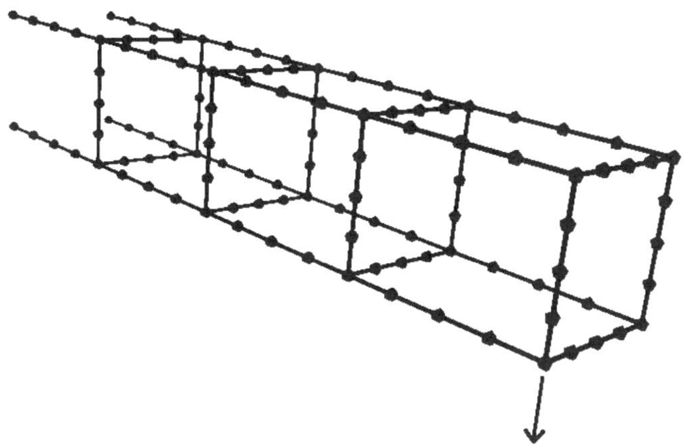

Figure 6.1. *Three-dimensional space structure and its finite element semi-discretization. Arrow designates the dof monitored in Figures* 6.2 *and* 6.3.

Both the new PITA proposed in this chapter and the parareal algorithm described in [15] are used for this purpose. Figure 6.2 reports the displacement and velocity results obtained for the dof designated by an arrow in Figure 6.1 using the parareal algorithm. These results highlight a numerical instability as well as a failure to converge to the sequential solution after three iterations. On the other hand, Figure 6.3, which reports the displacement and velocity time histories obtained for the same dof but using the new PITA framework, shows that convergence to the sequential solution is achieved in only two iterations.

In summary, the simple problem considered here illustrates the inability of the parareal [14, 15] and original PITA [1] frameworks to address second-order hyperbolic problems, while highlighting the excellent potential of the new PITA framework described in this chapter for time-parallelizing their solution.

6.6.2 Dynamic Responses of an F-16 Fighter Aircraft

Figure 6.4 displays an undamped ($D = 0$) finite element structural model of an F-16 fighter aircraft featuring bar, beam, solid, plate, shell, metallic and composite elements, and a total of 168,799 dofs. First, this model is set into free vibrations $\big(b(t) = 0\big)$ by specifying as initial displacement condition a linear combination of its first 16 mode shapes. The fine time step is set to $\Delta t = 1.1 \times 10^{-3}$ s (1.1 millisecond). This time step is typical for implicit nonlinear aeroelastic computations [21] and corresponds to sampling the period of the 16th mode by 60 points. The ratio of the coarse and fine time steps is set to $J = 20$ in order to consider a more difficult scenario than in the previous example. Figure 6.5 reports the displacement and velocity time histories predicted by the (new) PITA for a dof located on the aileron of the aircraft. After 2 iterations ($k = 1$), the results are accurate during the first 5 to 6 cycles of the dynamic response of the aircraft. After 3 iterations ($k = 2$), the predicted displacements are in excellent agreement with their sequential counterparts throughout all

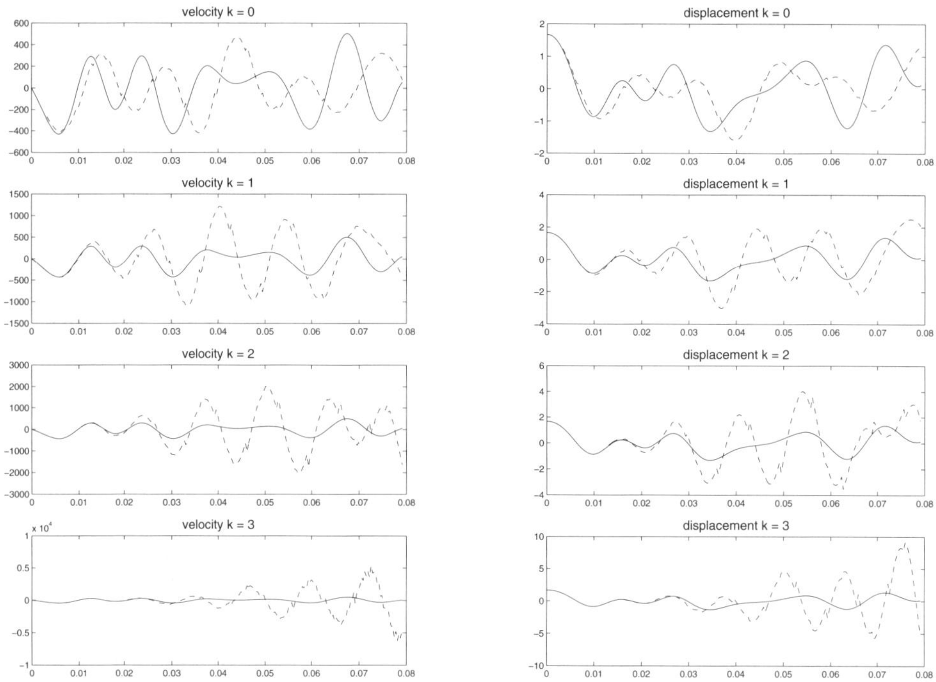

Figure 6.2. *Velocity (left) and displacement (right) time histories of the dof designated by an arrow in Figure* 6.1*: sequential solution (solid line) and parareal solution* [15] *(dashed line)—$J = 10$, $N_{ts} = 24$. Horizontal axis displays time, vertical axis displays displacement or velocity, and k denotes the iteration number.*

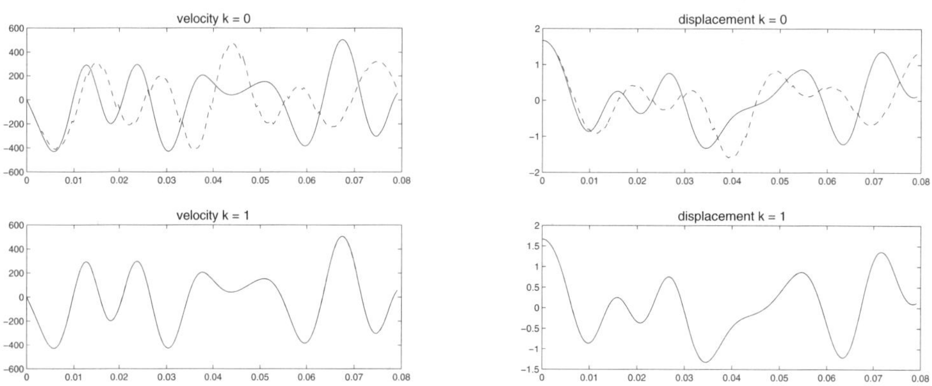

Figure 6.3. *Velocity (left) and displacement (right) time histories of the dof designated by an arrow in Figure* 6.1*: sequential solution (solid line) and PITA solution (dashed line)—$J = 10$, $N_{ts} = 24$. Horizontal axis displays time, vertical axis displays displacement or velocity, and k denotes the iteration number.*

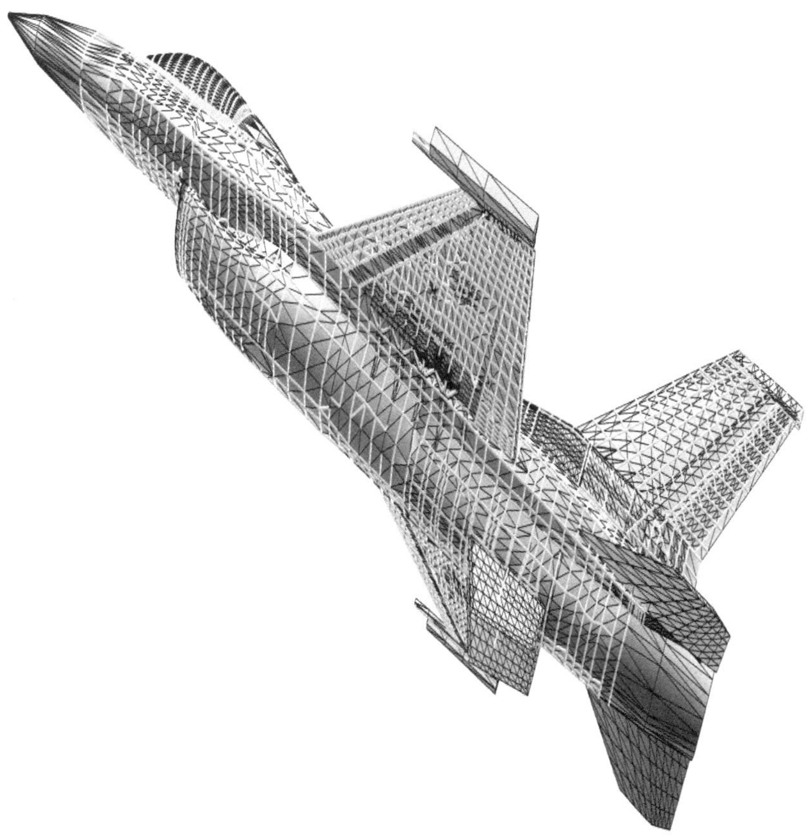

Figure 6.4. *Detailed finite element structural model of an F-16 fighter aircraft.*

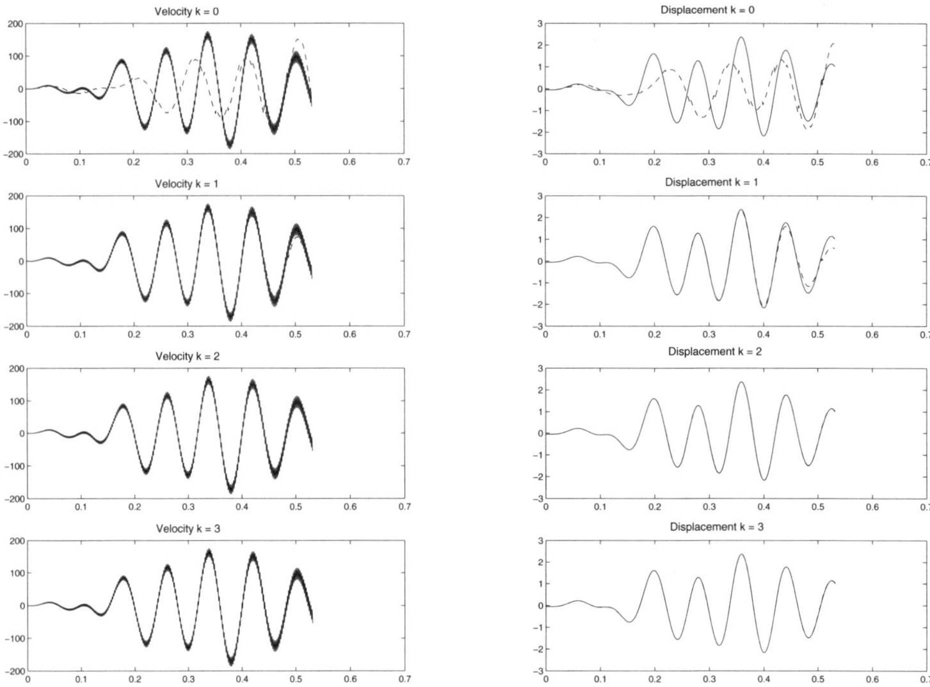

Figure 6.5. *Velocity (left) and displacement (right) time histories of a dof located on an aileron of the F-16 structural model set in free vibrations: sequential solution (solid line) and PITA solution (dashed line)—$J = 20$, $N_{ts} = 24$. Horizontal axis displays time, vertical axis displays displacement or velocity, and k denotes the iteration number.*

simulated time slices.

Next, the F-16 structural model is excited by gravity. Hence, the initial condition is set in this case to $b(t) = \mathcal{H}(t)$, where $\mathcal{H}(t)$ denotes a Heaviside function with an amplitude that varies at each dof of the model. The fine time step is set such that the response of the main eigenmode excited by this initial condition, which turns out to be mode 13, is well resolved. More specifically, the fine time step is set to $\Delta t = \frac{\mathcal{T}_{13}}{30}$, where $\mathcal{T}_{13}$ denotes the period of mode 13. The ratio of the coarse and fine time steps is set to $J = 10$ and 24 time slices are simulated using the new PITA framework of computations. Samples of the obtained displacement and velocity time histories are reported in Figures 6.6 and 6.7 for a vertical and a horizontal dof located on an aileron of the F-16, respectively. They reveal that the considered PITA demonstrates for this nonhomogeneous F-16 problem the same good performance as for the homogeneous one considered above.

6.7 Summary and conclusions

An ODE solver usually contains two main loops: an inner one which addresses a semidiscrete problem defined at a specific instance, and an outer one which advances the solution from a given instance to the next one. To a large extent, the parallelization of an ODE

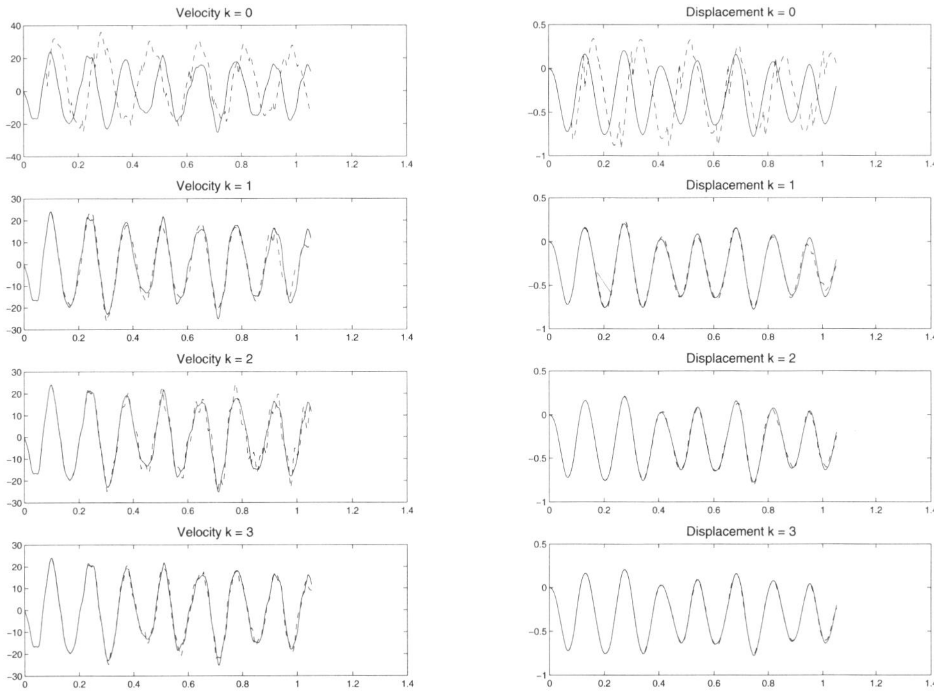

Figure 6.6. *Velocity (left) and displacement (right) time histories of a vertical dof located on an aileron of the F-16 structural model excited by gravity: sequential solution (solid line) and PITA solution (dashed line)—$J = 10$, $N_{ts} = 24$. Horizontal axis displays time, vertical axis displays displacement or velocity, and k denotes the iteration number.*

solver has essentially addressed the inner loop. This loop is also known as the space loop when the ODE (or system of ODEs) arises from the semidiscretization (or discretization in space) of a partial differential-evolution equation (PDE). By comparison, the parallelization of the outer loop, which in this case can be referred to as the time loop, has received little attention in the literature. This is because, in general, time parallelism, where the solution is computed simultaneously at different time instances, is harder to achieve than space parallelism, due to the inherently sequential nature of the time-integration process. Nevertheless, time parallelism can be of paramount importance to fast computations, for example, when space parallelism is unfeasible or cannot alone exploit all the processors available on a given massively parallel system, or when computing the solution of a system of ODEs in near-real time requires exploiting both space and time parallelisms. Two main approaches were proposed in the last decade for implementing time parallelism in an ODE solver. The first one is based on waveform relaxation [2, 3], and the second one on the classical principles of domain decomposition applied to the time-domain [1, 14, 15]. These approaches have met some success for systems of ODEs arising from the semidiscretization of parabolic and first-order hyperbolic problems but have failed to address those related to second-order hyperbolic problems (also known as oscillators). This chapter fills this gap as it proposes a new framework for designing time-parallel implicit time-integration algorithms (PITAs)

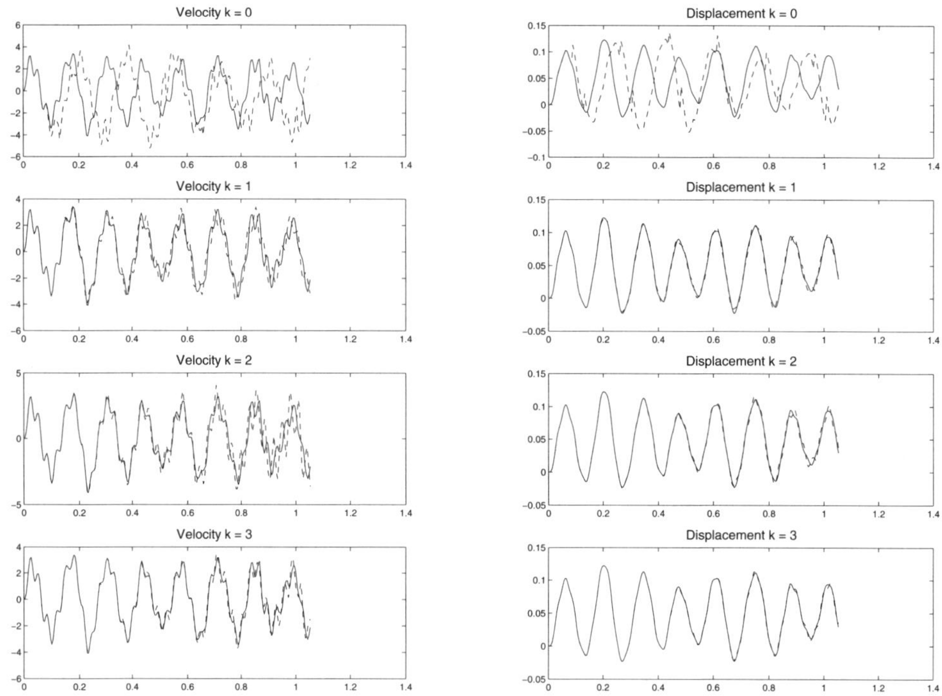

Figure 6.7. *Velocity (left) and displacement (right) time histories of a horizontal dof located on an aileron of the F-16 structural model excited by gravity: sequential solution (solid line) and PITA solution (dashed line)—$J = 10$, $N_{ts} = 24$. Horizontal axis displays time, vertical axis displays displacement or velocity, and k denotes the iteration number.*

that can handle linear second-order hyperbolic problems. The proposed framework is based on the PITA framework that was previously exposed in [1] and that is related to the parareal framework developed in [14, 15]. However, the parareal and original PITA frameworks for time-parallel computations fail to cope with systems of linear oscillators because they generate in this case a spurious beating phenomenon that severely restricts their stability time step. On the other hand, the new PITA framework proposed in this chapter incorporates a carefully designed projector that filters out the source of this spurious phenomenon and restores numerical stability for linear systems of oscillators. Sample linear structural dynamics applications discussed in this chapter, including the simulation of the transient response of an F-16 fighter aircraft to an initial perturbation, illustrate the potential of the new PITA framework for time-parallelizing the solution of systems of linear oscillators in view of achieving near-real-time processing. They also demonstrate the superior intrinsic performance of the proposed approach over the parareal and previously developed PITA frameworks.

Appendix A

Proposition 6.1. Given a problem of the form given in (6.1) with $b(t) = 0$ and a serial ITA for which G_{ITA}^C and G_{ITA}^F commute, the corresponding PITA satisfies

$$Y_k^i = \left[\sum_{l=0}^{k} \binom{i}{l} (\widetilde{G}_{ITA} - I)^l\right] Y_0^i, \tag{6.69a}$$

$$\Delta_k^i = \binom{i-1}{k} (\widetilde{G}_{ITA} - I)^{k+1} Y_0^i, \quad i > 0 \tag{6.69b}$$

$$C_k^i = \binom{i-1}{k+1} (\widetilde{G}_{ITA} - I)^{k+1} Y_0^i, \quad i > 0. \tag{6.69c}$$

Proof. The proof of the proposition is organized in two parts. First, it is shown that if the relationship (6.69a) holds, then the relationships (6.69b) and (6.69c) also hold. Then, the relationship (6.69a) is proved by induction.

Part 1. Assume that result (6.69a) holds. Consider first the case $i \leq k$. From the finite termination properties (6.26) it follows that $\Delta_k^i = 0$ and $C_k^i = 0$. From the convention (6.28) one deduces that $\binom{i-1}{k} = 0$ and $\binom{i}{k+1} = 0$, which shows that the relationships (6.69b) and (6.69c) also hold. Next, consider the case $i > k$. From the assumed result (6.69a) and the fact that G_{ITA}^C and G_{ITA}^F commute, it follows that (6.24) can be rewritten as

$$\begin{aligned}\Delta_k^i &= (G_{ITA}^F)^J \left[\sum_{l=0}^{k} \binom{i-1}{l} (\widetilde{G}_{ITA} - I)^l\right] Y_0^{i-1} - \left[\sum_{l=0}^{k} \binom{i}{l} (\widetilde{G}_{ITA} - I)^l\right] Y_0^i \\ &= \left[\widetilde{G}_{ITA} \sum_{l=0}^{k} \binom{i-1}{l} (\widetilde{G}_{ITA} - I)^l - \sum_{l=0}^{k} \binom{i}{l} (\widetilde{G}_{ITA} - I)^l\right] Y_0^i.\end{aligned} \tag{6.70}$$

Rewriting $\widetilde{G}_{ITA} = (\widetilde{G}_{ITA} - I) + I$ transforms the above expression into

$$\Delta_k^i = \left[\sum_{l=0}^{k} \binom{i-1}{l} (\widetilde{G}_{ITA} - I)^{l+1} + \sum_{l=0}^{k} \left(\binom{i-1}{l} - \binom{i}{l}\right) (\widetilde{G}_{ITA} - I)^l\right] Y_0^i. \tag{6.71}$$

Performing the change of variable $l' = l + 1$ in the first part of the above expression, that is, writing

$$\sum_{l=0}^{k} \binom{i-1}{l} (\widetilde{G}_{ITA} - I)^{l+1} = \sum_{l'=1}^{k+1} \binom{i-1}{l'-1} (\widetilde{G}_{ITA} - I)^{l'}, \tag{6.72}$$

then noting that $\binom{i-1}{0} = \binom{i}{0} = 1$ and therefore

$$\sum_{l=0}^{k} \left(\binom{i-1}{l} - \binom{i}{l}\right) (\widetilde{G}_{ITA} - I)^l = \sum_{l=1}^{k} \left(\binom{i-1}{l} - \binom{i}{l}\right) (\widetilde{G}_{ITA} - I)^l,$$

and finally, using the classical identity

$$\binom{i-1}{l-1} + \binom{i-1}{l} = \binom{i}{l}, \tag{6.73}$$

(6.71) can be transformed into

$$\Delta_k^i = \binom{i-1}{k}(\widetilde{G}_{ITA} - I)^{k+1} Y_0^i,$$

which proves the relationship (6.69b).

The relationship (6.69c) is proved by induction. First, it is noted that because of the first of the finite termination properties (6.26), this result holds for $i = k$ (recall convention (6.28)). Next, this result is assumed to hold for $i \geq k$ and then proved for $i+1$. Indeed, from (6.22), the identity (6.73), and the fact that G_{ITA}^C and G_{ITA}^F commute it follows that

$$\begin{aligned} C_k^{i+1} &= G_{ITA}^C(C_k^i + \Delta_k^i) \\ &= G_{ITA}^C\left[\binom{i-1}{k+1}(\widetilde{G}_{ITA} - I)^{k+1} Y_0^i + \binom{i-1}{k}(\widetilde{G}_{ITA} - I)^{k+1} Y_0^i\right] \\ &= G_{ITA}^C\left[\binom{i}{k+1}(\widetilde{G}_{ITA} - I)^{k+1}\right] Y_0^i \\ &= \left[\binom{i}{k+1}(\widetilde{G}_{ITA} - I)^{k+1}\right] Y_0^{i+1}, \end{aligned}$$

which completes the proof by induction of result (6.69c).

Part 2. For all $i \in \mathbb{N}$,

$$\sum_{l=0}^{0}\binom{i}{l}(\widetilde{G}_{ITA} - I)^l = \binom{i}{0} I = I,$$

which shows that result (6.69a) holds for $k = 0$. Next, suppose that this result holds for k. From Part 1 it follows that the relationships (6.69b) and (6.69c) also hold for k. Hence, from this observation and (6.14) it follows that

$$\begin{aligned} Y_{k+1}^i &= Y_k^i + (\Delta_k^i + C_k^i) \\ &= \left[\sum_{l=0}^{k}\binom{i}{l}(\widetilde{G}_{ITA} - I)^l + \binom{i}{k+1}(\widetilde{G}_{ITA} - I)^{k+1}\right] Y_0^i \\ &= \left[\sum_{l=0}^{k+1}\binom{i}{l}(\widetilde{G}_{ITA} - I)^l\right] Y_0^i, \end{aligned}$$

which completes the proof by induction of result (6.69a), and also completes the proof of Proposition 6.1. □

Appendix B

Proposition 6.3. If the amplification matrix G_{ITA} of the chosen ITA satisfies

$$\forall y_0 \in \mathbb{E}_{r(y_0)}, \quad \forall i \geq 0, \quad (G_{ITA})^i y_0 \in \mathbb{E}_{r(y_0)}, \tag{6.74}$$

the new PITA framework converges as soon as $\mathcal{S}_k = \mathbb{E}_{r(y_0)}$.

Proof. To begin, it is proved by induction on k that for an ITA that satisfies property (6.74) on both the coarse and fine time grids, the corresponding PITA generates updated seeds that verify

$$Y_k^i \in \mathbb{E}_{r(y_0)}, \; 0 \leq i \leq N_{ts} - 1, \; k \geq 0. \tag{6.75}$$

Indeed, consider first the case $k = 0$. The initialization on the coarse time grid yields $Y_0^i = (G_{ITA}^C)^i y_0$, $0 \leq i \leq N_{ts} - 1$. Hence, from property (6.68) it follows that result (6.75) holds for $k = 0$. Now, suppose that this result holds for k, that is,

$$Y_l^i \in \mathbb{E}_{r(y_0)}, \; 0 \leq i \leq N_{ts} - 1, \; l \leq k. \tag{6.76}$$

The objective becomes to prove that the above result also holds for $l = k + 1$. The proof is constructed here by induction on i. However, before proceeding with the proof by induction, two intermediate results are established. First, it is noted that definition (6.36) and assumption (6.76) imply that

$$\mathcal{S}_k \subset \mathbb{E}_{r(y_0)}. \tag{6.77}$$

Hence, for all $u \in \mathbb{E}_{r(y_0)}$, $P_k u \in \mathbb{E}_{r(y_0)}$, and $(I - P_k)u \in \mathbb{E}_{r(y_0)}$. Furthermore, exploiting property (6.68) on both the coarse and fine time grids allows us to conclude that $(G_{ITA}^F)^J P_k u \in \mathbb{E}_{r(y_0)}$ and $G_{ITA}^C(I - P_k)u \in \mathbb{E}_{r(y_0)}$. Therefore, from (6.52) one also concludes

$$\forall u \in \mathbb{E}_{r(y_0)}, \; \widehat{G}_{ITA,k}^C u \in \mathbb{E}_{r(y_0)}. \tag{6.78}$$

Second, it is also noted that substituting (6.54) written for i into (6.53) written for $i + 1$ gives

$$Y_{k+1}^{i+1} = (G_{ITA}^F)^J Y_k^i + \widehat{G}_{ITA,k}^C (Y_{k+1}^i - Y_k^i), \; 0 \leq i. \tag{6.79}$$

Now the proof by induction introduced above begins. For $i = 0$, $Y_{k+1}^0 = y_0$ and therefore result (6.76) holds for $l = k+1$ and $i = 0$. Suppose now that for $0 \leq i \leq N_{ts} - 2$, $Y_{k+1}^i \in \mathbb{E}_{r(y_0)}$. From this assumption, assumption (6.76), observation (6.78), property (6.74), and (6.79) it follows that $Y_{k+1}^{i+1} \in \mathbb{E}_{r(y_0)}$. This concludes the proof by induction on i that result (6.76) holds for $l = k + 1$, which in turn concludes the proof by induction on k of result (6.75).

Using the same reasoning as above, it is proved that

$$U_k^i \in \mathbb{E}_{r(y_0)}, \; 0 \leq i \leq N_{ts} - 1, \; 0 \leq k. \tag{6.80}$$

Finally, suppose now that for $k = k^\star$, $\mathcal{S}_{k^\star} \supset \mathbb{E}_{r(y_0)}$. From (6.77) which holds for any k it follows that $\mathcal{S}_{k^\star} = \mathbb{E}_{r(y_0)}$. Therefore, from this result and result (6.80), it follows that

$$P_k U_{k^\star}^i = U_{k^\star}^i, \quad 0 \leq i \leq N_{ts} - 1. \tag{6.81}$$

From (6.81) above and (6.56), it follows that

$$\Delta_{k^\star+1}^i = 0, \; 1 \leq i \leq N_{ts} - 1, \tag{6.82}$$

which implies the convergence of the PITA as soon as $\mathcal{S}_k = \mathbb{E}_{r(y_0)}$ and completes the proof of Proposition 6.3. □

Acknowledgments

This material is based upon work supported partially by the National Science Foundation under grant CNS-0451569-NCX, and partially by the Air Force Office of Scientific Research under grant FA9550-05-1-0038. Any opinions, findings, and conclusions or recommendations expressed in this material are those of the authors and do not necessarily reflect the views of the National Science Foundation or the Air Force Office of Scientific Research.

Bibliography

[1] C. Farhat and M. Chandesris, *Time-decomposed parallel time-integrators – part I: Theory and feasability studies for fluid, structure, and fluid-structure applications*, Internat. J. Numer. Methods Engrg., 58 (2003), pp. 1397–1434.

[2] J. White, A. Sangiovanni-Vincentelli, F. Odeh, and A. Ruehli, *Waveform relaxation: Theory and practice*, Trans. Soc. Computer Simulation, 2 (1985), pp. 95–133.

[3] A. Lumsdaine and D. Wu, *Krylov subspace acceleration of waveform relaxation*, SIAM J. Numer. Anal., 41 (2003), pp. 90–111.

[4] S. Vandewalle and R. Piessens, *Efficient parallel algorithms for solving initial-boundary value and time-periodic parabolic partial differential equations*, SIAM J. Sci. Stat. Comput., 13 (1992), pp. 1330–1346.

[5] K. Burrage, *Parallel methods for initial value problems*, Appl. Numer. Math., 11 (1993), pp. 5–25.

[6] S. Vandewalle and E. Van de Velde, *Space-time concurrent multigrid waveform relaxation*, Ann. Numer. Math., 1 (1994), pp. 347–360.

[7] G. Horton, S. Vandewalle, and P. Worley, *An algorithm with polylog parallel complexity for solving parabolic partial differential equations*, SIAM J. Sci. Comput., 16 (1994), pp. 531–541.

[8] G. Horton and S. Vandewalle, *A space-time multigrid method for parabolic differential PDEs*, SIAM J. Sci. Comput., 16 (1994), pp. 848–864.

[9] S. Ta'asan and H. Zhang, *Fourier-Laplace analysis of multigrid waveform relaxation method for hyperbolic equations*, BIT, 36 (1996), pp. 831–841.

[10] P. Amodio and L. Brugnano, *Parallel implementation of block boundary value methods for ODEs*, J. Comput. Appl. Math., 78 (1997), pp. 197–211.

[11] L. Brugnano and D. Trigiante, *Solving Differential Problems by Multistep Initial and Boundary Value Methods*, Stability and Control: Theory, Methods and Applications 6, Gordon and Breach Science Publishers, Amsterdam, 1998.

[12] L. F. Pavarino and A. Toselli, *Recent Developments in Domain Decomposition Methods*, Springer-Verlag, Berlin, 2002.

[13] P. Saha, J. Stadel, and S. Tremaine, *A parallel integration method for solar system dynamics*, Astron. J., 114 (1997), pp. 409–415.

[14] J. L. Lions, Y. Maday, and G. Turinici, *Résolution d'EDP par un schéma en temps "pararéel"*, C. R. Acad. Sci. Paris, Sér. I Math., 332 (2001), pp. 661–668.

[15] G. Bal and Y. Maday, *A "parareal" time discretization for non-linear PDEs with application to the pricing of an American put*, in Recent Developments in Domain Decomposition Methods, Lect. Notes Comput. Sci. Eng. 23, L. F. Pavarino and A. Toselli, eds., Springer-Verlag, Berlin, 2002, pp. 189–202.

[16] G. Staff and M. Rønquist, *Stability of the parareal algorithm*, in Domain Decomposition Methods in Science and Engineering, Lect. Notes Comput. Sci. Eng. 40, R. Kornhuber, R. H. W. Hoppe, D. E. Keyes, J. Périaux, O. Pironneau, and J. Xu, eds., Springer-Verlag, 2003, pp. 449–456.

[17] G. Bal, *On the convergence and the stability of the parareal algorithm to solve partial differential equations*, private communication, 2003.

[18] C. Farhat and P. S. Chen, *Tailoring domain decomposition methods for efficient parallel coarse grid solution and for systems with many right hand sides*, Contemp. Math., 180 (1994), pp. 401–406.

[19] C. Farhat, *Optimizing substructuring methods for repeated right hand sides, scalable parallel coarse solvers, and global/local analysis*, in Domain-Based Parallelism and Problem Decomposition Methods in Computational Science and Engineering, D. Keyes, Y. Saad, and D. G. Truhlar, eds., SIAM, Philadelphia, 1995, pp. 141–160.

[20] M. Bhardwaj, D. Day, C. Farhat, M. Lesoinne, K. Pierson, and D. Rixen, *Application of the FETI method to ASCI problems: Scalability results on one-thousand processors and discussion of highly heterogeneous problems*, Internat. J. Numer. Methods Engrg., 47 (2000), pp. 513–536.

[21] C. Farhat, P. Geuzaine, and G. Brown, *Application of a three-field nonlinear fluid-structure formulation to the prediction of the aeroelastic parameters of an F-16 fighter*, Comput. & Fluids, 32 (2003), pp. 3–29.

Chapter 7

Generalized SQP Methods with "Parareal" Time-Domain Decomposition for Time-Dependent PDE-Constrained Optimization

*Stefan Ulbrich**

7.1 Introduction

We consider a time-dependent PDE-constrained optimization problem of the form

$$\min_{y \in Y, u \in U} J(y, u) \quad \text{subject to } C(y, u) = 0. \tag{7.1}$$

Here $u \in U$ is the control living in a Banach space U, $y \in Y \subset C([0, T]; B)$ is a time-dependent state with Banach spaces B and Y, where $B \subset L^2(\Omega)$ with a domain $\Omega \subset \mathbb{R}^n$. The state equation $C(y, u) = 0$ is the appropriate formulation of a time-dependent PDE (or a system of time-dependent PDEs)

$$\begin{aligned} y_t + A(t, x, y, u) &= 0, \quad (t, x) \in (0, T) \times \Omega, \\ y(0, x) &= y_0(x), \quad x \in \Omega, \end{aligned} \tag{7.2}$$

with initial data $y_0 \in B$. For convenience we assume that boundary conditions are incorporated into the state space Y.

In many applications there are already robust and validated discretization schemes for (7.2) available. Lions, Maday, and Turinici [14] have observed that the time discretization can efficiently be parallelized by a time-domain decomposition technique, which the authors call the *parareal scheme*. The parareal scheme combines a coarse-grid predictor and a parallel fine-grid corrector. In this chapter we propose a generalized sequential quadratic programming (SQP) algorithm that allows the fast parallel solution of (7.1) by using the

*Fachbereich Mathematik, AG10, Technische Universität Darmstadt, Schlossgartenstr. 7, 64289 Darmstadt, Germany.

parareal technique for the state equation and the adjoint equation. The generalized SQP algorithm is inspired by the nonmonotone SQP method of Ulbrich and Ulbrich [18] that does—similar to filter methods [7, 8]—not use a penalty function as merit function. Our aim is to achieve high efficiency of the optimization method on parallel computers while requiring minimal changes of existing sequential state solvers by the user. Besides the parallelism provided by the time-domain decomposition the proposed method admits the application of highly efficient solvers inside the time domains, e.g., multigrid-based PDE solvers. These features allow the efficient solution of very large optimal control problems for time-dependent PDEs on parallel computers, while only minimal changes to existing state and adjoint solvers are necessary. In addition, the proposed approach is not restricted to special types of PDEs. It is only required that certain differentiability properties are satisfied, see Assumption 7.7, and that the state and adjoint solvers are convergent.

Recently, several approaches have been proposed for the efficient and parallel solution of optimal control problems for time-dependent PDEs. Time-domain decomposition techniques for the optimal control of the wave equation were considered by Lagnese and Leugering [12, 13] and for distributed linear quadratic optimal control problems by Heinkenschloss [9]. In these works the time-domain decomposition is carried out for the optimality system, and in contrast to the parareal technique no coarse-grid propagator is used. By using the parareal idea Maday and Turinici propose in [15] a preconditioned steepest descent method for the optimal control of time-dependent PDEs. Other recent approaches consider space-time multigrid methods for the optimal control of PDEs. In particular, Borzì proposes in [5] a space-time multigrid method for parabolic distributed optimal control problems. In contrast to the present chapter the approach in [5] does not allow one to apply user-provided solvers, since a specific multigrid algorithm is used.

This chapter is organized as follows. In Section 7.2 we recall the parareal time-domain decomposition technique and state known convergence results. In Section 7.3 we perform a time-domain decomposition of the optimal control problem (7.1) by applying the parareal technique to the state equation. Moreover, we derive optimality conditions for the decomposed problem and show that the parareal technique can also be applied to the adjoint system. In Section 7.4 we propose a generalized SQP method that allows the solution of (7.1) for arbitrary user-provided state solver and adjoint solver, where we use parareal solvers in the present chapter. We prove global convergence of the method. In Section 7.5 we use the generalized SQP method with the specific choice of parareal solvers for state and adjoint equations. We demonstrate the efficiency of the approach by numerical results for the optimal control of a semilinear parabolic equation in two dimensions.

7.2 Parareal Time-Domain Decomposition

The parareal algorithm was proposed by Lions, Maday, and Turinici [14] to speed up the numerical solution of time-dependent PDEs by using parallel computers with a sufficiently large number of processors.

7.2.1 Description of the Parareal Method

We consider a time-dependent PDE (or a system) of the general form

$$\begin{aligned} y_t + A(t, x, y) &= 0, \quad (t, x) \in (0, T) \times \Omega, \\ y(0, x) &= v_0(x), \quad x \in \Omega, \end{aligned} \tag{7.3}$$

where $y : [0, T] \to B$ maps time to a Banach space $B \subset L^2(\Omega)$, $v_0 \in B$ are initial data, and $A(t, x, y)$ is a possibly time-dependent partial differential operator in the variable $x \in \Omega$. The parareal technique, which was originally proposed in [14] and slightly modified in [2, 3], uses a decomposition

$$0 = T_0 < T_1 < \cdots < T_N = T$$

in time, which is uniform in the sense that

$$\eta_0 \Delta T \leq T_{n+1} - T_n \leq \Delta T, \quad \text{where } \Delta T := \max_{0 \leq n < N} T_{n+1} - T_n.$$

We assume that on each time domain $[T_n, T_{n+1}]$, $0 \leq n < N$, there exists a unique solution propagator

$$g(T_n, \cdot) : \quad v \in B \mapsto g(T_n, v) := y(T_{n+1}) \in B,$$

where y is the solution of

$$\begin{aligned} y_t + A(t, x, y) &= 0, \quad (t, x) \in (T_n, T_{n+1}) \times \Omega, \\ y(T_n, x) &= v(x), \quad x \in \Omega. \end{aligned} \tag{7.4}$$

Moreover, we assume that we have a coarse-grid approximation $g_\Delta(T_n, v)$ of the exact propagator $g(T_n, v)$ at our disposal.

Example 7.1. We will later use a backward Euler step

$$\frac{g_\Delta(T_n, v) - v}{T_{n+1} - T_n} + A(T_{n+1}, x, g_\Delta(T_n, v)) = 0$$

as coarse propagator. As we will see the dissipativity of the backward Euler discretization is useful to stabilize the parareal scheme.

The parareal technique is based on a multiple shooting reformulation of the initial value problem (7.3). It combines parallel fine-grid propagators with a coarse propagator for the iterative solution of the multiple shooting reformulation of (7.3).

More precisely, the parareal algorithm is defined as follows.

Algorithm 7.2 ("Parareal" Time Integration Method).
Compute approximations y_k^n of $y^n = y(T_n)$, $1 \leq n \leq N$, for the solution y of (7.3) as follows.

1. Initialize by coarse scheme: $y_1^0 = v_0$,

$$y_1^{n+1} = g_\Delta(T_n, y_1^n), \quad 0 \leq n < N.$$

2. For $k = 1, \dots, K-1$: $y_{k+1}^0 = v_0$

$$y_{k+1}^{n+1} = \underbrace{g_\Delta(T_n, y_{k+1}^n)}_{\text{predictor}} + \underbrace{\big(g(T_n, y_k^n) - g_\Delta(T_n, y_k^n)\big)}_{\substack{\text{corrector, computable} \\ \text{in parallel on time slabs}}}, \quad 0 \le n < N. \tag{7.5}$$

The sequential predictor step propagates the new approximation y_{k+1}^n by the coarse propagator. The error is corrected by using the approximation y_k^n of the previous iteration and can thus be computed in parallel.

Remark 7.3.

- In practice $g(T_n, y_k^n)$ is replaced by an approximation $g_\delta(T_n, y_k^n)$ obtained by an accurate fine-grid scheme.
- The exact solution $y^n = y(T_n)$ satisfies the parareal scheme, since $y^{n+1} = g(T_n, y^n)$.
- After N iterations the exact solution is obtained.
- The parareal scheme can directly be applied to nonlinear equations and is then a nonlinear iteration. Thus, user-provided nonlinear fine-grid solvers can be used directly. Besides the fact that nonlinear solvers can be more robust and efficient for some problems the nonlinear parareal algorithm has the advantage that the state in the time slabs does not have to be stored in contrast to, e.g., a Newton iteration with inner parareal solver for the linearized equation.

7.2.2 Convergence Properties of the Parareal Algorithm

In this subsection we collect recent convergence results for the parareal scheme. The first result yields a convergence order km after k parareal iterations provided the solution is smooth enough and the coarse propagator has order m and is Lipschitz.

Theorem 7.4 (Bal [2]). *Let $B_k \subset B_{k-1} \subset \cdots \subset B_0 = B$ be a scale of Banach spaces. Assume that*

- (7.3) *is stable in all spaces B_j, $0 \le j \le k$, i.e.,*

$$\|y(t)\|_{B_j} \le C\|v_0\|_{B_j} \quad \forall\, v_0 \in B_j, \quad t \in [0, T].$$

- *The coarse propagator g_Δ is Lipschitz in the sense that for $0 \le j < k$*

$$\max_{0 \le n < N} \|g_\Delta(T_n, v) - g_\Delta(T_n, w)\|_{B_j} \le (1 + C\Delta T)\|v - w\|_{B_j} \quad \forall\, u, v \in B_j.$$

- *The coarse propagator g_Δ has order m in the sense that for $0 \le j < k$ with $\delta g(T_n, v) := g(T_n, v) - g_\Delta(T_n, v)$ the estimate holds*

$$\max_{0 \le n < N} \|\delta g(T_n, v) - \delta g(T_n, w)\|_{B_j} \le C(\Delta T)^{m+1}\|v - w\|_{B_{j+1}} \quad \forall\, v, w \in B_{j+1}.$$

Then for initial data $v_0 \in B_k$ the parareal scheme after k iterations has an accuracy of order mk, i.e.,

$$\max_{1 \le n \le N} \|y(T_n) - y_k^n\|_{B_0} \le C(\Delta T)^{mk} \|v_0\|_{B_k}$$

with C independent of ΔT and v_0.

Proof. A proof can be found in [2, Theorem 1]. Our last assumption is a modification of assumption (H3) in [2], which is tailored directly to the proof of [2, Theorem 1]. □

This general result requires additional regularity of the solution in order to achieve order km. By exploiting dissipative effects of the coarse propagator, Bal [2] shows an improved result for certain linear partial differential operators by using Fourier analysis.

Theorem 7.5 (Bal [2]). *Consider the case of a linear Mth order one-dimensional spatial operator A with constant coefficients and symbol $P(\xi) = \alpha_0 + \sum_{k=1}^{M-1} \alpha_k \xi^k + |\xi|^M$, where $\alpha_k \in \mathbb{C}$ and $\alpha_0 \ge 0$ such that $P(\xi) \ge 0$ or $\Re(P(\xi)) > 0$ (e.g., heat equation, advection-diffusion equation). If for $T_n = n\Delta T$, $\Delta T = T/N$, the coarse propagator is given by the θ-scheme*

$$\frac{g_\Delta(T_n, v) - v}{\Delta T} + A(\theta g_\Delta(T_n, v) + (1-\theta)v) = 0,$$

then for $\theta \in (1/2, 1]$ the parareal scheme is stable and

$$\max_{1 \le n \le N} \|y(T_n) - y_k^n\|_{H^s(\mathbb{R})} \le C(\Delta T)^{mk} \|v_0\|_{H^s(\mathbb{R})}$$

for all $s \in \mathbb{R}$. Here $H^s(\mathbb{R})$ denotes the usual Sobolev space of order s.

Proof. As shown in [2], the result follows from Theorems 2 and 3 in [2]. □

7.2.3 Interpretation as Preconditioned Iteration

Rather than interpreting the parareal scheme as a scheme of higher order on the coarse grid, we prefer to view the parareal scheme as a preconditioned iteration. Consider for simplicity the case that (7.3) is linear. Then

$$g(T_n, y^n) = G_n y^n + c_n, \quad g_\Delta(T_n, y^n) = G_{\Delta,n} y^n + c_{\Delta,n}.$$

The relation

$$y^{n+1} = g(T_n, y^n), \quad 0 \le n < N, \quad y^0 = v_0,$$

can be written as

$$\begin{pmatrix} I & & & & \\ -G_0 & I & & & \\ & -G_1 & I & & \\ & & \ddots & \ddots & \\ & & & -G_{N-1} & I \end{pmatrix} \begin{pmatrix} y^0 \\ y^1 \\ \vdots \\ y^N \end{pmatrix} = \begin{pmatrix} v_0 \\ c_0 \\ \vdots \\ c_{N-1} \end{pmatrix}$$

which we write in short as

$$My = F.$$

Since the parareal update (7.5) can be written as

$$y_{k+1}^{n+1} - g_\Delta(T_n, y_{k+1}^n) = y_k^{n+1} - g_\Delta(T_n, y_k^n) + g(T_n, y_k^n) - y_k^{n+1}, \quad 0 \le n < N,$$

we see that with the coarse-grid approximation M_Δ of M

$$M_\Delta = \begin{pmatrix} I & & & & \\ -G_{\Delta,0} & I & & & \\ & -G_{\Delta,1} & I & & \\ & & \ddots & \ddots & \\ & & & -G_{\Delta,N-1} & I \end{pmatrix}$$

the parareal scheme (7.5) can be written as

$$y_{k+1} = y_k + M_\Delta^{-1}(F - My_k). \tag{7.6}$$

Theorems 7.4 and 7.5 show that under appropriate assumptions M_Δ^{-1} is a good preconditioner for the operator M.

To cover also the nonlinear case we introduce the operators

$$\mathcal{G}(y) = \begin{pmatrix} y^0 - v_0 \\ y^1 - g(T_0, y^0) \\ \vdots \\ y^N - g(T_{N-1}, y^{N-1}) \end{pmatrix}, \quad \mathcal{G}_\Delta(y) = \begin{pmatrix} y^0 - v_0 \\ y^1 - g_\Delta(T_0, y^0) \\ \vdots \\ y^N - g_\Delta(T_{N-1}, y^{N-1}) \end{pmatrix}.$$

Then the parareal scheme can be written as the preconditioned iteration

$$\mathcal{G}_\Delta(y_{k+1}) = \mathcal{G}_\Delta(y_k) - \mathcal{G}(y_k)$$

which is (7.6) in the linear case.

7.3 Time-Domain Decomposition of the Optimal Control Problem

We use a multiple shooting reformulation of the optimal control problem (7.1). The proposed optimization algorithm, however, is different from SQP-type optimization algorithms typically applied to multiple shooting reformulations of optimal control problems. In particular, our SQP algorithms are matrix free and, in the context of (7.1), use the parareal technique for the solution of the state equation and the adjoint equation.

Let as above $0 = T_0 < T_1 < \cdots < T_N = T$. We consider a PDE-constrained optimization problem (7.1) with state equation (7.2), where the cost functional J depends for simplicity only on the end states $y(T_n)$ of the time domains, i.e.,

$$J(y, u) = J^\Delta(y(T_0), \ldots, y(T_N), u). \tag{7.7}$$

Assume that for all $v \in B$, $u \in U$

$$\begin{aligned} &y_t + A(t, x, y(t), u(t)) = 0, \quad (t, x) \in (T_n, T_{n+1}) \times \Omega, \\ &y(T_n, x) = v(x), \quad x \in \Omega, \end{aligned}$$

has a unique solution and define the corresponding propagator

$$g(T_n, v; u) := y(T_{n+1}) \in B.$$

Using $y_\Delta = (y^0, \dots, y^N) \in B^{N+1} =: Y_\Delta$ as new state we can rewrite (7.1), (7.2), (7.7) as

$$\min_{y_\Delta=(y^0,\cdots,y^N)\in Y_\Delta, u\in U} J^\Delta(y, u) \quad \text{subject to } C^\Delta(y, u) = 0, \tag{7.8}$$

where $C^\Delta : Y_\Delta \times U \to Y_\Delta$,

$$C^\Delta(y_\Delta, u) = \begin{pmatrix} y^0 - v_0 \\ y^1 - g(T_0, y^0; u) \\ \vdots \\ y^N - g(T_{N-1}, y^{N-1}; u) \end{pmatrix}. \tag{7.9}$$

Obviously, the parareal algorithm can be used as an iterative algorithm to solve the decomposed state equation (7.9) in parallel. In the following we will develop an optimization algorithm for (7.8) that is motivated by the following basic concept:

- Use the parareal scheme as state solver in a generalized inexact SQP algorithm.
- Use the parareal scheme also for the adjoint equation to compute inexact gradients of the cost functional.

To develop the generalized SQP method we state first necessary optimality conditions for (7.8).

7.3.1 Optimality Conditions

The Lagrangian function for (7.8) is given by

$$L(y_\Delta, u, \lambda) = J^\Delta(y_\Delta, u) + \langle \lambda, C^\Delta(y^\Delta, u)\rangle_{Y^*_\Delta, Y_\Delta}.$$

By standard optimality theory we obtain the following result.

Proposition 7.6. *Assume that $J^\Delta : Y_\Delta \times U \to \mathbb{R}$, $C^\Delta : Y_\Delta \times U \to Y_\Delta$ are continuously differentiable. If $(\bar{y}_\Delta, \bar{u})$ is a local solution of (7.8), then there exists a Lagrange multiplier (adjoint state) $\bar{\lambda} \in Y^*_\Delta$ with*

$$L_y(\bar{y}_\Delta, \bar{u}, \bar{\lambda}) = J^\Delta_y(\bar{y}_\Delta, \bar{u}) + (C^\Delta_y)^*(\bar{y}_\Delta, \bar{u})\bar{\lambda} = 0 \quad \textit{(Adjoint equation)}, \tag{7.10}$$

$$L_u(\bar{y}_\Delta, \bar{u}, \bar{\lambda}) = J^\Delta_u(\bar{y}_\Delta, \bar{u}) + (C^\Delta_u)^*(\bar{y}_\Delta, \bar{u})\bar{\lambda} = 0 \quad \textit{(Stationarity)}, \tag{7.11}$$

$$C^\Delta(\bar{y}_\Delta, \bar{u}) = 0 \quad \textit{(State equation)}. \tag{7.12}$$

Proof. The structure of C^Δ implies immediately that $C_y^\Delta(\bar{y}_\Delta, \bar{u})$ is surjective. Hence, $C_{(y,u)}^\Delta(\bar{y}_\Delta, \bar{u})$ is surjective, which is a regularity condition for $(\bar{y}_\Delta, \bar{u})$. Therefore it is well known that the KKT conditions (7.10), (7.11), (7.12) hold at a local minimum $(\bar{y}_\Delta, \bar{u})$ of (7.8). □

7.3.2 Structure of the Adjoint Equation

We have with $G_n(y^n, u) := g_{y^n}(T_n, y^n; u)$, which is the propagator for the linearized state equation,

$$C_y^\Delta(y, u) = \begin{pmatrix} I & & & & \\ -G_0 & I & & & \\ & -G_1 & I & & \\ & & \ddots & \ddots & \\ & & & -G_{N-1} & I \end{pmatrix}.$$

Thus, we obtain the following structure of the adjoint equation:

$$(C_y^\Delta)^*(y_\Delta, u)\lambda + J_y^\Delta(y_\Delta, u) = \begin{pmatrix} I & -G_0^* & & & \\ & I & -G_1^* & & \\ & & \ddots & \ddots & \\ & & & I & -G_{N-1}^* \\ & & & & I \end{pmatrix} \begin{pmatrix} \lambda^0 \\ \vdots \\ \lambda^N \end{pmatrix} + J_y^\Delta(y_\Delta, u) = 0. \tag{7.13}$$

Here G_n^* is the propagator for the adjoint equation on time domain n backwards in time. Therefore, the parareal scheme is directly applicable backwards in time to the adjoint equation.

In the rest of the chapter we will assume that we have a parareal solver for the state equation (7.9) and a parareal solver for the adjoint equation (7.13) at our disposal.

7.4 A Generalized SQP Method

We propose in this section a generalized SQP algorithm for the solution of the problem

$$\min_{y \in Y, u \in U} \quad J(y, u) \quad \text{subject to} \quad C(y, u) = 0. \tag{7.1}$$

Motivated by the time-dependent PDE-constrained problem (7.8) the design of the proposed optimization algorithm offers the following features:

- The algorithm solves the state equation and the optimization problem simultaneously.
- Arbitrary user-provided iterative solvers for the state equation and the adjoint equation can be used, in our case parallel parareal solvers.
- The optimization algorithm requires only very few iterations of the solvers in each step by controlling the inexactness.

- The algorithm ensures convergence to a stationary point.

For the considered time-dependent problem we have in addition the following properties:

- A nonlinear parareal solver can be used for nonlinear state equations.
- Only the state on the time-domain interfaces has to be stored; the states in the interior of the time domains can be handled locally by the local solver of the time domain.

7.4.1 Basic Assumptions

We will work under the following assumptions.

Assumption 7.7. *With $X := Y \times U$ the following hold for a given open convex set $X_0 = Y_0 \times U_0 \subset X$:*

- *The mappings*

$$\begin{aligned}(y,u) &\in X \mapsto J(y,u),\\ (y,u) &\in X \mapsto C(y,u) \in Z\end{aligned}$$

are continuously differentiable and the derivatives are uniformly bounded and Lipschitz on X_0.

- *For any $u \in U_0$ there exists a unique solution $y(u) \in Y_0$ of $C(y(u),u) = 0$.*
- *The derivative $C_y(y,u) \in \mathcal{L}(Y,Z)$ has an inverse that is uniformly bounded for all $(y,u) \in X_0$.*

7.4.2 Review of Classical Trust-Region SQP Methods

Classical SQP Method

It is well known that the local SQP method is given by the iteration

$$x_{k+1} = x_k + s_k,$$

where s_k solves the *SQP problem* at $(x_k, \lambda_k) \in X \times Z^*$

$$\begin{aligned}&\min_{s \in X}\ q_k(s) := L(x_k,\lambda_k) + \langle L_x(x_k,\lambda_k), s\rangle_{X^*,X} + \frac{1}{2}\langle s, H_k s\rangle_{X,X^*}\\ &\text{subject to } C(x_k,\lambda_k) + C_x(x_k,\lambda_k)s = 0\end{aligned}$$

with $x_k = (y_k, u_k)$ and $H_k = L_{xx}(x_k,\lambda_k)$ (or an approximation). In *reduced SQP methods* the quadratic model $q_k(s_y, s_u)$ is reduced to the control component s_u by choosing λ_k according to the adjoint equation

$$L_y(x_k,\lambda_k) = 0 \tag{7.14}$$

and by replacing H_k by

$$B_k := \begin{pmatrix} 0 & 0 \\ 0 & \hat{B}_k \end{pmatrix},$$

where $\hat{B}_k$ is an approximation of the reduced Hessian

$$\hat{H}_k = W_k^* H_k W_k, \quad W_k = \begin{pmatrix} -(C_y^{-1}C_u)(x_k) \\ I_U \end{pmatrix}.$$

Here I_U denotes the identity on U. This leads to the *reduced SQP problem* at (x_k, λ_k)

$$\min_{s=(s_y,s_u)\in X} \quad \hat{q}_k(s_u) := L(x_k,\lambda_k) + \langle L_u(x_k,\lambda_k), s_u\rangle_{U^*,U} + \frac{1}{2}\langle s_u, \hat{B}_k s_u\rangle_{U,U^*}$$

$$\text{subject to } \ C(x_k,\lambda_k) + C_y(x_k,\lambda_k)s_y + C_u(x_k,\lambda_k)s_u = 0.$$

Trust-Region Globalization

Following the composite-step approach by Byrd and Omojokun (see [6, 17]), a trust-region globalization of SQP methods can be obtained by a step decomposition

$$s_k = s_k^n + s_k^t$$

with *quasi-normal step* s_k^n and a *tangential step* s_k^t.Here, the quasi-normal step $s_k^n = (s_{y,k}^n, 0)$ updates only the state and is obtained as an approximate solution of the quasi-normal problem

$$\min_{s_y^n\in Y} \|C(x_k) + C_y(x_k)s_y^n\|_Z^2 \quad \text{subject to} \quad \|s_y^n\|_Y \le \delta_k, \tag{7.15}$$

where $\delta_k > 0$ is the current trust-region radius. The tangential step s_k^t is an approximate solution of the *tangential problem*

$$\min_{s^t=(s_y^t,s_u^t)\in X} q_k(s_k^n + s^t) \ \text{ subject to } \ C_x(x_k)s^t = 0, \ \ \|s_u^t\|_U \le \delta_k.$$

For reduced SQP methods the problem simplifies to the following: Compute an approximate solution $s_{u,k}^t$,

$$\min_{s_u^t\in U} \hat{q}_k(s_u^t) \ \text{ subject to } \ \|s_u^t\|_U \le \delta_k, \tag{7.16}$$

and then compute $s_{y,k}^t$ by solving

$$C_y(x_k)s_{y,k}^t = -C_u(x_k)s_{u,k}^t.$$

The evaluation of the step $s_k = s_k^n + s_k^t$ and the adaption of the trust-region radius can now be performed in various ways, e.g., by using a merit function [6] or by using techniques without merit function such as filter techniques [7, 8] or nonmonotone trust-region techniques [18].

7.4.3 Development of the Generalized SQP Algorithm

The classical SQP method has several drawbacks that limit the flexibility of the method.

- The SQP method uses a Newton iteration for the simultaneous solution of the state equation. But often the user provides already his own efficient and robust solver (sometimes nonlinear).
- For nonlinear state equations nonlinear solvers can be just as efficient as the solver for the linearized equation (e.g., nonlinear parareal solvers).
- The linearized state equation can be inappropriate as approximation of the state equation (for example for flows with shocks [19, 20]).

We circumvent these limitations by using nonmonotone trust-region techniques similar to [18] together with a generalized quasi-normal step. Again, we use a step decomposition

$$s_k = s_k^n + s_k^t.$$

The generalized tangential step $s_k^t = (s_{k,y}^t, s_{k,u}^t)$ is obtained as follows.

Generalized Tangential Step

Compute $s_{k,u}^t$ as an approximate solution of the reduced tangential problem

$$\min_{s_u^t \in U} \hat{q}_k(s_u^t) \quad \text{subject to} \quad \|s_u^t\|_U \le \delta_k \tag{7.16}$$

satisfying a fraction of Cauchy decrease condition (FCD)

$$\hat{q}_k(0) - \hat{q}_k(s_{k,u}^t) \ge \kappa_q \|L_u(x_k, \lambda_k)\|_{U^*} \min\{\|L_u(x_k, \lambda_k)\|_{U^*}, \delta_k\} \tag{7.17}$$

with a constant $\kappa_q > 0$. Then we set

$$s_k^t = (s_{k,y}^t, s_{k,u}^t), \tag{7.18}$$

where we require only that

$$\|s_{k,y}^t\|_Y \le \kappa_t \|s_{k,u}^t\|_U \tag{7.19}$$

with a fixed constant $\kappa_t > 0$. The choice $s_{k,y}^t = 0$ is allowed.

Remark 7.8. It is well known that FCD holds if $\|\hat{B}_k\|_{U,U^*}$ is uniformly bounded and if s_k^t yields at least a fixed fraction of the Cauchy decrease, which is the decrease along the steepest descent direction $-\nabla \hat{q}_k(0)$ inside the trust region.

Inexact Adjoint State

Instead of choosing the multiplier λ_k as the exact solution of the adjoint equation (7.14) we admit inexact adjoint states λ_k satisfying

$$\|L_y(x_k, \lambda_k)\|_{Y^*} \le \kappa_\lambda \min\{\|L_u(x_k, \lambda_k)\|_{U^*}, \delta_k\} \tag{7.20}$$

with a constant $\kappa_\lambda > 0$. This requirement is similar to the inexactness framework of Heinkenschloss and Vicente [10].

Generalized Quasi-Normal Step

Instead of the Newton-like quasi-normal step (7.15) we invoke a user-provided iterative solver, in our case the parareal state solver, to compute a generalized quasi-normal step

$$s_k^n = (s_{k,y}^n, 0) \tag{7.21}$$

such that the following nonmonotone decrease condition holds:

$$\max\left\{R_k, \|C(x_k)\|_Z^2\right\} - \|C(x_k + s_k)\|_Z^2 \geq \kappa_c \|C(x_k)\|_Z^2, \quad \kappa_c \in (0, 1), \tag{7.22}$$

where we require that with constants $\delta_c > 0$, $\kappa_n > 0$ it holds that

$$\|s_{k,y}^n\|_Y \leq \kappa_n(\|C(x_k)\|_Z + \delta_k) \quad \text{if } \|C(x_k)\|_Z + \delta_k \leq \delta_c. \tag{7.23}$$

Here $R_k \geq \|C(x_k)\|_Z^2$, $R_k > 0$, is a relaxation term which is chosen as $R_k = \|C(x_k)\|_Z^2$ unless feasibility is much better than stationarity, i.e., $\|C(x_k)\|_Z \ll \|\nabla_u L(x_k, \lambda_k)\|_{U^*}$.

Remark 7.9. If Assumption 7.7 holds with a convex ε-neighborhood X_0 of all x_k, $x_k + s_k^t$, then (7.23) can always be ensured. In fact, under this assumption the Newton–Kantorovich theorem ensures the existence of some $\delta > 0$ such that the following holds: in the case $\|C(x_k + s_k^t)\|_Z \leq \delta$ Newton's method for the equation

$$C(y_k + s_{k,y}^t + z, u_k + s_{k,u}^t) = 0 \tag{7.24}$$

starting with $z_0 = 0$ generates iterates z_j that converge with linear rate $1/2$ to a point z^* with $C(y_k + s_{k,y}^t + z^*, u_k + s_{k,u}^t) = 0$ and the iteration stays in a ball of radius $\leq 2M\|C(x_k + s_k^t)\|_Z \leq 2M\delta$, where M is a bound for $\|C_y^{-1}\|_{Z,Y}$ on X_0 (Assumption 7.7 allows a uniform choice for δ and M). But since C is Lipschitz on X_0 with a constant L_C by Assumption 7.7 and since $\|s_k^t\|_X \leq (1 + \kappa_t)\delta_k$, we have

$$\|C(x_k + s_k^t)\|_Z \leq \|C(x_k)\|_Z + L_C(1 + \kappa_t)\delta_k \leq (1 + L_C)(1 + \kappa_t)(\|C(x_k)\|_Z + \delta_k),$$

and hence the condition $\|C(x_k + s_k^t)\|_Z \leq \delta$ is satisfied whenever $\|C(x_k)\|_Z + \delta_k \leq \delta_c$ with $\delta_c = \delta(1 + L_C)^{-1}(1 + \kappa_t)^{-1}$. Moreover, the iteration stays in a ball of radius $\leq \kappa_n(\|C(x_k)\|_Z + \delta_k)$ with $\kappa_n = (1 + L_C)(1 + \kappa_t)2M$. Since $R_k > 0$ and $R_k \geq \|C(x_k)\|_Z^2$, (7.22) will eventually hold with $x_k + s_k = (y_k + s_{k,y}^t + z_j, u_k + s_{k,u}^t)$ and the resulting $s_{k,y}^n = z_j$ satisfies (7.23) for the above choices of δ_c and κ_n.

If $\|s_k^t\|_X \leq \kappa_s \|C(x_k)\|_Z$ with a constant $\kappa_s > 0$, then there exists $\delta > 0$ small enough, such that in the case $\|C(x_k + s_k^t)\|_Z \leq \delta$ the decrease condition (7.22) holds after a single Newton step by the local quadratic convergence.

The above considerations show that also any solver for (7.24) which converges for the starting point $z_0 = 0$ to the closest point z^* with $C(y_k + s_{k,y}^t + z^*, u_k + s_{k,u}^t) = 0$ will eventually generate an iterate $s_{k,y}^n = z_j$ satisfying (7.22), (7.23).

The purpose of the relaxation term R_k is to allow a controlled increase of the constraint violation if the current point is almost feasible but the stationarity is unsatisfactory. This is essential to obtain an efficient all-at-once method, since otherwise the algorithm would produce almost feasible iterates as soon as an almost feasible point was generated.

Similar as in [18] we use the following algorithm to adjust R_k.

Algorithm 7.10 (Computation of the Relaxation Parameter).
Input: $k,\ j_k,\ \|L_u(x_k,\lambda_k)\|_{U^*},\ \|C(x_k)\|_Z$.

Output: $R_k,\ j_{k+1}$.

Given: Tolerance sequence $a_j \searrow 0,\ \frac{a_{j+1}}{a_j} \geq \alpha_0 > 0,\ \alpha, \beta \in]0, \frac{1}{2}[$.

If $\|C(x_k)\|_Z \geq \min\left\{\alpha a_{j_k}, \beta\|L_u(x_k,\lambda_k)\|_{U^*}\right\}$: set $R_k := \|C(x_k)\|_Z^2,\ j_{k+1} := j_k$.

Else: set $R_k := \min\left\{a_{j_k}^2, \|\nabla_u L_u(x_k,\lambda_k)\|_{U^*}^2\right\},\ j_{k+1} := j_k + 1$.

Remark 7.11. In [18] it is in addition assumed that the tolerance sequence decays slowly enough that

$$\sum_{j=0}^{\infty} a_j^{\eta} < \infty$$

with a fixed constant $\eta > 4/3$. This is necessary to allow transition to fast local convergence when classical tangential step and quasi-normal steps are used.

Evaluation of Steps

For the evaluation of steps we adapt the merit-function-free strategy in [18] that provides a similar acceptance behavior as filter methods [7, 8] and is convenient to integrate in our generalized SQP framework. The basic idea is to compare the model reduction

$$\Delta q_k(s_k^t) := \hat{q}_k(0) - \hat{q}_k(s_{k,u}^t)$$

obtained in the quasi-tangential step with the actual reduction

$$\Delta L_k(s_k) := L_k(x_k,\lambda_k) - L_k(x_k + s_k, \lambda_k)$$

of the Lagrangian function if the model reduction $\Delta q_k(s_k^t)$ is sufficiently large compared to the constraint violation. More precisely, if

$$\Delta q_k(s_k^t) \geq (\nu\|C(x_k)\|_Z^2)^{\mu} \tag{7.25}$$

with constants $\nu > 0$, $\mu \in (2/3, 1)$, then we accept the step if in addition

$$\Delta L_k(s_k) \geq \eta_1 \Delta q_k(s_k^t)$$

with a constant $\eta_1 \in (0, 1)$.

If (7.25) is violated, then we always accept the step. In this case (7.22) ensures a nonmonotone decrease of the constraint violation.

Remark 7.12. In this chapter we could allow $\mu \in (0, 1)$ in (7.25). The range $\mu \in (2/3, 1)$ allows transition to fast local convergence when classical tangential step and quasi-normal steps are used; see [18].

The generalized SQP Method

Combining all ingredients we obtain the following algorithm.

Algorithm 7.13 (Generalized SQP Method without Merit Function).
Let $\eta_1 \in (0,1)$, $0 < \gamma_1 < 1 < \gamma_2$, $\nu > 0$, $\mu \in]\frac{2}{3}, 1[$.
Choose $x_0 \in X$, $\delta_0 \geq \delta_{min} > 0$, $\hat{B}_0 \in \mathcal{L}(U, U^*)$, and set $j_0 := 0$, $k := 0$.

1. Compute $C(x_k)$, $J(x_k)$.
2. Compute an approximate adjoint state λ_k satisfying (7.20).
3. If $\|C(x_k)\|_Z + \|L_x(x_k)\|_{X^*} = 0$: STOP ($x_k$ is KKT point).
4. Compute the generalized tangential step s_k^t according to (7.16), (7.18) satisfying (7.17), (7.19).
5. Determine the relaxation parameter R_k and j_{k+1} according to Algorithm 7.10.
6. Starting from $x_k + s_k^t$ compute a generalized quasi-normal step s_k^n satisfying (7.21), (7.22).
7. If $\Delta q_k(s_k^t) \geq (\nu \|C(x_k)\|_Z^2)^\mu$ goto 8 else goto 9.
8. If $\Delta L_k(s_k) < \eta_1 \, \Delta q_k(s_k^t)$ then set $x_{k+1} = x_k$, $j_{k+1} = j_k$. (step rejected)
   Set $\delta_{k+1} = \gamma_1 \delta_k$, $k := k+1$ and goto 2.
9. Set $x_{k+1} = x_k + s_k^t + s_k^n$, compute new Hessian update $\hat{B}_{k+1}$. (step successful)
   Choose $\delta_{k+1} \in [\max\{\delta_{min}, \delta_k\}, \max\{\delta_{min}, \gamma_2 \delta_k\}]$.
   Set $k := k+1$ and goto 1.

7.4.4 Convergence Result

We will prove global convergence under the following assumption.

Assumption 7.14.

- *Assumption 7.7 holds for an open convex ε-neighborhood $X_0 = Y_0 \times U_0$ containing all trial points x_k, $x_k + s_k$, $x_k + s_k^t$.*
- *There is a constant $M_B > 0$ with*

$$\|\hat{B}_k\|_{U,U^*} \leq M_B \quad \forall k.$$

- *λ_k remains uniformly bounded.*
- *$J(x_k)$ is bounded from below.*

Well Definedness

Lemma 7.15. *Let Assumption* 7.14 *hold. If Algorithm* 7.13 *does not terminate finitely, then it generates an infinite sequence of successful steps.*

Proof. We have already mentioned in Remarks 7.8 and 7.9 that the requirements on s_k^t and s_k^n can be met.

Assume that there is $K > 0$ such that all steps s_k with $k \geq K$ are not successful. Since the algorithm does not terminate, we have with a constant $\varepsilon > 0$

$$\|C(x_k)\|_Z + \|L_x(x_k, \lambda_k)\|_{X^*} \geq 2\varepsilon.$$

We will derive a contradiction.

By the mechanism of the algorithm all iterations $k \geq K$ can only be unsuccessful if

$$\delta_{k+1} = \gamma_1 \delta_k \to 0 \quad \text{and} \quad \Delta q_k(s_k^t) \geq (\nu \|C(x_k)\|_Z^2)^\mu \quad \forall\, k \geq K. \tag{7.26}$$

Since

$$\Delta q_k(s_k^t) \leq \|L_u(x_k, \lambda_k)\|_{U^*} \delta_k + \frac{1}{2} \|\hat{B}_k\|_{U,U^*} \delta_k^2 \leq c_1(\delta_k + \delta_k^2), \tag{7.27}$$

we conclude with (7.26) that

$$0 \leq (\nu \|C(x_k)\|_Z^2)^\mu \leq \Delta q_k(s_k^t) \to 0 \quad \text{for } k \to \infty$$

and therefore

$$\|C(x_k)\|_Z \to 0, \quad \delta_k \to 0 \tag{7.28}$$

(we even have $\|C(x_k)\|_Z = \|C(x_K)\|_Z = 0$ for $k \geq K$, but for later reference it is useful to use only (7.28)). Hence, we find $K' \geq K > 0$ with

$$\|L_x(x_k, \lambda_k)\|_{X^*} \geq \varepsilon \quad \forall\, k \geq K'.$$

Then we have either $\|L_u(x_k, \lambda_k)\|_{U^*} \geq \varepsilon/2$ or $\|L_y(x_k, \lambda_k)\|_{X^*} \geq \varepsilon/2$, and thus by (7.20)

$$\|L_u(x_k, \lambda_k)\|_{U^*} \geq \frac{\varepsilon}{2 + 2\kappa_\lambda} =: \kappa_1 \varepsilon \quad \forall\, k \geq K'. \tag{7.29}$$

Hence, the quasi-tangential step leads by (7.17) and (7.26) to a model decrease with

$$\begin{aligned} \Delta q_k(s_k^t) &\geq \kappa_q \kappa_1 \varepsilon \min\{\kappa_1 \varepsilon, \delta_k\} \quad \forall\, k \geq K', \\ \Delta q_k(s_k^t) &\geq \nu^\mu \|C(x_k)\|_Z^{2\mu} \quad \forall\, k \geq K. \end{aligned} \tag{7.30}$$

We show that this eventually implies

$$\Delta L_k(s_k) \geq \eta_1 \Delta q_k(s_k^t),$$

and thus the step would be accepted in contradiction to our assumption. In fact, we find by (7.28) some $K'' \geq K'$ with

$$\|C(x_k)\|_Z + \delta_k \leq \delta_c \quad \forall\, k \geq K''.$$

Thus (7.23) yields

$$\|s_k^n\|_X \le \kappa_n(\|C(x_k)\|_Z + \delta_k) \quad \forall k \ge K'',$$

and we have with some $\xi \in [0, 1]$

$$\begin{aligned}|\Delta L_k(s_k) - \Delta q_k(s_k^t)| &\le \|L_x(x_k + \xi s_k, \lambda_k) - L_x(x_k, \lambda_k)\|_{X^*}\|s_k\|_X \\ &\quad + \|L_y(x_k, \lambda_k)\|_{Y^*}\|s_{k,y}\|_Y + \frac{1}{2}\|\hat{B}_k\|_{U,U^*}\|s_{k,u}^t\|_U^2 \\ &\le c_2\|s_k\|_X^2 \le c_2(1 + \kappa_t + \kappa_n)^2(\|C(x_k)\|_Z + \delta_k)^2.\end{aligned}$$

Since $\mu \in (2/3, 1)$, a comparison with (7.30) shows that there is $0 < \delta \le \delta_c$ with

$$|\Delta L_k(s_k) - \Delta q_k(s_k^t)| \le (1 - \eta_1)\Delta q_k(s_k^t) \quad \forall k \ge K' \text{ with } \|C(x_k)\|_Z + \delta_k \le \delta,$$

and thus

$$\Delta L_k(s_k) \ge \eta_1 \Delta q_k(s_k^t) \quad \forall k \ge K' \text{ with } \|C(x_k)\|_Z + \delta_k \le \delta.$$

Since $\|C(x_k)\|_Z + \delta_k \to 0$ by (7.28) this shows that the step would eventually be successful, which contradicts our assumption. □

Convergence to Feasible Points

Lemma 7.16. *If $R_k = \|C(x_k)\|_Z^2$ for all successful iterations $k \ge K$, then*

$$\lim_{k\to\infty} \|C(x_k)\|_Z = 0.$$

Proof. In this case we have

$$\|C(x_k)\|_Z^2 - \|C(x_{k+1})\|_Z^2 \ge \kappa_c\|C(x_k)\|_Z^2 \quad \forall k \ge K \text{ with } x_{k+1} \ne x_k,$$

and thus $\|C(x_{k+1})\|_Z^2 \le (1 - \kappa_c)\|C(x_k)\|_Z^2$ for all successful iterations $k \ge K$. Since $C(x_{k+1}) = C(x_k)$ for rejected steps and since we have infinitely many successful steps by Lemma 7.15, we conclude that $\|C(x_k)\|_Z \to 0$. □

Otherwise, the following lemma applies.

Lemma 7.17. *If k is a successful iteration and $j_{k+1} = j_k + 1$ in Algorithm 7.10, then*

$$\|C(x_l)\|_Z \le a_{j_k} \quad \forall l \ge k.$$

Moreover, we have

$$\|C(x_k)\|_Z \le a_{j_k} \quad \forall k \quad \text{with} \quad j_k \ge 1.$$

Proof. Consider Algorithm 7.10 for a successful iteration k. If $j_{k+1} = j_k + 1$, then we must have

$$\|C(x_k)\|_Z < \min\{\alpha a_{j_k}, \beta\|L_u(x_k)\|_{U^*}\} \le \alpha a_{j_k}.$$

We now show by induction

$$\|C(x_l)\|_Z \le a_{j_k} \quad \forall\, l \ge k. \tag{7.31}$$

Since x_k and j_k change only for successful iterations it is sufficient to show (7.31) only for successful iterations l. For $l = k$ (7.31) is already shown. Now we have by using that $R_l \le \max\{\|C(x_l)\|_Z^2, a_{j_l}^2\}$ together with (7.22)

$$\|C(x_{l+1})\|_Z^2 \le \max\left\{R_l, \|C(x_l)\|_Z^2\right\} \le \max\left\{a_{j_l}^2, \|C(x_l)\|_Z^2\right\} \le a_{j_k}^2.$$

To show the second assertion we observe that $\|C(x_k)\|_Z \le a_{j_k}$ is already proven for every successful iterations k with $j_{k+1} = j_k + 1$ (and this includes the first iteration with $j_{k+1} = 1$). Now it follows also for all $l > k$ with $j_l = j_k$ that

$$\|C(x_l)\|_Z \le a_{j_k} = a_{j_l}. \quad \square$$

Together we obtain the following result.

Theorem 7.18. *We have*

$$\lim_{k\to\infty} \|C(x_k)\|_Z = 0.$$

Proof. If $R_k = \|C(x_k)\|_Z^2$ for all successful iterations $k \ge K$, this follows from Lemma 7.16. Otherwise we have by Algorithm 7.10 that $j_{k+1} = j_k + 1$ for an infinite sequence of successful iterations. Thus, $j_k \to \infty$ and we find $K > 0$ with $j_k \ge 1$ for all $k \ge K$. The previous lemma now yields

$$\|C(x_k)\|_Z \le a_{j_k} \quad \forall\, k \ge K,$$

and thus $\|C(x_k)\|_Z \to 0$, since $a_{j_k} \to 0$. $\square$

Global Convergence Result

We have the following convergence result.

Theorem 7.19. *Let Assumption* 7.14 *hold. If Algorithm* 7.13 *does not terminate finitely, then it generates an infinite sequence of iterates with*

$$\lim_{k\to\infty} \|C(x_k)\|_Z = 0, \quad \liminf_{k\to\infty} \|\nabla_x L_k\|_{X^*} = 0.$$

We start with the following auxiliary result.

Lemma 7.20. *Let Assumption* 7.14 *hold and assume that*

$$\|\nabla_x L_k\|_{X^*} \ge \varepsilon > 0 \quad \forall\, k \ge K'.$$

Then there is $K \geq K'$ and a constant $\delta > 0$ such that

$$\delta_k \geq \gamma_1 \min\{\delta_{min}, \delta/2\} =: \delta' \quad \textit{for all successful iterations } k \geq K. \tag{7.32}$$

Proof. We know already by Theorem 7.18 that $\|C(x_k)\|_Z \to 0$. After each successful step the trust-region radius is $\geq \delta_{min}$. Unsuccessful steps and a reduction of δ_k occur only in the case $\Delta q_k(s_k^t) \geq (\nu\|C(x_k)\|_Z^2)^\mu$.

In the latter case we deduce (7.29) and (7.30) exactly as in the proof of Lemma 7.15. As in the proof of Lemma 7.15 we obtain a constant $0 < \delta \leq \delta_c$ with

$$\Delta L_k(s_k) \geq \eta_1 \Delta q_k(s_k^t) \quad \forall k \geq K' \text{ with } \|C(x_k)\|_Z + \delta_k \leq \delta.$$

Now choose $K \geq K'$ so large that $\|C(x_k)\|_Z \leq \delta/2$ for all $k \geq K$. Then all steps $k \geq K$ are accepted as soon as $\delta_k \leq \delta/2$, which yields the lower bound in (7.32). □

***Proof of Theorem* 7.19.** We know by Theorem 7.18 that $\|C(x_k)\|_Z \to 0$. Assume that $\liminf_{k\to\infty} \|\nabla_x L_k\|_{X^*} > 0$. Then we find $K' \geq 0$ with

$$\|\nabla_x L_k\|_{X^*} \geq \varepsilon > 0 \quad \forall k \geq K'.$$

Now Lemma 7.20 yields $K \geq K'$ such that

$$\delta_k \geq \gamma_1 \min\{\delta_{min}, \delta/2\} =: \delta' \quad \text{for all successful iterations } k \geq K. \tag{7.32}$$

We show next that there is $K'' \geq K$ with

$$\Delta q_k(s_k^t) \geq (\nu\|C(x_k)\|_Z^2)^\mu \quad \text{for all successful iterations } k \geq K''. \tag{7.33}$$

As in (7.29) we obtain

$$\|L_u(x_k, \lambda_k)\|_{U^*} \geq \kappa_1 \varepsilon \quad \forall k \geq K',$$

and thus by the Cauchy decrease condition (7.17) and (7.32)

$$\Delta q_k(s_k^t) \geq \kappa_q \kappa_1 \varepsilon \min\{\kappa_1\varepsilon, \delta_k\} \geq \kappa_q \kappa_1 \varepsilon \min\left\{\kappa_1\varepsilon, \delta'\right\} =: \varepsilon'$$

for all successful iterations $k \geq K$. Since $\|C(x_k)\|_Z \to 0$ we deduce (7.33) with $K'' \geq K$ large enough.

Now (7.33) and the mechanism of the algorithm yield

$$\Delta L_k(s_k) \geq \eta_1 \Delta q_k(s_k^t) \geq \eta_1 \varepsilon' \quad \text{for all successful iterations } k \geq K''.$$

Finally, $\|C(x_k)\|_Z \to 0$ yields

$$L(x_k, \lambda_k) - L(x_{k+1}, \lambda_{k+1}) \geq \Delta L_k(s_k) - \|\lambda_{k+1} - \lambda_k\|_{Z^*}\|C(x_{k+1})\|_Z \geq \frac{\eta_1\varepsilon'}{2}$$

for all successful iterations $k \geq K'''$ with some $K''' \geq K''$. We deduce that $L(x_k, \lambda_k) \to -\infty$. This is a contradiction, since $J(x_k, \lambda_k)$ is bounded from below, $\|\lambda_k\|_{Z^*}$ and $\|C(x_k)\|_Z$ are uniformly bounded, and thus $L(x_k, \lambda_k)$ is bounded from below. □

7.5 Application of the Generalized SQP Method with Parareal Solvers

In this section we apply the generalized SQP method in Algorithm 7.13 to the specific decomposed time-dependent problem (7.8) for a semilinear parabolic equation in two dimensions and use parareal solvers for the state equation (7.9) and the adjoint equation (7.13).

7.5.1 An Optimal Control Problem for a Semilinear Parabolic Equation

We consider the problem

$$\min_{y \in W(0,T), u \in L^2((0,T)\times\Omega)} \quad J(y,u) := \frac{1}{2}\int_\Omega (y(T) - y_d)^2\, dx + \frac{\alpha}{2}\int_0^T \int_\Omega u^2\, dx\, dt$$

subject to

$$\begin{aligned} & y_t - \Delta y + \beta y^3 = u \quad \text{on } (0, T \times \Omega, \\ & y|_{(0,T)\times\partial\Omega} = 0, \\ & y(0,\cdot) = v_0 \quad \text{on } \Omega, \end{aligned} \tag{7.34}$$

where $\beta \geq 0$ is a constant. As in [4] the state and control spaces are given by

$$\begin{aligned} & U = L^2((0,T)\times\Omega), \\ & Y = W(0,T) = \left\{ y : y \in L^2(0,T; H_0^1(\Omega)),\ y_t \in L^2(0,T; H^{-1}(\Omega)) \right\}. \end{aligned}$$

Since $W(0,T) \hookrightarrow C([0,T]; L^2(\Omega))$, it is natural to choose in the decomposed problem (7.8), (7.9) the state space $Y_\Delta = B^{N+1}$ with $B = L^2(\Omega)$. Thus

$$C^\Delta : (L^2(\Omega))^{N+1} \times L^2(\Omega \times (0,T)) \to (L^2(\Omega))^{N+1},$$

$$C^\Delta(y_\Delta, u) = \begin{pmatrix} y^0 - v_0 \\ y^1 - g(T_0, y^0; u) \\ \vdots \\ y^N - g(T_{N-1}, y^{N-1}; u) \end{pmatrix}.$$

For the linear case $\beta = 0$ it is easy to check that Assumption 7.7 is satisfied for all convex open and bounded sets $X_0 = Y_0 \times U_0 \subset Y \times U$. For the semilinear case $\beta > 0$ it follows from the theory of semilinear parabolic equations that (7.34) admits for all $u \in U$ and initial data $v_0 \in W_0^{1,p}$, $p > 1$, a unique solution $y \in Y$; see [4, 16]. It is beyond the scope of this chapter to analyze (7.34) in detail. For sufficiently smooth Lipschitz continuous monotone increasing nonlinearities $f(y)$ instead of βy^3 results can be found in [4]. Semilinear elliptic equations are considered in [1, 11].

7.5.2 Propagators in the Parareal Scheme

For the implementation of the parareal state solver we use the equidistant time decomposition

$$T_n = n\Delta T, \quad \Delta T = \frac{T}{N}.$$

Coarse Propagator

For the coarse propagator $g_\Delta(T_n, v; u)$ we apply one backward Euler step in time and use a standard 5-point stencil for the Laplacian; i.e., $y^{n+1} = g_\Delta(T_n, v; u)$ is given by

$$\frac{y^{n+1} - v}{\Delta T} - Ay^{n+1} + v^3 = u^n.$$

Fine Propagator

To approximate the exact propagator $g(T_n, v; u)$ we apply N_f steps of a Crank–Nicolson scheme with time step $\delta T = \Delta T / N_f$ and use a standard 5-point stencil for the Laplacian. Hence, $y_f^{n+1} = g_\delta(T_n, v; u)$ is given by

$$\begin{aligned} &v^0 = v, \\ &\frac{v^{j+1} - v^j}{\delta T} - A\frac{v^{j+1} + v^j}{2} + \frac{(v^{j+1})^3 + (v^j)^3}{2} = u_{n,j}, \quad j = 0, \dots, N_f - 1, \\ &y_f^{n+1} = v^{N_f}. \end{aligned}$$

The implicit equation in each time step is solved by 1–3 Newton iterations in each time step.

7.5.3 Implementation of the Generalized SQP Method

We use the following solvers and approximation of the reduced Hessian:

- Nonlinear parareal solver for the state equation starting with the current state, parareal solver for the adjoint equation.
- Limited memory BFGS update for the reduced Hessian. $\hat{B}_0^{-1}$ is chosen as the inverse of the Hessian of the L^2-regularization. The update is skipped if the positive definiteness would be destroyed.
- As approximate solution of (7.16) we minimize along the Newton step in the trust region if it satisfies a sufficient decrease condition. Otherwise we minimize along the steepest descent direction in the trust region.

Remark 7.21. For the solution of the timestep equations in the propagators, standard multigrid solvers can directly be used. It would be possible to control the inexactness of the multigrid solvers by the generalized SQP methods.

In addition, a coarser space grid could be used for the coarse propagator.

7.5.4 Numerical Results

We consider the following specific problem:

$$\begin{aligned} &T = 1, \quad \alpha = 10^{-3}, \quad \beta = 1, \\ &y_d(x) = (2 + \sin(3T + 3x_1 - x_2^2))\, x_1 x_2 (1 - x_1)(1 - x_2), \\ &v_0(x) = (2 + \sin(3x_1 - x_2^2))\, x_1 x_2 (1 - x_1)(1 - x_2). \end{aligned}$$

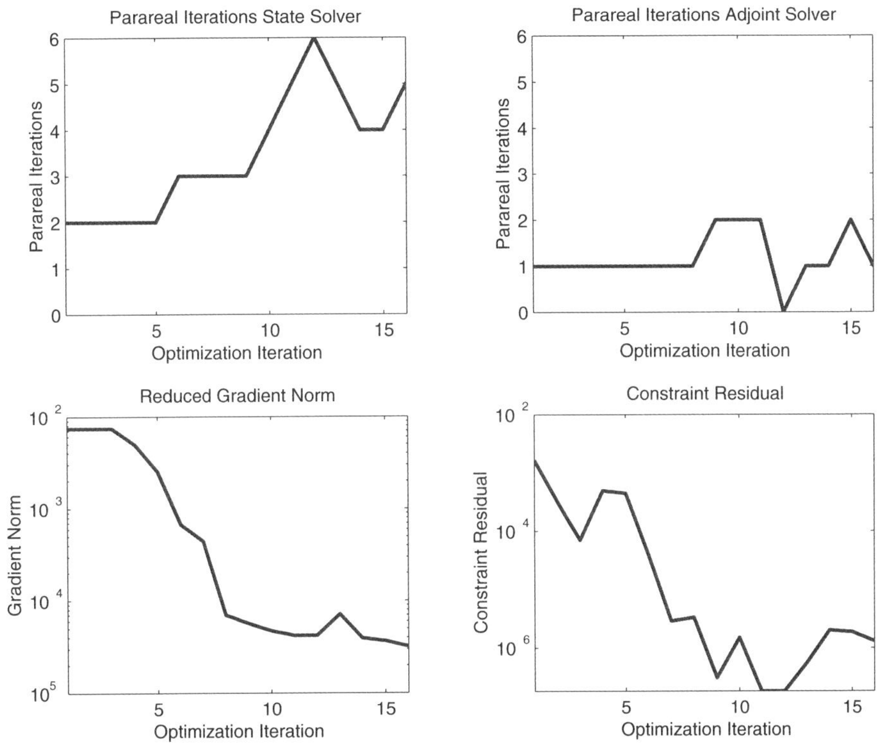

Figure 7.1. *Iteration history for the generalized SQP parareal method.*

For the discretization we use the following mesh sizes:

$$N = 20 \text{ time domains}, \quad \Delta T = \frac{1}{N}, \quad \delta T = \frac{\Delta T}{40}, \quad 40 \times 40 \text{ equidistant space grid.}$$

For conditions (7.22) and (7.20) we use the constants

$$\kappa_c = 10^{-1}, \quad \kappa_\lambda = 10^{-2}.$$

The iteration history is shown in Figure 7.1. The generalized SQP parareal method takes 16 optimization iterations. The upper plots show the number of parareal iterations for the state equation (upper left) and for the adjoint equation (upper right) versus the optimization iteration. We see that in the mean 3.5 parareal iterations for the state equation and 1.2 parareal iterations for the adjoint equation are necessary in each optimization iteration. The lower figures plot the norm of the reduced gradient (lower left) and the constraint residual (lower right) versus the optimization iteration. Note that one state solve requires 5–6 parareal iterations to achieve a comparable constraint residual of 10^{-6}. This yields the following cost ratio between optimization and one state solve:

$$\frac{\text{time(solution of optimization problem)}}{\text{time(solution of state equation)}} = \frac{16 \times (3.5 + 1.2) \text{ parareal its.}}{5 \text{ parareal its.}} \approx 15.$$

Parallel Efficiency

With $N = 20$ processors we have

$$\frac{\text{time(parallel parareal it.)}}{\text{time(serial state solve)}} = \frac{20 \text{ coarse} + 40 \text{ fine steps}}{800 \text{ fine steps}} = 0.075.$$

Since the optimization with serial state and adjoint solver takes also about 16 iterations, this yields

$$\frac{\text{time(parallel optimization)}}{\text{time(serial optimization)}} \approx 0.075 \times \frac{3.5 + 1.2}{1 + 1} \approx \frac{1}{5.7},$$

and thus the parallel efficiency is

$$\frac{5.7}{20} = 28.5\%.$$

We believe that a comparable parallel efficiency can be obtained for very large problems with many time steps even if many processors are used. Moreover, the parallel optimization allows us to treat much larger problems than a serial implementation on one processor, since it automatically distributes the problem to the resources of the available processors.

7.6 Conclusions

We have proposed a generalized SQP method for the solution of PDE-constrained optimization problems that works with any user-provided iterative solvers for the state equation and the adjoint equation. The algorithm uses—similar to filter methods [7, 8]—an approach without merit function and is a true all-at-once method that achieves feasibility and optimality simultaneously. The inaccuracy in the user-provided solvers is controlled by the SQP method in such a way that global convergence is ensured. While the proposed method is applicable to general problems of the form (7.1), we have shown that the generalized SQP method can efficiently be applied to time-dependent PDE-constrained optimization problems by using highly parallel parareal state and adjoint solvers. This approach provides a modular and flexible framework to obtain a parallel optimization solver for large-scale transient PDE-constrained optimization problems. The user has only to provide fine and coarse propagators for the considered PDE and its adjoint. We have shown that the approach yields promising numerical results for the optimal control of a semilinear parabolic problem in two dimensions. The application to other problems, in particular to the optimal control of the Navier–Stokes equations, is a topic for future research.

While it is quite straightforward to handle additional control constraints, we plan to extend the approach to problems with state constraints, which is important for applications. Moreover, the parareal technique could also be applied after a time-domain decomposition of the optimality system. We plan to investigate this approach. Finally, the proposed generalized SQP method is in particular well suited for the optimal control of flows with shocks, since nonlinear solvers for the state equation can directly be used, which avoids the difficulty of linearizing the state equation at shocks. Results for the optimal control of discontinuous flows will be presented elsewhere.

Bibliography

[1] N. Arada, J.-P. Raymond, and F. Tröltzsch, *On an augmented Lagrangian SQP method for a class of optimal control problems in Banach spaces*, Comput. Optim. Appl., 22 (2002), pp. 369–398.

[2] G. Bal, *On the convergence and the stability of the parareal algorithm to solve partial differential equations*, in Proceedings of DD15, Berlin, Lect. Notes Comput. Sci. Eng. 40, Springer-Verlag, Berlin, 2004, pp. 425–432.

[3] G. Bal and Y. Maday, *A "parareal" time discretization for non-linear PDE's with application to the pricing of an American put*, in Recent Developments in Domain Decomposition Methods, Zürich, 2001, Lect. Notes Comput. Sci. Eng. 23, Springer-Verlag, Berlin, 2002, pp. 189–202.

[4] M. Bergounioux and F. Tröltzsch, *Optimality conditions and generalized bang-bang principle for a state-constrained semilinear parabolic problem*, Numer. Funct. Anal. Optim., 17 (1996), pp. 517–536.

[5] A. Borzì, *Multigrid methods for parabolic distributed optimal control problems*, J. Comput. Appl. Math., 157 (2003), pp. 365–382.

[6] J. E. Dennis, Jr., M. El-Alem, and M. C. Maciel, *A global convergence theory for general trust-region-based algorithms for equality constrained optimization*, SIAM J. Optim., 7 (1997), pp. 177–207.

[7] R. Fletcher, N. I. M. Gould, S. Leyffer, P. L. Toint, and A. Wächter, *Global convergence of a trust-region SQP-filter algorithm for general nonlinear programming*, SIAM J. Optim., 13 (2002), pp. 635–659.

[8] R. Fletcher and S. Leyffer, *Nonlinear programming without a penalty function*, Math. Program., 91 (2002), pp. 239–269.

[9] M. Heinkenschloss, *A time-domain decomposition iterative method for the solution of distributed linear quadratic optimal control problems*, J. Comput. Appl. Math., 173 (2005), pp. 169–198.

[10] M. Heinkenschloss and L. N. Vicente, *Analysis of inexact trust-region SQP algorithms*, SIAM J. Optim., 12 (2001), pp. 283–302.

[11] M. Hintermüller and M. Ulbrich, *A mesh-independence result for semismooth Newton methods*, Math. Program., 101 (2004), pp. 151–184.

[12] J. E. Lagnese and G. Leugering, *A posteriori error estimates in time-domain decomposition of final value optimal control of the acoustic wave equation*, Appl. Math. Optim., 46 (2002), pp. 263–290.

[13] J. E. Lagnese and G. Leugering, *Time-domain decomposition of optimal control problems for the wave equation*, Systems Control Lett., 48 (2003), pp. 229–242.

[14] J.-L. LIONS, Y. MADAY, AND G. TURINICI, *Résolution d'EDP par un schéma en temps "pararéel"*, C. R. Acad. Sci. Paris Sér. I Math., 332 (2001), pp. 661–668.

[15] Y. MADAY AND G. TURINICI, *A parareal in time procedure for the control of partial differential equations*, C. R. Math. Acad. Sci. Paris, 335 (2002), pp. 387–392.

[16] P. NEITTAANMÄKI AND D. TIBA, *Optimal Control of Nonlinear Parabolic Systems*, Monogr. Textbooks Pure Appl. Math. 179, Marcel Dekker, New York, 1994.

[17] E. O. OMOJOKUN, *Trust Region Algorithms for Optimization with Nonlinear Equality and Inequality Constraints*, Ph.D. thesis, University of Colorado, Boulder, CO, 1989.

[18] M. ULBRICH AND S. ULBRICH, *Non-monotone trust region methods for nonlinear equality constrained optimization without a penalty function*, Math. Program., 95 (2003), pp. 103–135.

[19] S. ULBRICH, *A sensitivity and adjoint calculus for discontinuous solutions of hyperbolic conservation laws with source terms*, SIAM J. Control Optim., 41 (2002), pp. 740–797.

[20] S. ULBRICH, *Adjoint-based derivative computations for the optimal control of discontinuous solutions of hyperbolic conservation laws*, Systems Control Lett., 48 (2003), pp. 309–324.

Chapter 8

Simultaneous Pseudo-Timestepping for State-Constrained Optimization Problems in Aerodynamics

Subhendu B. Hazra and Volker Schulz**

8.1 Introduction

With the increase in efficiency of computer technology and availability of high-fidelity CFD codes, simulation-based optimization is becoming a more and more reliable tool for preliminary design methodology. Among various optimization strategies used for this method, gradient-based methods are used in most of the practical applications. For the computation of sensitivities, the adjoint method has advantages over finite difference methods and is applied in most of the practical applications as well.

In this chapter we present a method for solving aerodynamic shape optimization problems involving state constraints. It is based on simultaneous pseudo-timestepping. Instead of solving the KKT system using some iterative techniques, we consider the pseudo-time embedded nonstationary system of equations. The advantages of this method are that it requires no additional globalization technique in the design space, that it blends in nicely with a previously existing pseudo-timestepping method for the states only, and that a block preconditioner can be used for convergence acceleration which stems from reduced SQP methods. Also for the problem class that we consider in our applications, the state and costate solutions are obtained using pseudo-timestepping on the corresponding unsteady equations. Therefore, it seems natural to use pseudo-time embedding for the design equation as well.

Optimization in aerodynamics is an active area of current research and many possibilities are being considered. Among various methods, evolutionary genetic algorithms are receiving considerable attention, since they promise to lead to the global optimum. However, their application is still limited due to the enormous costs involved. The other methods

*Department of Mathematics, University of Trier, D-54286 Trier, Germany.

used in most of the practical applications are gradient-based algorithms using adjoint sensitivities. These methods require lesser computational cost than evolutionary algorithms, but the overall cost is still quite high. This is due to the fact that typically in each design update of the method, the state and costate equations are to be solved quite accurately. Even though efficient CFD solvers are available, repeating the process several times adds up, so that the cost can be quite high. Computational results based on these methodologies have been presented in [8, 10, 21, 22, 32, 33] on structured grids. An application of this method on unstructured grids has been presented in [1]. This approach with a less accurate state and costate solution has been performed in Iollo, Kuruvila, and Ta'asan [19].

In [14], we proposed a new method for solving such problems using simultaneous pseudo-timestepping. It is well known that there is a strong correlation between iterative methods and pseudo-timestepping [5, 24] which has been exploited for the construction of a timestepping method in the spirit of reduced SQP (RSQP) methods. This formulation is advantageous, since the steady state flow is obtained by integrating the pseudo-unsteady Euler (or Navier–Stokes) equations in this problem class. Therefore, one can use the same timestepping scheme for the whole set of equations and preconditioners can be used to accelerate the convergence. Certain preconditioners are equivalent to SQP methods, whose mathematical background is well studied. In [15, 16] we have implemented that method for the shape design example using Euler equations. In [17] we extended the same method for problems involving additional state constraints. The idea of partially RSQP methods [6, 12], which is a generalization of standard RSQP methods, e.g., as in [13, 27], is used for this extension. This chapter is supplemented with two application examples for drag reduction with constant lift for RAE2822 and RAE5225 airfoils. The number of iterations required for the full optimization problem is less than 7 times that of the analysis problem. This implies a drastic reduction of the computational cost compared to usual gradient methods that solve the state and adjoint PDEs exactly at each optimization iteration.

The chapter is organized as follows. For the sake of completeness, we explain briefly the simultaneous pseudo-timestepping method of [17] in the next section. Section 8.3 presents the back projection method into the additional state constraint. Numerical results are presented in Section 8.4. We draw our conclusions in Section 8.5.

8.2 Simultaneous Pseudo-Timestepping for Optimization Problems

The optimization problem that we are dealing with in this study is of the class of optimal control problems including also the subclass of shape design problems. These problems can be written in abstract form as

$$\begin{aligned} &\min \ I(w,q) \\ &\text{subject to} \quad c(w,q) = 0, \\ &\qquad\qquad h(w,q) = 0, \end{aligned} \tag{8.1}$$

where $(w,q) \in X \times P$ (X, P are appropriate Hilbert spaces), $I : X \times P \to \mathbb{R}$, and $c : X \times P \to Y$ are twice Frechet-differentiable (with Y an appropriate Banach space). The Jacobian, $J = \frac{\partial c}{\partial w}$, is assumed to be invertible. Here the equation $c(w,q) = 0$ represents the steady state flow equations (in our case Euler equations) together with boundary conditions, where w is the vector of state variables and q is the vector of design variables. The objective $I(w, q)$ is the drag and the scalar-valued function $h(w, q)$ is the lift of an airfoil for the purposes of this chapter. Typically, in practical applications, additional constraints of the form

$$h(w, q) \geq 0$$

often pose severe restrictions on the validity region of the model or for the design construction. For the sake of simplicity in presentation, we discuss only the equality constraints, which is the case in our application. The inequality constraints can be handled by an active set strategy. For the scalar constraint of the airfoil design problem, an active set strategy is trivial.

The necessary optimality conditions can be formulated using the Lagrangian functional

$$L(w, q, \lambda) = I(w, q) - \lambda^\top c(w, q) - \mu^\top h(w, q), \tag{8.2}$$

where λ and μ are the Lagrange multipliers or the adjoint variables from the dual Hilbert space. If $\hat{z} = (\hat{w}, \hat{q})$ is a minimum, then there exist a $\hat{\lambda}$ and a $\hat{\mu}$ such that

$$\nabla_z L(\hat{z}, \hat{\lambda}, \hat{\mu}) = \nabla_z I(\hat{z}) - \hat{\lambda}^\top \nabla_z c(\hat{z}) - \hat{\mu}^\top \nabla_z h(\hat{z}) = 0. \tag{8.3}$$

Hence, the necessary optimality conditions are

$$c(w, q) = 0, \tag{8.4a}$$
$$h(w, q) = 0, \tag{8.4b}$$
$$\nabla_w L(w, q, \lambda, \mu) = 0, \tag{8.4c}$$
$$\nabla_q L(w, q, \lambda, \mu) = 0. \tag{8.4d}$$

This set of nonlinear equations can be solved using an iterative method, for example, using an approximate RSQP method. A detailed discussion of this approach can be found in [34, 35]. A step of this method can be interpreted as an approximate Newton step for the necessary conditions of finding the extremum of problem (8.1), since the updates of the variables are computed according to the linear system

$$\begin{pmatrix} 0 & 0 & \left(\frac{\partial h}{\partial w}\right)^\top & A^\top \\ 0 & B & \left(\frac{\partial h}{\partial q}\right)^\top & \left(\frac{\partial c}{\partial q}\right)^\top \\ \frac{\partial h}{\partial w} & \frac{\partial h}{\partial q} & 0 & 0 \\ A & \frac{\partial c}{\partial q} & 0 & 0 \end{pmatrix} \begin{pmatrix} \Delta w \\ \Delta q \\ \Delta \mu \\ \Delta \lambda \end{pmatrix} = \begin{pmatrix} -\nabla_w L \\ -\nabla_q L \\ -h \\ -c \end{pmatrix}, \tag{8.5}$$

where A is some approximation of the Jacobian $J = \frac{\partial c}{\partial w}$ and B is the reduced Hessian.

As a special iterative method to determine the solution of (8.4), we use simultaneous pseudo-timestepping to reach the steady states of the following pseudo-time embedded evolution equations:

$$\begin{aligned} \frac{dw}{dt} + c(w,q) &= 0, \\ \frac{d\mu}{dt} + h(w,q) &= 0, \\ \frac{d\lambda}{dt} + \nabla_w L(w,q,\lambda) &= 0, \\ \frac{dq}{dt} + \nabla_q L(w,q,\lambda) &= 0. \end{aligned} \tag{8.6}$$

This approach has the advantage that it blends nicely with a pseudo-timestepping flow solver.

In [19], Iollo, Kuruvila, and Ta'asan proposed an approach in which pseudo-time embedding is suggested for the state and costate equations and the design equation is solved as an additional algebraic equation, in particular, in the form of an additional boundary condition for boundary control problems. An additional state constraint of the form $h(w,q) = 0$ is not considered there. In that case, the pseudo-time embedded system would be written in our notation as

$$\begin{aligned} \frac{dw}{dt} + c(w,q) &= 0, \\ \frac{d\lambda}{dt} + \nabla_w L(w,q,\lambda) &= 0, \\ \nabla_q L(w,q,\lambda) &= 0. \end{aligned} \tag{8.7}$$

This is a differential algebraic system of equations, where one has to provide some means to solve the design equation alone and no preconditioner is used for this system.

The pseudo-time embedded system (8.6) is usually a stiff system of ODEs . Therefore explicit timestepping schemes may converge very slowly or may even diverge. In order to accelerate convergence, this system needs some preconditioning . In our implementation, we use the inverse of the matrix in (8.5) as a preconditioner for the timestepping process. The pseudo-time embedded system of ODEs that we consider is

$$\begin{pmatrix} 0 & 0 & \left(\frac{\partial h}{\partial w}\right)^\top & A^\top \\ 0 & B & \left(\frac{\partial h}{\partial q}\right)^\top & \left(\frac{\partial c}{\partial q}\right)^\top \\ \frac{\partial h}{\partial w} & \frac{\partial h}{\partial q} & 0 & 0 \\ A & \frac{\partial c}{\partial q} & 0 & 0 \end{pmatrix} \begin{pmatrix} \dot{w} \\ \dot{q} \\ \dot{\mu} \\ \dot{\lambda} \end{pmatrix} = \begin{pmatrix} -\nabla_w L \\ -\nabla_q L \\ -h \\ -c \end{pmatrix}. \tag{8.8}$$

This seems natural, since (8.5) can be considered as an explicit Euler discretization for the corresponding timestepping that we envision. Also, due to its block structure, it is computationally inexpensive. The preconditioner employed is similar to the preconditioners for KKT systems discussed in [2, 3] in the context of Krylov subspace methods and in [4] in the context of Lagrange–Newton–Krylov–Schur methods.

Within the inexact reduced SQP preconditioner, one has to look for an appropriate approximation of the reduced Hessian, B. In particular, when dealing with PDEs constituting the state equations, the reduced Hessian can often be expressed as a pseudo-differential operator, the symbol of which can be computed and exploited for preconditioning purposes as in [14].

Using Gaussian elimination for partial reduction, the above system (8.8) can be reduced to a system

$$\begin{pmatrix} B & g_h \\ g_h^\top & 0 \end{pmatrix} \begin{pmatrix} \dot{q} \\ \dot{\mu} \end{pmatrix} = \begin{pmatrix} -\nabla_q L + \left(\frac{\partial c}{\partial q}\right)^\top A^{-\top} \nabla_w L \\ -h + \frac{\partial h}{\partial w} A^{-1} c \end{pmatrix}, \tag{8.9}$$

where

$$g_h := \left(\frac{\partial h}{\partial q} - \frac{\partial h}{\partial w} A^{-1} \frac{\partial c}{\partial q} \right)^\top = \begin{bmatrix} -A^{-1} \frac{\partial c}{\partial q} \\ I \end{bmatrix}^\top \begin{bmatrix} \nabla_w h \\ \nabla_q h \end{bmatrix}$$

is the reduced gradient of $h(w, q)$.

We use this reduction strategy in the following to solve problem (8.1) using simultaneous pseudo-timestepping. First, we consider the system without the additional state constraint $h(w, q)$ as discussed in [15]. Then the linearized constraint h leads to the QP

$$\begin{aligned} \min \; & \frac{1}{2} \dot{q}^\top B \dot{q} + g^\top \dot{q} \\ \text{subject to } \; & g_h^\top \dot{q} = -h(w, q) + \frac{\partial h}{\partial w} A^{-1} c, \end{aligned} \tag{8.10}$$

where $g = (\nabla_q L - (\frac{\partial c}{\partial q})^\top A^{-\top} \nabla_w L)$ is the reduced gradient of the objective function. Construction of each of the reduced gradients g and g_h requires (approximate) solution of one adjoint problem.

Therefore, this reduction can be interpreted as a projection of the design velocity of the corresponding equation (8.8) without the state constraint towards the linearized lift constraint and thus resembling dynamic projection strategies onto invariants as in [36].

8.3 Back Projection

Due to the nonlinearity of the problem there is some deviation from the additional state constraint. Therefore we use the following correction strategy in each optimization step to avoid this deviation. We minimize the distance between the point q_0 and the manifold $\mathcal{S}$ of constant lift $\mathcal{S} = \{q \mid h(w(q)) = l_0\}$ (Figure 8.1), where l_0 is the given constant value of the state constraint that we wish to achieve. That is to solve ideally the problem

$$\begin{aligned} \min \; & \tfrac{1}{2} \|q - q_0\|^2 \\ \text{subject to } \; & h(w(q)) - l_0 = 0. \end{aligned} \tag{8.11}$$

We use one step of a generalized Gauss–Newton technique to solve this problem. Since the stiffness matrix of the flow equations is approximated by A when forming the reduced

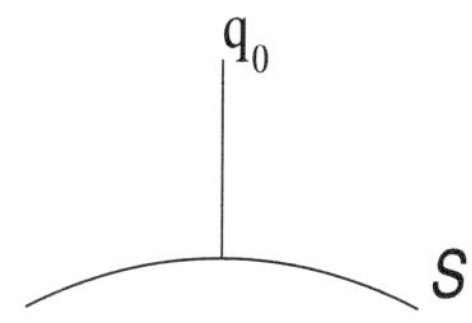

Figure 8.1. *Projection.*

gradient g_h, we compute the step Δq from

$$\begin{aligned} &\min \ \tfrac{1}{2}||\Delta q||^2 \\ &\text{subject to} \ \ g_h^\top \Delta q = -h(w,q) + \frac{\partial h}{\partial w} A^{-1} c. \end{aligned} \tag{8.12}$$

The necessary optimality conditions lead to the system of equations

$$\begin{pmatrix} I & g_h \\ g_h^\top & 0 \end{pmatrix} \begin{pmatrix} \Delta q \\ \nu \end{pmatrix} = \begin{pmatrix} 0 \\ -(h(w,q) - \frac{\partial h}{\partial w} A^{-1} c) \end{pmatrix},$$

where ν is the Lagrange multiplier for the system (8.12). Solving this system gives

$$\Delta q = \frac{-\left(h(w,q) - \frac{\partial h}{\partial w} A^{-1} c\right)}{g_h^\top g_h} g_h.$$

Hence, the corrected step is given by

$$q^{k+1} = q_0^{k+1} - \Delta q. \tag{8.13}$$

The overall algorithm reads as follows.

Algorithm 8.1.
The simultaneous pseudo-timestepping for the preconditioned system.

(0) Set $k := 0$; start at some initial guess w_0, λ_0, and q_0.

(1) Compute λ^{k+1} marching one step in time for the adjoint equations.

(2) Compute sensitivities using state and adjoint solutions.

(3) Determine some approximation B_k of the projected Hessian of the Lagrangian.

(4) Solve the quadratic subproblem (8.10) to get $\dot{q}$.

(5) March in time one step for the design equation as follows:
$q^{k+1} = q^k + \Delta t \cdot \dot{q}$.

(6) Use the correction step (8.13) for the new step.

(7) Compute w^{k+1} marching one step in time for the state equations.

(8) Set $k := k + 1$; go to (1) until convergence.

In the current implementation, $A^{-1}c$ in the right-hand side of the constraint in (8.10) is approximated by $\dot{w}$ value from the previous iteration as it is updated only at step (7) of the above algorithm. It is a "one-shot" method, since we perform one time step for each design update.

The details of governing equations, discretization, surface parameterization, gradient computation, and grid perturbation strategies can be found in [15, 16, 17].

8.4 Numerical Results and Discussion

The optimization method is applied to two test cases of RAE2822 and RAE 5225 airfoils for drag reduction with constant lift. For RAE2822 airfoil the physical domain is discretized using an algebraically generated (193 × 33) C-grid. The camberline representation of the airfoil is parameterized by 21 Hicks–Henne parameters. A complete optimization cycle is performed under the optimization platform SynapsPointerPro [7]. We start the optimization iteration with the solution obtained after 500 time steps of the state equations (w_0) and 800 time steps of the costate equations (λ_0). The FLOWer code [25, 26] of the German Aerospace Center (DLR) is used for solving the forward and adjoint equations.

The design equation is integrated in time using an explicit Euler scheme, and the state and costate equations are integrated in time using a 5-stage Runge–Kutta scheme. Therefore the time steps used for the three sets of equations are not the same. In the current implementation of FLOWer the time steps are not the same even in each discretization cell as they are determined independently according to the local stability. However, this has no effect on the final solution at steady state.

One of the main issues of using this kind of preconditioned pseudo-timestepping is the approximation of the reduced Hessian. In [11, 15] it is demonstrated that the reduced Hessian preconditioner should be consistent with the approximate solvers for the forward and adjoint system. The implementation of the reduced Hessian approximation is based on second-order information similar to the case of BFGS optimization methods. We define $s_k := (q_{k+1} - q_k)$ and $z_k := (\nabla I_{k+1} - \nabla I_k)$, where k represents the iteration number and the gradients are formed by the current approximation. Then the curvature in the direction s_k is obtained from the product $(z_k^T s_k)$. If the curvature is positive, the reduced Hessian is approximated by

$$B_k = \bar{\beta} \frac{z_k^T s_k}{z_k^T z_k} \delta_{ij},$$

where $\bar{\beta}$ is a constant. Otherwise, it is approximated by $\beta \delta_{ij}$, where β is a different constant.

The optimization iteration is stopped when $||\dot{q}||_\infty < 0.0005$. This requires 1900 optimization iterations. Figure 8.2 presents the optimization convergence history of this case. After the convergence of the optimization iteration, another 600 time steps are performed for the state equation on the optimized geometry to reduce its residual further so that the force coefficients and surface pressure distribution can be compared with those obtained by other methods.

Figure 8.3 presents the comparisons of initial and final airfoils (upper left), camber lines (upper right), and surface pressure distribution. Since the major amount of drag in

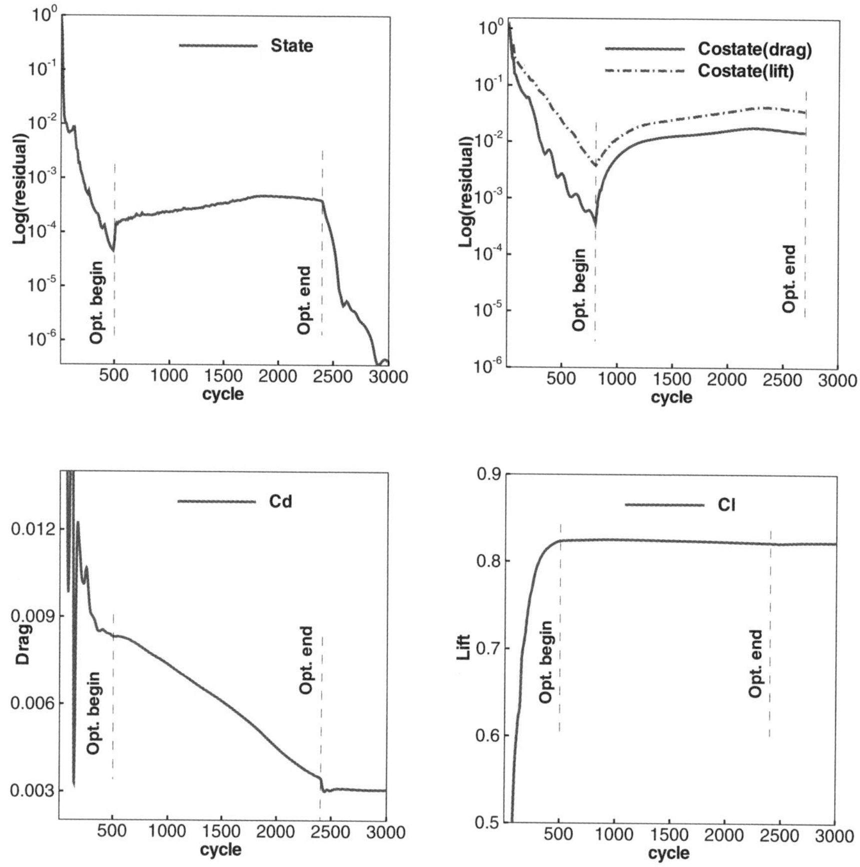

Figure 8.2. *Convergence history of the optimization iterations.*

the transonic regime is caused by the shock, therefore the optimization for drag reduction results in a shockless airfoil.

Each forward and adjoint run requires approximately 2500 iterations in time to reach steady state. The total number of time steps required in this case is $(500 + 1900 + 600)$ for state solvers, and two times $(800 + 1900)$ for adjoint solvers. This together is approximately 5.2 times that of the forward run. The effort to solve the design equation is almost negligible in comparison to state or costate solutions. Also in every optimization iteration the gradients are to be computed twice (once for the drag and once for the lift). The time required for this is also negligible in comparison to the state or costate solutions. Additionally, the simultaneous pseudo-time method needs a new grid, obtained by grid perturbation, after each optimization iteration. Additional time is required to write the output after every iteration and read the same before each iteration, as the iterations start with solution values

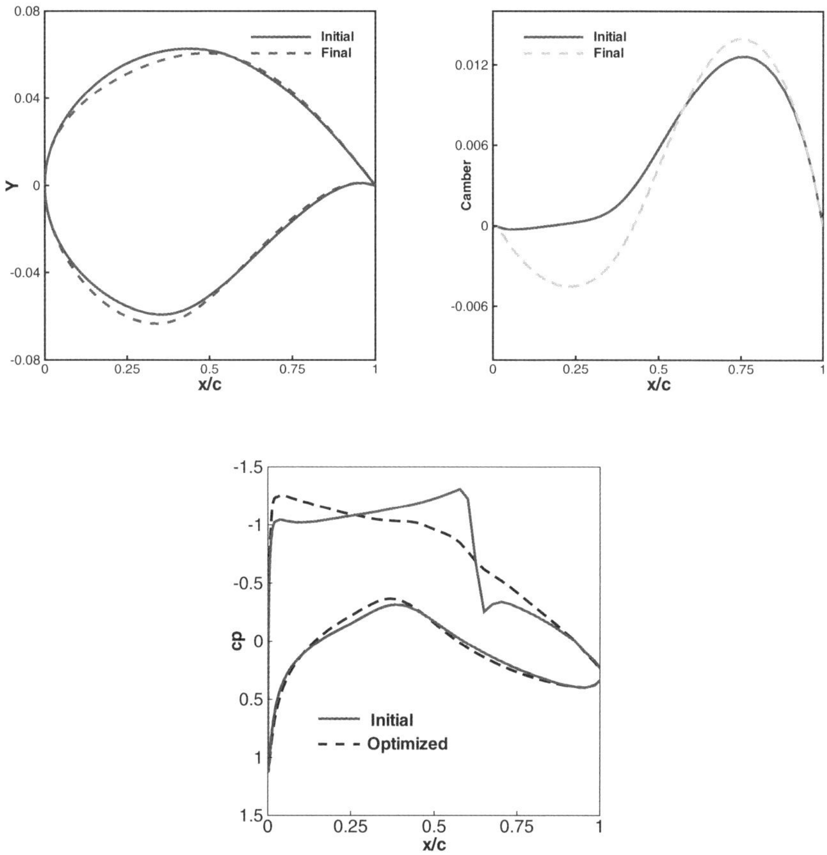

Figure 8.3. *Comparison of initial and final quantities.*

from the previous iteration. However, the total time required for this overhead is less than one complete forward run. If we add up everything, the total effort is less than 7 times that of a forward simulation run. In terms of CPU time, the complete optimization cycle needs about 80 minutes on an Intel Xeon CPU 1700MHz machine.

In the pseudo-time optimization iteration, the initial drag of 0.0081012 is reduced to 0.003070 in the optimization process which is a reduction of more than 62%. We imposed a constraint of constant lift. This constraint is quite well satisfied as the final reduction of the lift is less than 0.18%.

In the second test case of the RAE5225 airfoil, the physical domain is divided into a (256 × 61) C-grid. The airfoil is parameterized by 21 Hicks–Henne parameters as in the earlier case. The initial state and costate values for the optimization iteration are taken from the solution obtained after 2500 Runge–Kutta time steps. The convergence history

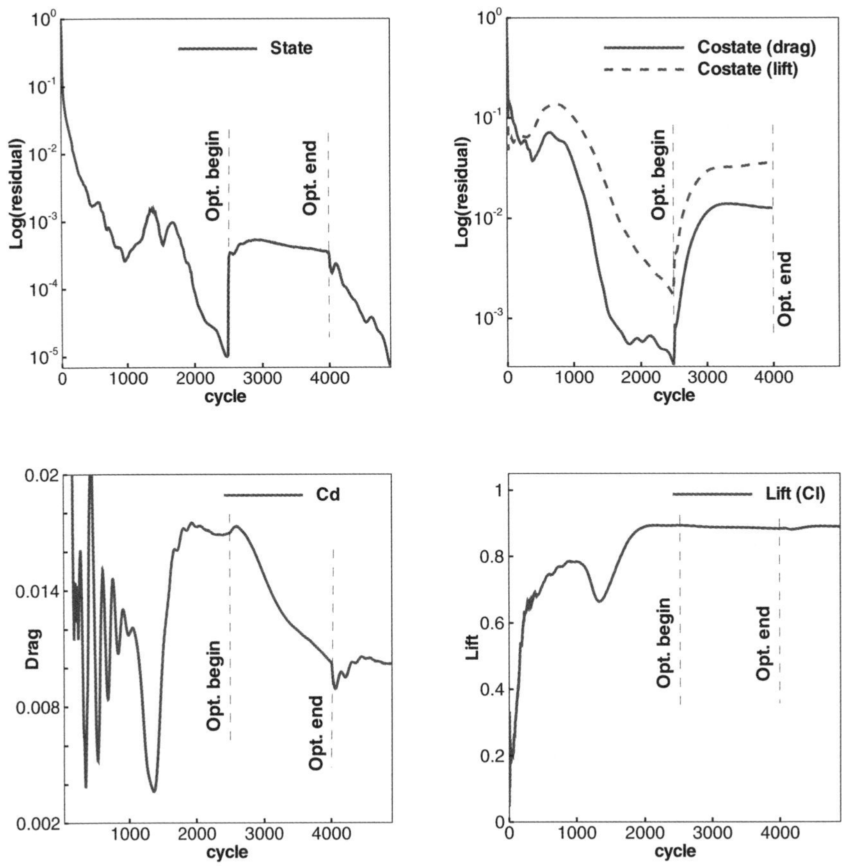

Figure 8.4. *Convergence history of the optimization iterations.*

is presented in Figure 8.4. The initial and optimized airfoil and camberline is presented in Figure 8.5 (top) and initial and optimized surface pressure is presented in Figure 8.5 (bottom). In this case the optimization requires 1500 iterations. The drag is reduced by 37% and the change in lift is less than 0.2%.

8.5 Conclusions

A new method has been proposed for aerodynamic shape optimization with state constraints which is based on simultaneous pseudo-timestepping. The solution of the KKT system is obtained in terms of steady states of the pseudo-unsteady state, costate, and design equations. For convergence acceleration of the timestepping process a preconditioner is used which is motivated by a continuous reinterpretation of RSQP methods. The overall cost

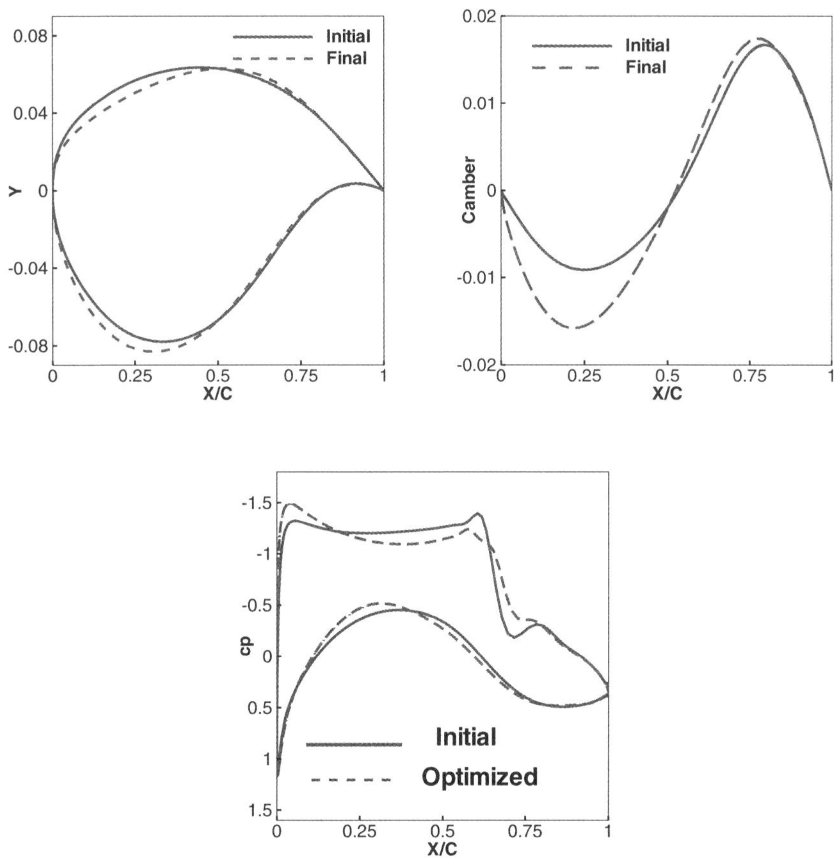

Figure 8.5. *Comparison of initial and final quantities.*

of computation is approximately 7 times that of a forward simulation run. Application to three-dimensional problems is our next goal.

Acknowledgments

We would like to thank N. Gauger and J. Brezillon for discussions and helpful hints for implementations with the code FLOWer.

Bibliography

[1] W. K. ANDERSON AND V. VENKATAKRISHNAN, *Aerodynamic design optimization on unstructured grids with a continuous adjoint formulation*, AIAA 97-0643, 1997.

[2] A. Battermann and M. Heinkenschloss, *Preconditioners for Karush-Kuhn-Tucker systems arising in the optimal control of distributed systems*, in Optimal Control of PDE, Vorau 1996, W. Desch, F. Kappel, and K. Kunisch, eds., Birkhäuser-Verlag, Basel, 1996, pp. 15–32.

[3] A. Battermann and E. W. Sachs, *Block preconditioners for KKT systems in PDE-governed optimal control problems*, in Fast Solution of Discretized Optimization Problems, K. H. Hoffmann, R. H. W. Hoppe, and V. Schulz, eds., Birkhäuser-Verlag, Basel, 2001, pp. 1–18.

[4] G. Biros and O. Ghattas, *Parallel Lagrange–Newton–Krylov–Schur methods for PDE-constrained optimization. Part* I: *The Krylov–Schur solver*, SIAM J. Sci. Comput., 27 (2005), pp. 687–713.

[5] T. S. Coffey, C. T. Kelley, and D. E. Keyes, *Pseudotransient continuation and differential-algebraic equations*, SIAM J. Sci. Comput., 25 (2003), pp. 553–569.

[6] R. Fletcher, *Practical Methods of Optimization*, John Wiley & Sons, New York, 1987.

[7] O. Fromman, *SynapsPointerPro v*2.50, Synaps Ingenieure Gesellschaft mbH, Bremen, Germany, 2002.

[8] N. R. Gauger, *Aerodynamic shape optimization using the adjoint Euler equations*, in Proceedings of the GAMM workshop on Discrete Modeling and Discrete Algorithms in Continuum Mechanics, Logos-Verlag, Berlin, 2001, pp. 87–96.

[9] N. R. Gauger, *Das Adjungiertenverfahren in der aerodynamischen Formoptimierung*, DLR-Report DLR-FB-2003-05, TU Braunschweig, Braunschweig, Germany, 2003.

[10] N. R. Gauger and J. Brezillon, *Aerodynamic shape optimization using adjoint method*, J. Aero. Soc. India, 54 (2002), pp. 247–254.

[11] I. Gherman and V. Schulz, *Preconditioning of one-shot pseudo-timestepping methods for shape optimization*, PAMM, 5 (2005), pp. 741–742.

[12] P. E. Gill, W. Murry, and M. H. Wright, *Practical Optimization*, Academic Press, London, 1981.

[13] C. B. Gurwitz and M. L. Overton, *Sequential quadratic programming methods based on approximating a projected Hessian matrix*, SIAM J. Sci. Stat. Comput., 10 (1989), pp. 631–653.

[14] S. B. Hazra and V. Schulz, *Simultaneous pseudo-timestepping for PDE-model based optimization problems*, BIT, 44 (2004), pp. 457–472.

[15] S. B. Hazra, V. Schulz, J. Brezillon, and N. R. Gauger, *Aerodynamic shape optimization using simultaneous pseudo-timestepping*, J. Comput. Phys., 204 (2005), pp. 46–64.

[16] S. B. Hazra, *An efficient method for aerodynamic shape optimization problems*, AIAA 4628-2004, 2004.

[17] S. B. Hazra and V. Schulz, *Simultaneous Pseudo-Timestepping for Aerodynamic Shape Optimization Problems with State Constraint*, Forschungsbericht Nr.04-7, Department of Mathematics/Computer Science, University of Trier, Trier, Germany, 2004.

[18] R. M. Hicks and P. A. Henne, *Wing design by numerical optimization*, J. Aircraft, 15 (1978), pp. 407–412.

[19] A. Iollo, G. Kuruvila, and S. Ta'asan, *Pseudo-Time Method for Optimal Shape Design Using Euler Equations*, ICASE Report 95-59, NASA Langley Research Center, Hampton, VA, 1995.

[20] A. Jameson, *Aerodynamic design via control theory*, J. Sci. Comput., 3 (1988), pp. 23–260.

[21] A. Jameson, *Automatic Design of Transonic Airfoils to Reduce Shock Induced Pressure Drag*, MAE report 1881, presented at the 31st Israel Annual Conference on Aviation and Aeronautics, 1990.

[22] A. Jameson, *Optimum aerodynamic design using CFD and control theory*, in Proceedings of the AIAA 12th Computational Fluid Dynamics Conference, AIAA 95-1729-CP, 1995.

[23] A. Jameson and J. Reuther, *Control theory based airfoil design using the Euler equations*, AIAA Proceedings 94-4272-CP, 1994.

[24] C. T. Kelley and D. E. Keyes, *Convergence analysis of pseudo-transient continuation*, SIAM J. Numer. Anal., 35 (1998), pp. 508–523.

[25] N. Kroll, C. C. Rossow, K. Becker, and F. Thiele, *The MEGAFLOW - a numerical flow simulation system*, in Proceedings of the 21st ICAS Symposium, Paper 98-2.7.4, Melbourne, Australia, 1998.

[26] N. Kroll, C. C. Rossow, K. Becker, and F. Thiele, *The MEGAFLOW project*, Aerosp. Sci. Technol., 4 (2000), pp. 223–237.

[27] F.-S. Kupfer, *An infinite-dimensional convergence theory for reduced SQP methods in Hilbert space*, SIAM J. Optim., 6 (1996), pp. 126–163.

[28] J. L. Lions, *Optimal Control of Systems Governed by Partial Differential Equations*, Springer-Verlag, New York, 1971.

[29] O. Pironneau, *On optimum design in fluid mechanics*, J. Fluid Mech., 64 (1974), pp. 97–110.

[30] O. Pironneau, *On optimum profiles in Stokes flow*, J. Fluid Mech., 59 (1973), pp. 117–128.

[31] O. PIRONNEAU, *Optimal Shape Design for Elliptic Systems*, Springer-Verlag, New York, 1982.

[32] J. REUTHER AND A. JAMESON, *Aerodynamic shape optimization of wing and wing-body configurations using control theory*, AIAA 95-0123, 1995.

[33] J. REUTHER, A. JAMESON, J. FARMER, L. MARTINELLI, AND D. SAUNDERS, *Aerodynamic shape optimization of complex aircraft configurations via an adjoint formulation*, AIAA 96-0094, 1996.

[34] V. H. SCHULZ, *Reduced SQP Methods for Large-Scale Optimal Control Problems in DAE with Application to Path Planning Problems for Satellite Mounted Robots*, Ph.D. thesis, Universität Heidelberg, Heidelberg, Germany, 1996.

[35] V. H. SCHULZ, *Solving discretized optimization problem by partially reduced SQP methods*, Comput. Vis. Sci., 1 (1998), pp. 83–96.

[36] V. H. SCHULZ, H. G. BOCK, AND M. C. STEINBACH, *Exploiting invariants in the numerical solution of multipoint boundary value problems for DAEs*, SIAM J. Sci. Comput., 19 (1998), pp. 440–467.

Chapter 9

Digital Filter Stepsize Control in DASPK and Its Effect on Control Optimization Performance

Kirsten Meeker, Chris Homescu†, Linda Petzold†, Hana El-Samad‡, Mustafa Khammash‡, and Gustaf Söderlind§*

9.1 Introduction

It has long been known that the solutions produced by adaptive solvers for ODE and DAE systems, while generally reliable, are not smooth with respect to perturbations in initial conditions or other problem parameters. This property seems particularly important in the control and optimization of dynamical systems, where the optimizer is generally expecting the DAE solver to return solutions that vary smoothly with respect to the parameters.

Derivative-based optimizers, while very efficient and robust, assume a substantial amount of smoothness in the problem. If the objective function depends on the numerical solution of a differential equation, it is possible that small discontinuities resulting from adaptive stepsize control decisions (in the ODE or DAE solver) could lead to a less than optimal performance of the optimizer. Hence it seems likely that the performance of the optimizer might be improved by selecting the stepsizes in such a way that the performance of the solver is less sensitive to perturbations in the problem.

Söderlind [12] and Söderlind and Wang [13] have recently developed a digital filter stepsize controller that has a theoretical basis from control and appears to result in a much smoother dependence of the solution on problem parameters. We have implemented this

*Department of Computer Science, University of California, Santa Barbara, CA, 93106.

†Department of Computer Science and Department of Mechanical Engineering, University of California, Santa Barbara, CA, 93106.

‡Department of Mechanical Engineering, University of California, Santa Barbara, CA, 93106.

§Centre for Mathematical Sciences, Lund University, Lund, Sweden.

digital filter stepsize controller in the DAE solver *DASPK*3.1 and used the new solver for the optimization of dynamical systems. We have obtained a markedly improved performance of the optimizer, as a result of the new stepsize controller.

The problem is given by

$$\min_{\mathbf{u}} \Phi(\mathbf{u}) = \int_0^T \Psi(\mathbf{y}, \mathbf{u}, t)\, dt \tag{9.1}$$
$$\text{subject to} \quad \mathbf{F}(t, \mathbf{y}, \mathbf{y}', \mathbf{u}) = 0, \quad \mathbf{y}(0) = \mathbf{y}_0(\mathbf{u}),$$
$$g(\mathbf{u}) \leq 0,$$

where Ψ is the objective function, $\mathbf{F}$ is the DAE system which is a function of state $\mathbf{y}$ and control $\mathbf{u}$, and g($\mathbf{u}$) are constraints. In the formulation considered here, the problem is solved by a *shooting*-type method [1]: whenever the objective function needs to be evaluated for a given $\mathbf{u}$, we first solve the DAEs for $\mathbf{y}$. We selected a derivative-based optimization method for two reasons: (a) the optimizer is fast and reliable, and b) the DAE solver *DASPK*3.1 is able to efficiently compute the required derivatives via sensitivity analysis [9].

We implemented the new digital controller in the large-scale DAE solver *DASPK*3.1 [9], which has a facility for sensitivity analysis. The resulting version of DASPK, denoted *DASPKmod*, was tested on several simulation and sensitivity problems [11]. It was found that its speed is generally comparable to that of *DASPK*3.1, but its performance is more predictable for small perturbations in the problem or code parameters.

To determine whether the new stepsize controller could lead to faster convergence in the control of dynamical systems, we tested the new solver in the optimization of an interesting problem from systems biology. We employed *DASPKmod* together with the derivative-based optimizer KNITRO [14] to solve this optimization problem. The numerical results demonstrate that the optimization process (combined with *DASPKmod*) took substantially fewer iterations compared with the case when KNITRO was used in conjunction with *DASPK*3.1.

The remainder of the chapter is organized as follows. Section 9.2 contains a short description of the DAE solver *DASPK*3.1, including the computation of sensitivities (which are needed for the gradient-based optimization). In Section 9.3 we compare the characteristics of the original stepsize controller and the new digital filter stepsize controller. The test problem, which involves the optimization of the heat shock response in *E. coli*, is briefly presented in Section 9.5. Cast as a multiobjective optimization problem, it is solved using a powerful hybrid interior method, KNITRO, summarized in Section 9.4. The numerical results are provided in Section 9.6. Our results are summarized in Section 9.7.

9.2 The DAE Solver DASPK

*DASPK*3.1 [9] solves a differential algebraic system of the form

$$\begin{aligned} \mathbf{F}(t, \mathbf{y}, \mathbf{y}', \mathbf{u}) &= 0\,, \\ \mathbf{y}(t_0, \mathbf{u}) &= \mathbf{y}_0\,, \end{aligned} \tag{9.2}$$

where $\mathbf{u}$ are parameters of the problem. The k-step backward differentiation formula (BDF) method employed by *DASPK*3.1 approximates the derivative $\mathbf{y}'$ using k past values of the solution **y**:

$$\mathbf{F}\left(t_{n+1}, \mathbf{y}_{n+1}, \frac{1}{h\beta_0}\sum_{i=0}^{k}\alpha_i \mathbf{y}_{n+1-i}, \mathbf{u}\right) = 0\,. \tag{9.3}$$

The order of the approximation can be varied by changing the number of past solution values used. A modified Newton method is used to solve the implicit equation for y_n at each time step. Selection of the time step is the focus of the next section, which describes two approaches for the stepsize controller. The linear systems may be solved by direct methods, or by preconditioned Krylov iteration.

The sensitivities are the derivatives of the state variables **y** with respect to the problem parameters **u**. They will be employed as gradient components, required by the derivative-based optimization approach. These derivatives are computed by the addition of sensitivity equations to the system being solved. The number of additional equations is $n_s = n_y \cdot n_u$ [10]. Using the notation $s_i = \frac{\partial \mathbf{y}}{\partial u_i}$, the new system is given by

$$\mathbf{F}(t, \mathbf{y}, \mathbf{y}', \mathbf{u}) = 0\,, \tag{9.4}$$

$$\frac{\partial \mathbf{F}}{\partial \mathbf{y}} s_i + \frac{\partial \mathbf{F}}{\partial \mathbf{y}'} s_i' + \frac{\partial \mathbf{F}}{\partial \mathbf{u}} = 0, \quad i = 1, \ldots, n_u\,. \tag{9.5}$$

The sensitivity equations are generated using the automatic differentiation software ADIFOR [2]. The staggered corrector method in *DASPK*3.1 solves the entire system in two steps [6]. First, it computes the approximation to the solution **y** to (9.4) at the next time step using the BDF and a Newton iteration. Then, it solves the sensitivity equations (9.5) over the same time step using the BDF discretization and a second Newton iteration. The Jacobian matrix used to solve the original system and the sensitivity system are dependent only on the DAE solution, so they need only be computed once following the solution of the DAE on each time step.

9.3 Stepsize Controller

9.3.1 Original DASPK Approach

*DASPK*3.1 selects the stepsize h_{n+1} by

$$h_{n+1} = \left(\frac{\epsilon}{\hat{r}_n}\right)^{\frac{1}{k+1}} h_n\,, \tag{9.6}$$

where h_{n+1} is the next stepsize, ϵ is a fraction of the desired error tolerance, $\hat{r}_n$ is the estimated local error, and k is the order of BDF employed by *DASPK*3.1.

The dependence of the local error on the stepsize is described by

$$\hat{r}_n = \hat{\Phi}_n h_n^{k+1}\,, \tag{9.7}$$

where $\hat{\Phi}_n$ is the norm of the principle error function. It has been observed in practice that the output stepsize sequence and the resulting error $\hat{r}$ of the solver can exhibit large variations due to even very small perturbations in the input. For simulation alone, this is not generally an issue, as long as the solution is still within the error tolerance. But it becomes a critical matter of concern in problems such as optimization of DAE systems, where smoothness of the computed DAE solution is needed by the optimizer.

9.3.2 Digital Filter Stepsize Controller

Söderlind and Wang observed that the frequency response of the simple stepsize controller used in *DASPK*, as well as in most other popular DAE and ODE solvers, emphasizes high frequencies. They proposed in [12, 13] a stepsize low-pass filter to produce better frequency response while preserving stability. We have replaced the original stepsize controller in *DASPK*3.1 by their filter.

With this controller, the stepsize is updated according to

$$h_{n+1} = \left(\frac{\epsilon}{\hat{r}_n}\right)^{\beta_1} \left(\frac{\epsilon}{\hat{r}_{n-1}}\right)^{\beta_2} \left(\frac{h_n}{h_{n+1}}\right)^{-\alpha_2} h_n\,, \tag{9.8}$$

where $k\beta_1 = k\beta_2 = \alpha_2 = \frac{1}{4}$, $k = \hat{p} + 1$, and $\hat{p}$ is the order of convergence.

The filter is named for the performance characteristics that it is controlling: the stepsize h, the *orders of dynamics, adaptivity*, and *filter* (according to this naming scheme, the original stepsize controller in *DASPK*3.1 is H110, while the new filter is denoted as H211). The H110 and H211 controllers differ in order of dynamics and filter order but have the same order of adaptivity.

Definition 9.1. *The order of dynamics p_D of the closed-loop system is defined to be the degree of q in the denominator of the stepsize transfer function $H_{\hat{\Phi}}(q)$.*

Definition 9.2. *Letting the error transfer function $R_{\hat{\Phi}}(q)$ have all its poles strictly inside the unit circle, if $|R_{\hat{\Phi}}(q)| = O(|q-1|^{p_A})$ as $q \to 1$, the order of adaptivity p_A of the controller is defined to be p_A.*

Definition 9.3. *Letting the stepsize transfer function $H_{\hat{\Phi}}(q)$ have all its poles strictly inside the unit circle, if $|H_{\hat{\Phi}}(q)| = O(|q+1|^{p_F})$ as $q \to -1$, the stepsize filter order at $q = -1$ is p_F.*

The *stepsize* and *error transfer functions* are equations describing the stepsize and error response to controller input, and can be derived from the stepsize recursion relations (9.8) and (9.6), and the assumption that the local error is proportional to the stepsize

$$\hat{r}_n = \hat{\Phi}_n h_n^k, \tag{9.9}$$

where $\hat{\Phi}_n$ is the norm of the principal error function. Taking logarithms of (9.8)–(9.9), solving for $\log \hat{r}$ and $\log h$, and setting $\log \epsilon$ (a constant) to zero yields the stepsize and error transfer functions for the H110 controller

$$H_{\hat{\Phi}}(q) = -\frac{1}{kq}, \quad R_{\hat{\Phi}}(q) = \frac{q-1}{q},$$

and the stepsize and error transfer functions for the H211 controller

$$H_{\hat{\Phi}}(q) = -\frac{\beta_1 q + \beta_2}{(q-1)(q+\alpha_2) + k(\beta_1 q + \beta_2)}, \quad R_{\hat{\Phi}}(q) = \frac{(q-1)(q+\alpha_2)}{(q-1)(q+\alpha_2) + k(\beta_1 q + \beta_2)}.$$

A controller is stable if all poles (roots of the denominator of $H_{\hat{\Phi}}(q)$) are strictly inside the unit circle. The single pole $q = 0$ of the elementary controller's stepsize transfer function $H_{\hat{\Phi}}(q)$ is located at the origin, and so the H110 controller is highly stable. The order of dynamics quantifies how quickly and with what function the stepsize response decays to zero from an initial value, given no input disturbance (the homogeneous solution). The elementary controller has dynamic order one, and so it decays linearly to a stepsize dependent on the initial value. The filter order determines the order of repeated averaging which smooths the stepsize. This elementary controller has filter order zero; therefore it does no smoothing of successive stepsizes. Given that the controller is stable, has a linear decay function, and has a reasonably fast response, the next property to examine is the controller's frequency response in both the stepsize and error outputs.

The stepsize frequency response $|H_{\hat{\Phi}}(e^{i\omega})| \equiv 1$ for $\omega \in [0, \pi]$, and so it is independent of the frequency ω. The output stepsize h will have the same spectral content as the input disturbance $\hat{\Phi}$ because all frequencies have equal weighting. The error frequency response of a stable controller is $|R_{\hat{\Phi}}(e^{i\omega})| = O(\omega^{p_a})$ as $\omega \to 0$. The output error $\hat{r}$ will emphasize high frequencies of the input disturbance $\hat{\Phi}$. The combined result is the undesired emphasis on high frequencies, making the output stepsize and error rougher than the input disturbance, and prompting the design of a digital filter stepsize controller.

The stability of the H211 controller is not as straightforward, and must be guaranteed by careful filter parameter selection [12]. The digital filter stepsize controller has a higher order of dynamics $p_D = 2$ and a higher filter order $p_F = 1$. The effect of these improvements is to give it a faster and smoother response. The order of adaptivity is unchanged from the elementary controller, and so the error frequency response again will emphasize high frequencies of the input disturbance. However, the stepsize controller has been designed to shape the frequency response to be more uniform across all frequencies. This is done by placing the zeros (roots of the numerator of the stepsize transfer function) at frequencies we desire to suppress.

9.4 The Optimization Solver KNITRO

The heat shock optimization problem solved in this chapter is a multiobjective optimization problem. In single-objective optimization, the goal is to find the best solution, which corresponds to the minimum or maximum value of the objective function. In multiobjective optimization with conflicting objectives, there is no single optimal solution. The interaction among different objectives gives rise to a set of compromise solutions in which no objective can be improved without degrading another objective. Such a solution is called "Pareto optimal," after the economist who first analyzed income distribution in this way. Moreover, the underlying DAE system is very stiff.

Thus we needed to select a very robust algorithm from the class of gradient-based methods. Our choice was KNITRO, an optimization solver presented in [3, 14, 15]. Its excellent performance is due to the fact that it incorporates within the interior point method two powerful tools for solving nonlinear problems: sequential quadratic programming (SQP) and trust-region techniques. SQP efficiently handles nonlinearities in the constraints. Trust-region strategies allow the algorithm to treat convex and nonconvex problems uniformly, permit the direct use of second derivative information, and provide a safeguard in the presence of nearly dependent constraint gradients.

KNITRO is a hybrid interior method (also known as barrier method), where the original problem is replaced by a series of barrier subproblems controlled by a barrier parameter. The algorithm [3, 15] uses trust regions and a merit function to promote convergence. It performs one or more minimization steps on each barrier problem and then decreases the barrier parameter. The step computed at every iteration is decomposed into a normal step, whose goal is to improve feasibility, and a tangent step, toward optimality.

A typical iteration computes a primary step by solving primal-dual equations (using direct linear algebra) and performs a line search to ensure decrease in a merit function. In order to obtain global convergence in the presence of nonconvexity and Hessian or Jacobian singularities, the primary step is replaced, under certain circumstances, by a safeguarding trust-region step. The second derivatives of the objective function and constraints are approximated using quasi-Newton updating.

The DAE solver is employed for two purposes in the optimization. First, it computes the underlying numerical solution, evaluates the integrand in the objective function, and through a quadrature equation computes the integral appearing in the optimization objective function. Second, it calculates the sensitivities of the objective function, which form the gradients needed for the optimization.

9.5 Heat Shock Model

Survival of organisms such as *E. coli* in extreme conditions, like exposure to high temperatures, has necessitated the evolution of stress response networks that detect and respond to environmental changes. The consequences of unmediated heat at the cellular level are the unfolding, misfolding, or aggregation of cell proteins, which threaten the life of the cell. Cells respond to heat stress by initiating the production of heat shock proteins (HSPs), whose function is to refold denatured proteins into their native states. The heat shock response system is among the most important and challenging protective systems of the cell. This system consists of intricate control mechanisms for detecting the presence of heat-related protein damage, and initiating an appropriate response through the synthesis of HSPs. In the cytoplasmic heat shock response system of the bacterium *E. coli*, HSPs at 30°C represent less than 2% of the total protein content in the cytoplasm, whereas at 46°C, close to the growth cutoff for the organism, HSPs account for 20–25% of the total protein content [7].

HSP levels are regulated through an intricate architecture of feedback loops centered around the σ^{32}-factor. The σ^{32}-factor regulates the transcription of the HSPs under normal and stress conditions. The enzyme RNA polymerase (RNAP), bound to σ^{32}, recognizes the heat shock gene promoters and transcribes specific heat shock genes. This transcription process is controlled through the tight regulation of σ^{32}, whose synthesis, activity, and stability are modulated by intricate feedback and feedforward loops that incorporate information about the temperature and the folding state of the cell.

Building on the model of El-Samad et al. [4] and Kurata et al. [8], we formulated an optimization problem which asked whether the heat shock response is naturally balancing the cost of producing HSPs, versus the benefits in protein folding resulting from their presence in large numbers in the cell. Our motivation for addressing this problem is the following. A first glance at the heat shock response system would suggest that overproducing HSPs in anticipation of heat is the best strategy. However, this would only be true if the sole objective considered was the efficient folding of proteins. On the other hand, when one considers the HSPs' use of the cell's limited resources, it becomes apparent that an excess of HSPs can be expensive. In light of this analysis, the following questions arise naturally: *How does the heat shock response balance these two conflicting performance objectives? Is it solving an optimization problem whose result is a fine balance between the protective effect of HSPs and the metabolic burden of overexpressing them?*

We sought an answer to these questions by formulating and solving a multiobjective optimization problem where both the cost of producing and maintaining HSPs, in addition to the cost of unfolded proteins, are considered as optimization criteria. We used the mathematical model of the heat shock response reported in [4, 8]. The model employs first-order kinetics (law of mass action) to describe the various components of the heat shock system, namely the synthesis of HSPs and their regulator σ^{32}, in addition to protein folding and the association/dissociation activity of molecules. Reactions involving association/dissociation of molecules are often faster than synthesis and degradation of new molecules. Therefore, the model yields itself naturally to a separation of time scales and hence numerical stiffness. This stiffness imposed the transformation of the differential equations that describe the fast states into algebraic constraints through a partial equilibrium approximation, yielding an index-one system of DAEs. The DAE describes the evolution of 11 differential variables, coupled to the evolution of 20 algebraic variables, and includes 32 parameters. This structure of the heat shock model depicts a real biological system with a substantial amount of complexity and high dimensionality, and makes an ideal testbed for the application of our new numerical techniques.

9.5.1 Pareto Optimality

Solving our problem involves the simultaneous optimization of multiple objectives. The principle of multiobjective optimization is different from that of single-objective optimization. In single-objective optimization, the goal is to find the best solution, which corresponds to the minimum or maximum value of the objective function. But there is no single optimal solution in multiobjective optimization (due to conflicting objectives). Instead, we obtain a set of "compromise" solutions, largely known as the Pareto optimal solutions.

Many other examples of multiobjective optimization can be found among real-world engineering design or decision-making problems. For example, in bridge construction, a good design is characterized by low total mass and high stiffness; in aircraft design, simultaneous optimization of fuel efficiency, payload, and weight is required; in chemical plant design, or in design of a groundwater remediation facility, total investment and net operating costs are considered; in the traditional portfolio optimization problem, minimum risk and maximum fiscal return are attempted.

Each solution of the Pareto optimal set is not dominated by any other solution. In going from one solution to another, it is not possible to improve on one objective without making at least one of the other objectives worse. Using multiobjective analysis, decision makers can better assess the trade-offs between different objectives. Although they cannot identify a solution that is clearly the best, they can discover a set of near-optimal solutions, and have reasonable grounds to make sensible decisions and avoid clearly poor alternatives.

9.5.2 Multiobjective Optimization Formulation for Heat Shock

Using the framework of multiobjective optimization, the heat shock problem can be formulated as

$$\begin{aligned} &\min_{\mathbf{u}} \mathbf{F}_\alpha(\mathbf{u}) \equiv \int_0^T y_1^2(t,\mathbf{u})\,dt + \alpha \int_0^T y_2^2(t,\mathbf{u})\,dt \\ &\text{subject to} \quad g(\mathbf{u}) \le 0\,, \end{aligned} \tag{9.10}$$

where y_1 and y_2 are specific components of $\mathbf{y}$, the solution of the DAE describing the heat shock model.

The variables y_1 and y_2 correspond to the levels of unfolded proteins and, respectively, chaperons. For every value of the weight α, the parameters of the model, lumped in the control vector $\mathbf{u}$, should be determined in order to minimize the cost function $F_\alpha(\mathbf{u})$. The value of α represents the relative weight (importance) for one of the objectives versus the other. It should be mentioned that different values of α yield different optimal solutions. The control vector $\mathbf{u}$ is composed of 32 model parameters for the heat shock model, e.g., binding and degradation coefficients or synthesis rates for different proteins, σ factors, and chaperons [8].

We denote by $\mathbf{u}_\alpha^{OPTIM}$ the value of the parameter vector that generates the optimal solution for the corresponding weight α. By plotting

$$\int_0^T y_1^2(t,\mathbf{u}_\alpha^{OPTIM})\,dt \quad \text{versus} \quad \int_0^T y_2^2(t,\mathbf{u}_\alpha^{OPTIM})\,dt \tag{9.11}$$

we obtain a Pareto optimal curve. Using this curve, we can then answer our questions about the optimality of the heat shock system by inspecting where the point of operation of the wild-type system falls with respect to that curve. Figure 9.1 shows the Pareto optimal curve (red) generated by the solution of our optimization problem. The plot also shows the cost of chaperons and unfolded proteins for wild-type heat shock in *E. coli*. The closeness of this point to the optimal curve indicates that the heat shock response is operating in a near-optimal regime, perhaps as a result of a long evolutionary past that converged to a near-optimal solution. To check that our finding is substantial and to rule out the possibility that the structure of the model itself is generating near-optimal solutions for all parameter values, we randomly generated combinations of the parameters and computed their costs. These randomly chosen parameter combinations mostly yielded operating points well inside the nonoptimal region, therefore substantiating our conclusions. Three illustrative points are shown in Figure 9.1.

Since the main objective of this chapter is to present the improvement in the optimizer's performance as a result of the new stepsize controller, we provide only a brief discussion of the heat shock response and its numerical Pareto curve. For an in-depth description the reader is referred to El-Samad et al. [5].

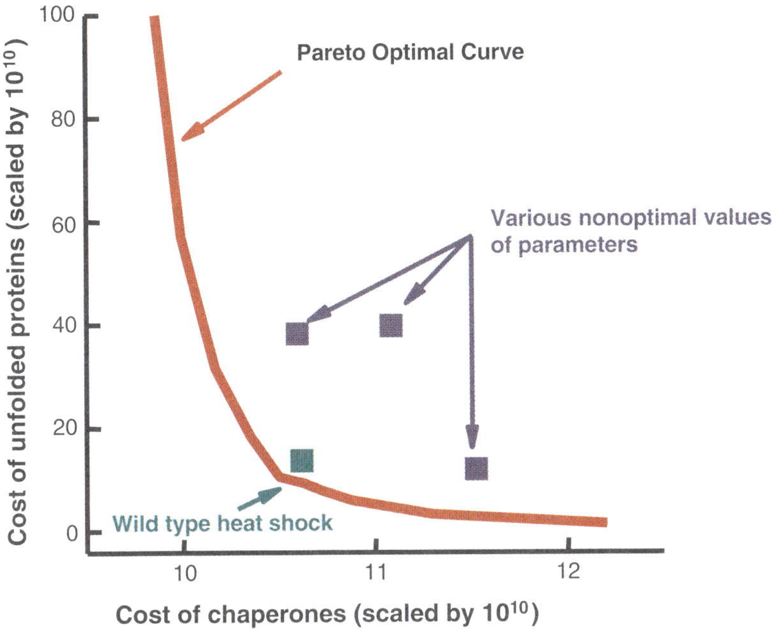

Figure 9.1. *Pareto optimal design of the heat shock response.*

9.6 Numerical Efficiency Comparison

To reduce the influence of the unsteady part of the heat shock model on the optimal solution, we solved the heat shock problem in two optimization stages. The goal of the first stage was to reach a steady state solution for the DAE system. The values of the parameters $\mathbf{u}$ and state variables $\mathbf{y}$ for the *wild-type heat shock* were given as initial conditions for the control parameter $\mathbf{u}$, respectively for the state variables for the DAE system describing the underlying process. The number of iterations was fixed at 1000 for each weight α in the objective function (9.10). The number was selected based on numerical experiments which showed that for a larger number of iterations the cost functional decreases by an extremely small amount (less than 0.0001%) at each additional iteration and thus not promising enough improvement in the objective function to warrant the extra computational effort.

For the second stage of the optimization the initial conditions for the control vector $\mathbf{u}$ were taken to be the optimal values obtained at the end of the first optimization run. The initial conditions for the state variables $\mathbf{y}$ were obtained from the steady state of the heat shock model, which was found via a run of the model using the optimal values of $\mathbf{u}$ from stage 1. For the second optimization stage, the number of iterations was bounded but not fixed. By allowing this number to vary, we were able to assess the improvements in the optimizer due to the digital stepsize controller.

We compare the optimal solutions and the computational efficiency with the solution computed with *DASPK*3.1 and, respectively, *DASPKmod*. As mentioned before, for a multiobjective optimization problem it is not expected to have a single optimal solution but rather to identify a set of good (near-optimal) solutions that reduce the value of the objective functional while satisfying the constraints (the DAE system). The value of the objective function obtained using *DASPKmod* was similar (within the tolerance error) to the value of the objective function after optimization with *DASPK*3.1 as the DAE solver. Next, we turn our attention to the computational costs of the two approaches. There is a clear advantage in both the computational time and the number of iterations for the optimization based on *DASPKmod*.

Figure 9.2 compares the efficiency of the two stepsize controllers for three values of the parameter α. In the first stage the number of DASPK steps was smaller with the digital filter stepsize controller than with the original stepsize controller, indicating that the average stepsize was larger with the new filter. However, the processor time taken to solve the problem was not consistently reduced. In the second stage the number of optimization steps and the corresponding time used to solve the problem were dramatically reduced. This was true even when the DASPK number of steps was not reduced when $\alpha = 20$. We conjecture that the performance improvement is due to the smoother behavior of the solution components with respect to perturbation of the problem parameters.

9.7 Conclusion

Many optimization problems consist of the minimization of a cost functional which is constructed from solutions of underlying differential systems, obtained using adaptive (DAE or ODE) solvers. Thus the iterative process is influenced by the numerical approach chosen to solve these differential systems, in particular by the smoothness of the selection of the time stepsize for the DAE or ODE solver.

We have obtained numerical results which show that, by choosing a smoother stepsize controller, the performance of the optimizer is greatly improved. Our methodology was tested on a real-life systems biology problem, the heat shock response of *E. coli*. The number of optimization iterations was reduced, while maintaining the same accuracy or better for the numerical optimal solution.

Acknowledgments

We would like to thank Dr. Lina Wang of the Centre for Mathematical Sciences at Lund University, Sweden, for providing us with her digital stepsize controller implemented in DASSL. This work was supported by DOE DE-FG03-00ER25430, NSF/ITR ACI-0086061, NSF CTS-0205584, and U.S. Army DAAD19-03-D-0004.

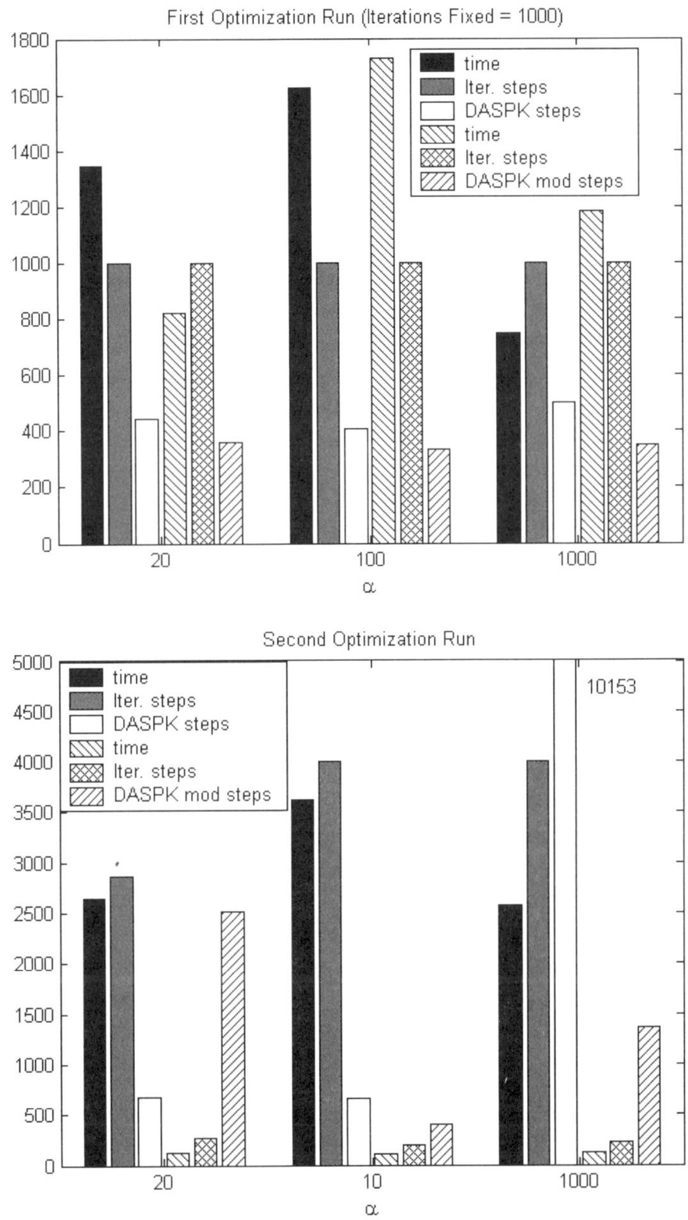

Figure 9.2. *Performance of the optimizer versus the weight parameter* α. *The underlying DAE system* (9.2) *for the heat shock problem is solved using either the original controller (DASPK*3.1*) or the digital filter stepsize controller (DASPKmod). The solid bars are the statistics for DASPK*3.1 *and the striped bars denote DASPKmod results. The statistics describe the CPU time (in seconds), the number of iteration steps, and the number of DASPK steps.*

Bibliography

[1] U. M. Ascher, R. M. M. Mattheij, and R. D. Russell, *Numerical Solution of Boundary Value Problems for Ordinary Differential Equations*, SIAM, Philadelphia, 1995.

[2] C. Bischof, A. Carle, G. Corliss, A. Griewank, and P. Hovland, *ADIFOR-Generating derivative codes from Fortran programs*, Sci. Program., 1 (1992), pp. 11–29.

[3] R. H. Byrd, M. E. Hribar, and J. Nocedal, *An interior point algorithm for large-scale nonlinear programming*, SIAM J. Optim., 9 (1999), pp. 877–900.

[4] H. El-Samad, H. Kurata, J. C. Doyle, C. A. Gross, and M. Khammash, *Surviving heat shock: Control strategies for robustness and performance*, Proc. Natl. Acad. Sci. USA, 102 (2005), pp. 2736–2741.

[5] H. El-Samad, C. Homescu, M. Khammash, and L. Petzold, *The heat shock response: Optimization solved by evolution?*, in Proceedings of the Fifth International Conference on Systems Biology, ICSB 2004, Heidelberg, Germany, 2004.

[6] W. Feehery, J. Tolsma, and P. Barton, *Efficient sensitivity analysis of large-scale differential-algebraic systems*, Appl. Numer. Math., 25 (1997), pp. 41–54.

[7] C. A. Gross, *Function and regulation of the heat shock proteins*, in Escherichia coli and Salmonella: Cellular and Molecular Biology, ASM Press, F. C. Neidhart, ed., 1996, pp. 1384–1394.

[8] H. Kurata, H. El-Samad, T. M. Yi, M. Khammash, and J. C. Doyle, *Feedback regulation of the heat shock response in E. Coli*, in Proceedings of the 40th IEEE Conference on Decision and Control, 2001, pp. 837–842.

[9] S. Li and L. R. Petzold, *Design of new DASPK for Sensitivity Analysis*, Technical report TRCS99-28, UCSB, Santa Barbara, CA, 1999.

[10] S. Li and L. Petzold, *Software and algorithms for sensitivity analysis of large-scale differential algebraic systems*, J. Comput. Appl. Math., 125 (2000), pp. 131–145.

[11] K. Meeker, *Digital filter stepsize control in DASPK and its effect on control optimization performance*, M.Sc. thesis, UCSB, Santa Barbara, CA, 2004.

[12] G. Söderlind, *Digital filters in adaptive time stepping*, ACM Trans. Math. Software, 29 (2003), pp. 1–26.

[13] G. Söderlind and L. Wang, *Adaptive time-stepping and computational stability*, ACM Trans. Comput. Log., 5 (2002), pp. 1–14.

[14] R. A. Waltz and J. Nocedal, *KNITRO User's Manual*, Technical report OTC 2003/05, Optimization Technology Center, Northwestern University, Evanston, IL, 2003.

[15] R. A. Waltz, J. L. Morales, J. Nocedal, and D. Orban, *KNITRO-Direct: A Hybrid Interior Algorithm for Nonlinear Optimization*, Technical report OTC 2003/10, Optimization Technology Center, Northwestern University, Evanston, IL, 2003.

Part III

Reduced-Order Modeling

Chapter 10

Certified Rapid Solution of Partial Differential Equations for Real-Time Parameter Estimation and Optimization

Martin A. Grepl, Ngoc C. Nguyen†, Karen Veroy†, Anthony T. Patera†, and Gui R. Liu†*

10.1 Introduction

Engineering analysis requires the prediction of selected "outputs" s relevant to ultimate component and system performance; typical outputs include critical stresses or strains, flow rates or pressure drops, and various measures of temperature and heat flux. These outputs are functions of "inputs" μ that serve to identify a particular configuration of the component or system; typical inputs reflect geometry, properties, and boundary conditions and loads.

In many cases, the input-output function is best articulated as a (say) linear functional ℓ of a field variable $u(\mu)$ that is the solution to an input-parameterized partial differential equation (PDE); typical field variables and associated PDEs include temperature and steady/unsteady conduction, displacement and equilibrium elasticity/Helmholtz, and velocity and steady incompressible Navier–Stokes. System behavior is thus described by an input-output relation $s(\mu) = \ell(u(\mu))$, the evaluation of which requires solution of the underlying PDE.

Our focus is on "deployed" systems—components or processes *in operation* in the field—and associated "Assess-Act" scenarios. In the Assess stage we pursue robust parameter estimation (inverse) procedures that map measured-observable outputs to (all) possible system-characteristic and environment-state inputs. In the subsequent Act stage we then pursue adaptive design (optimization) procedures that map mission-objective outputs to best control-variable inputs. The computational requirements on the PDE-induced evaluation

*Massachusetts Institute of Technology, Room 3-264, 77 Massachusetts Avene, Cambridge, MA 02139.
†National University of Singapore, 10 Kent Ridge Crescent, Singapore 117576.

$\mu \to s$ are formidable: the response must be *real-time*—we must "Assess-Act" *immediately*; and the outputs must be rigorously *certified*—we must "Assess-Act" *safely* and *feasibly* [24].

We describe here a method for real-time certified evaluation of PDE input-output relations; the two ingredients are reduced-basis (RB) approximation [2, 8, 10, 13, 19, 23, 25, 27] and a posteriori error estimation [19, 22, 29, 36, 37, 38]. We first describe the approach for elliptic linear second-order PDEs—Sections 10.2–10.5; we then consider extensions to certain nonlinear (incompressible Navier–Stokes) and parabolic (heat) equations—Sections 10.6 and 10.7, respectively.

10.2 Abstract Statement: Elliptic Linear Equations

We first consider the following "exact" (superscript e) problem: given $\mu \in \mathcal{D} \subset \mathbb{R}^P$, we evaluate $s^e(\mu) = \ell(u^e(\mu))$, where $u^e(\mu)$ satisfies the weak form of our μ-parameterized PDE, $a(u^e(\mu), v; \mu) = f(v)$, for all $v \in X^e$. Here μ and $\mathcal{D}$ are the input and (closed) input domain, respectively; $u^e(\mu)$ is our field variable; X^e is a Hilbert space with inner product (w, v) and associated norm $\|w\| = \sqrt{(w, w)}$; and $a(\cdot, \cdot; \mu)$ and $f(\cdot), \ell(\cdot)$ are X^e-continuous bilinear and linear functionals, respectively.

Our interest here is in second-order PDEs, and thus $(H_0^1(\Omega))^\nu \subset X^e \subset (H^1(\Omega))^\nu$: here $\Omega \subset \mathbb{R}^d$ is our spatial domain; $\nu = 1$ for a scalar field variable and $\nu = d$ for a vector field variable; and $H^1(\Omega)$ (respectively, $H_0^1(\Omega)$) is the usual Hilbert space of derivative square-integrable functions (respectively, derivative square-integrable functions that vanish on the domain boundary, $\partial\Omega$) [30]. The associated inner product $(\cdot, \cdot)$ is a μ-independent continuous coercive symmetric bilinear form over X^e that perforce induces an $(H^1(\Omega))^\nu$-equivalent norm $\|\cdot\|$.

We next introduce X (typically, $X \subset X^e$), a reference finite element approximation space of finite dimension $\mathcal{N}$. Our reference (or "truth") finite element approximation $u(\mu) \in X$ is then defined by $a(u(\mu), v; \mu) = f(v)$, for all $v \in X$: $u(\mu) \in X$ is a calculable surrogate for $u^e(\mu)$ upon which we will build our RB approximation and with respect to which we will evaluate the RB error; $u(\mu)$ also serves as the "classical alternative" relative to which we will assess the efficiency of our approach. We assume that $\|u^e(\mu) - u(\mu)\|$ is suitably small and hence that $\mathcal{N}$ is typically very large; our formulation must be both *stable* and *efficient* as $\mathcal{N} \to \infty$.

We shall make two crucial hypotheses. The first hypothesis is related to well-posedness and is often verified only a posteriori. We assume that the inf-sup parameter, $\beta(\mu) \equiv \inf_{w \in X} \sup_{v \in X}[a(w, v; \mu)/(\|w\|\|v\|)]$, is strictly positive: $\beta(\mu) \geq \beta_0 > 0$, for all $\mu \in \mathcal{D}$. The second hypothesis is related primarily to numerical efficiency and is typically verified a priori. We assume that a is *affine* in the parameter μ: $a(w, v; \mu) = \sum_{q=1}^{Q} \Theta^q(\mu) a^q(w, v)$, for $q = 1, \ldots, Q$ parameter-*dependent* functions $\Theta^q(\mu) : \mathcal{D} \to \mathbb{R}$ and parameter-*independent* continuous bilinear forms $a^q(w, v)$. The affine assumption may in fact be relaxed [5].

10.3 Reduced-Basis Approximation

The reduced-basis (RB) approximation was first introduced in the late 1970s in the context of nonlinear structural analysis [2, 23] and subsequently abstracted and analyzed [8, 27] and extended [10, 13, 25] to a much larger class of parameterized PDEs. We first introduce nested samples $S_N \equiv \{\mu_1 \in \mathcal{D}, \ldots, \mu_N \in \mathcal{D}\}$, $1 \le N \le N_{\max}$, and associated nested "Lagrangian" RB spaces $W_N \equiv \text{span}\,\{\zeta_n(\mu_n) \equiv u(\mu_n),\ 1 \le n \le N\}$, $1 \le N \le N_{\max}$. Our RB approximation is then the following: given $\mu \in \mathcal{D}$, evaluate $s_N(\mu) = \ell(u_N(\mu))$, where $u_N(\mu)$ satisfies $a(u_N(\mu), v; \mu) = f(v)$, for all $v \in W_N$. We consider here only Galerkin projection.

In essence, W_N comprises "snapshots" on the parametrically induced manifold $\mathcal{M} \equiv \{u(\mu) \mid \mu \in \mathcal{D}\} \subset X$. It is clear that $\mathcal{M}$ is very *low dimensional*; furthermore, it can be shown under our hypotheses—we consider the equations for the sensitivity derivatives and invoke stability and continuity—that $\mathcal{M}$ is very *smooth*. We thus anticipate that $u_N(\mu) \to u(\mu)$ very rapidly, and hence—at least for modest P—we may choose $N \ll \mathcal{N}$. Many numerical examples justify this expectation (see Sections 10.5, 10.6, and 10.7); and, in certain simple cases, exponential convergence can be proven [20]. We emphasize that the deployed context requires global RB approximations that are *uniformly* (rapidly) convergent over the entire parameter domain $\mathcal{D}$; proper choice of the parameter samples S_N is thus crucial (see Section 10.4).

We now represent $u_N(\mu)$ as $u_N(\mu) = \sum_{j \in \mathbb{N}} u_{N\,j}(\mu)\zeta_j$, where $\mathbb{N} \equiv \{1, \ldots, N\}$, and $\mathbb{N}_{\max} \equiv \{1, \ldots, N_{\max}\}$. Our RB output may then be expressed as $s_N(\mu) = \sum_{j \in \mathbb{N}} u_{N\,j}(\mu)\ell(\zeta_j)$, where—we now invoke our affine assumption—the $u_{N\,j}(\mu)$, $1 \le j \le N$, satisfy the $N \times N$ linear algebraic system

$$\sum_{j \in \mathbb{N}} \left\{ \sum_{q \in \mathbb{Q}} \Theta^q(\mu) a^q(\zeta_j, \zeta_i) \right\} u_{N\,j}(\mu) = f(\zeta_i) \quad \forall\, i \in \mathbb{N}\,, \tag{10.1}$$

where $\mathbb{Q} \equiv \{1, \ldots, Q\}$. (In practice we replace the ζ_j, $1 \le j \le N$, with a $(\cdot,\cdot)$-orthonormalized system; the algebraic stiffness matrix is then well conditioned.) It is clear from (10.1) that we may pursue an offline-online computational strategy [3, 13, 19, 29] ideally suited to the deployed real-time context.

In the *offline* stage—performed *once*—we first solve for the ζ_i, for all $i \in \mathbb{N}_{\max}$; we then form *and store* $f(\zeta_i)$, $\ell(\zeta_i)$, for all $i \in \mathbb{N}_{\max}$, and $a^q(\zeta_j, \zeta_i)$, for all $(i, j) \in \mathbb{N}^2_{\max}$, for all $q \in \mathbb{Q}$. In the online stage—performed many times, for each new μ "in the field"—we first assemble and subsequently invert the (full) $N \times N$ "stiffness" matrix $\sum_{q \in \mathbb{Q}} \Theta^q(\mu) a^q(\zeta_j, \zeta_i)$ to obtain the $u_{N\,j}(\mu)$, $1 \le j \le N$—at cost $O(QN^2) + O(N^3)$; we then evaluate the sum $\sum_{j \in \mathbb{N}} u_{N\,j}(\mu)\ell(\zeta_j)$ to obtain $s_N(\mu)$—at cost $O(N)$. The online complexity is *independent* of $\mathcal{N}$, and hence—given that $N \ll \mathcal{N}$—we shall realize extremely rapid "deployed" response.

10.4 A Posteriori Error Estimation

We first "presume" $\tilde{\beta}(\mu)$, a (to-be-constructed) positive lower bound for the inf-sup parameter, $\beta(\mu)$: $\beta(\mu) \ge \tilde{\beta}(\mu) \ge \tilde{\beta}_0 > 0$, for all $\mu \in \mathcal{D}$. We next introduce the dual norm of the residual $\varepsilon_N(\mu) \equiv \sup_{v \in X}[R(v; \mu)/\|v\|\,]$, where $R(v; \mu) \equiv f(v) - a(u_N(\mu), v; \mu)$, for all $v \in X$.

We may now define our error estimator, $\Delta_N(\mu) \equiv \varepsilon_N(\mu)/\tilde{\beta}(\mu)$, and associated effectivity, $\eta_N(\mu) \equiv [\Delta_N(\mu)/\|u(\mu) - u_N(\mu)\|]$. We can then readily demonstrate [29, 38] that

$$1 \leq \eta_N(\mu) \leq \gamma(\mu)/\tilde{\beta}(\mu) \quad \forall\, \mu \in \mathcal{D},\ \forall\, N \in \mathbb{N}_{\max}, \tag{10.2}$$

where $\gamma(\mu) \equiv \sup_{w\in X} \sup_{v\in X}[a(w, v; \mu)/(\|w\|\|v\|)]$ is our continuity "constant." The left inequality states that $\Delta_N(\mu)$ is a *rigorous* upper bound for $\|u(\mu) - u_N(\mu)\|$; the right inequality states that $\Delta_N(\mu)$ is a (reasonably) *sharp* upper bound.

We may also develop bounds for the error in the output; we consider here the special "compliance" case in which $\ell = f$ and a is symmetric—more general functionals ℓ and nonsymmetric a require adjoint techniques [29]. We first define our output error estimator, $\Delta_N^s(\mu) \equiv \varepsilon_N^2(\mu)/\tilde{\beta}(\mu)$, which scales as the *square* of the dual norm of the residual, $\varepsilon_N(\mu)$. We can then demonstrate [22, 29, 38] that $1 \leq \Delta_N^s(\mu)/|s(\mu) - s_N(\mu)|$, for all $\mu \in \mathcal{D}$, for all $N \in \mathbb{N}_{\max}$—$\Delta_N^s(\mu)$ is a rigorous upper bound for $|s(\mu) - s_N(\mu)|$; we may further prove in the coercive case [29] that $\Delta_N^s(\mu)/|s(\mu) - s_N(\mu)| \leq \gamma(\mu)/\tilde{\beta}(\mu)$—$\Delta_N^s(\mu)$ is a (reasonably) sharp upper bound.

It remains to develop appropriate constructions and associated offline-online computational procedures for the efficient calculation of $\varepsilon_N(\mu)$ and $\tilde{\beta}(\mu)$. To begin, we consider the former [19, 22, 29]: we invoke duality, our RB expansion, the affine parametric dependence of a, and linear superposition to express

$$\varepsilon_N^2(\mu) = (\mathcal{C}, \mathcal{C}) + \sum_{q\in\mathbb{Q}} \sum_{n\in\mathbb{N}} \Theta^q(\mu) u_{N\,n}(\mu) \left\{ 2(\mathcal{C}, \mathcal{L}_n^q) + \sum_{q'\in\mathbb{Q}} \sum_{n'\in\mathbb{N}} \Theta^{q'}(\mu) u_{N\,n'}(\mu) (\mathcal{L}_n^q, \mathcal{L}_{n'}^{q'}) \right\},$$

where $\mathcal{C} \in X$ and $\mathcal{L}_n^q \in X$, for all $n \in \mathbb{N}$, for all $q \in \mathbb{Q}$ satisfy the *parameter-independent* Poisson(-like) problems $(\mathcal{C}, v) = f(v)$, for all $v \in X$ and $(\mathcal{L}_n^q, v) = -a^q(\zeta_n, v)$, for all $v \in X$.

An efficient offline-online decomposition may now be identified. In the offline stage—performed only once—we first solve for $\mathcal{C}$ and $\mathcal{L}_n^q$, for all $n \in \mathbb{N}_{\max}$, for all $q \in \mathbb{Q}$; we then form *and store* the associated parameter-independent inner products $(\mathcal{C}, \mathcal{C})$, $(\mathcal{C}, \mathcal{L}_n^q)$, $(\mathcal{L}_n^q, \mathcal{L}_{n'}^{q'})$, for all $(n, n') \in \mathbb{N}_{\max}^2$, for all $(q, q') \in \mathbb{Q}^2$. In the online stage—performed many times, for each new value of μ "in the field"—we simply evaluate the $\varepsilon_N^2(\mu)$ sum in terms of $\Theta^q(\mu)$, $u_{N\,n}(\mu)$, and the precomputed inner products—at cost $O(Q^2 N^2)$. The online cost is *independent* of $\mathcal{N}$ and, for Q not too large, commensurate with the online cost to evaluate $s_N(\mu)$.

Finally, we turn to the development of our lower bound $\tilde{\beta}(\mu)$ for the inf-sup "constant" $\beta(\mu)$. For simplicity, we consider here the particular case $P = 1$, $Q = 2$, $\Theta^1(\mu) = 1$, and $\Theta^2(\mu) = \mu$, such that $a(w, v; \mu) \equiv a^1(w, v) + \mu a^2(w, v)$; we further suppose that $\mathcal{D}$ is convex. The more difficult general case, in which $P > 1$ and the $\Theta^q(\mu)$, $1 \leq q \leq Q$, are general functions of μ, is considered in [22] and illustrated in subsequent sections of the present chapter. As our point of departure, we note that $\beta(\mu) \equiv \inf_{v\in X} \sqrt{b(v, v; \mu)/\|v\|^2}$, where $b(w, v; \mu) = (T^\mu w, T^\mu v)$, for all $w, v \in X$, and $w \in X \to T^\mu w \in X$ is defined as $(T^\mu w, v) = a(w, v; \mu)$, for all $v \in X$.

Next, given any $\overline{\mu} \in \mathcal{D}$, we introduce $t(w, v; \mu; \overline{\mu}) \equiv b(w, v; \overline{\mu}) + (\mu - \overline{\mu})[a^2(w, T^{\overline{\mu}} v) + a^2(v, T^{\overline{\mu}} w)]$ and $\mathcal{D}^{\overline{\mu}} \equiv \{\mu \in \mathcal{D} \,|\, t(v, v; \mu; \overline{\mu}) \geq 0\}$. We may then define $\tau(\mu; \overline{\mu}) \equiv \inf_{v\in X} \sqrt{t(v, v; \mu; \overline{\mu})/\|v\|^2}$, for all $\mu \in \mathcal{D}^{\overline{\mu}}$. Our function $\tau(\mu; \overline{\mu})$ enjoys three properties: (i) $\beta(\mu) \geq \tau(\mu; \overline{\mu}) \geq 0$, for all $\mu \in \mathcal{D}^{\overline{\mu}}$; (ii) $\tau(\mu; \overline{\mu})$ is concave in μ over the convex domain $\mathcal{D}^{\overline{\mu}}$; and (iii) $\tau(\mu; \overline{\mu})$ is "tangent" to $\beta(\mu)$ at $\mu = \overline{\mu}$. (To make property (iii) rigorous we must in general consider nonsmooth analysis and also possibly a continuous spectrum as $\mathcal{N} \to \infty$.)

We can now develop our inf-sup lower bound $\tilde{\beta} : \mathcal{D} \to \mathbb{R}$. We first specifiy a constant $\overline{\epsilon} \in]0, 1[$. We then introduce a sample $E_J \equiv \{\overline{\mu}_1 \in \mathcal{D}, \ldots, \overline{\mu}_J \in \mathcal{D}\}$ and associated set of polytopes $C_J \equiv \{\mathcal{P}_1 \subset \mathcal{D}^{\overline{\mu}_1}, \ldots, \mathcal{P}_J \subset \mathcal{D}^{\overline{\mu}_J}\}$ that satisfy (a) a "positivity condition," $\tau(\mu; \overline{\mu}_j) \geq \overline{\epsilon}\, \beta(\overline{\mu}_j)$, for all $\mu \in \mathcal{P}_j$, $1 \leq j \leq J$, and (b) a "coverage condition," $\mathcal{D} \subset \cup_{j=1}^{J} \mathcal{P}_j$; we may now define (and compute) our lower bound as

$$\tilde{\beta}(\mu) \equiv \max_{\{j \in \{1,\ldots,J\} \,|\, \mu \in \mathcal{P}_j\}} \overline{\epsilon}\, \beta(\overline{\mu}_j). \tag{10.3}$$

(We can also develop piecewise *linear* approximations, though—as discussed further in Section 10.5—our inf-sup lower bound need not be highly accurate.) It is readily demonstrated that $\tilde{\beta}(\mu)$ has the requisite theoretical and computational attributes: $\beta(\mu) \geq \tilde{\beta}(\mu) \geq \overline{\epsilon}\, \beta_0 > 0$, for all $\mu \in \mathcal{D}$; and the online complexity $\mu \to \tilde{\beta}(\mu)$ depends only on J—we need only find the maximum of the pretabulated $\beta(\overline{\mu}_j)$, $1 \leq j \leq J$, and multiply by $\overline{\epsilon}$—which in turn depends only on P and Q and *not* on $\mathcal{N}$. Properties (i), (ii), and (iii) permit us to parlay relatively few expensive (offline) evaluations into a very inexpensive global (online) lower bound.

We further elaborate on two points: verification of the "positivity condition"; and choice of $\overline{\epsilon}$. As regards the former, the key ingredient is the concavity of $\tau(\mu; \overline{\mu}_j)$ in μ: we need only confirm that $\tau(\cdot; \overline{\mu}_j) \geq \overline{\epsilon}\beta(\overline{\mu}_j)$ at the two end points of $\mathcal{P}_j$ to conclude that $\tau(\mu; \overline{\mu}_j) \geq \overline{\epsilon}\beta(\overline{\mu}_j)$, for all $\mu \in \mathcal{P}_j$. (It thus follows, since $\beta(\mu) \geq \tau(\mu; \overline{\mu}_j)$, for all $\mu \in \mathcal{P}_j$, that $\beta(\mu) \geq \overline{\epsilon}\beta(\overline{\mu}_j)$, for all $\mu \in \mathcal{P}_j$—this proves the lower bound, (10.3).) As regards the choice of $\overline{\epsilon}$, there is clearly a trade-off between good effectivity and computational effort: the larger we choose $\overline{\epsilon}$, the better our lower bound and hence the better (lower) our effectivity, $\eta_N(\mu)$—but at the expense of more regions and hence larger J.

As an illustrative example of our inf-sup lower bound construction we consider the Helmholtz-elasticity crack problem of the next section for $\mu = (\omega^2 \in [2.5, 5.0],\ z = 1.0,\ L = 0.2)$—the crack location, z, and crack length, L, are fixed, and only the frequency squared, ω^2, is permitted to vary—and material damping coefficient $d_m = 0.1$. We find that a sample $E_{J=3}$ suffices to satisfy our positivity and coverage conditions for $\overline{\epsilon} = 0.4$. We present in Figure 10.1(a) $\beta(\mu)$; $\tau(\mu; \overline{\mu}_j)$ for $\mu \in \mathcal{D}^{\overline{\mu}_j}$, $1 \leq j \leq J$; and our lower bound (10.3). We note that $\beta(\mu)$ is not concave (or convex) or even quasi-concave, and hence $\tau(\mu; \overline{\mu})$ is a necessary intermediary in the construction of our lower bound.

In conclusion, we can calculate a rigorous and sharp upper bound for $|s(\mu) - s_N(\mu)|$, $\Delta_N^s(\mu) \equiv \varepsilon_N^2(\mu)/\tilde{\beta}(\mu)$, with online complexity *independent* of $\mathcal{N}$. These inexpensive error bounds serve most crucially in the deployed stage—to choose optimal N, to confirm the desired accuracy, to establish strict feasibility, and to control suboptimality. However, the bounds may also be gainfully enlisted in the predeployed stage—to construct optimal samples S_N [22, 38]: Given Ξ^F, a very fine random sample over the parameter space $\mathcal{D}$ of size $n_F \gg 1$, and an initial sample $S_1^{\mathrm{opt}} = \mu_1^*$, we [DO $N = 2, \ldots, N_{\max}$; $\mu_N^* = \arg\max_{\mu \in \Xi^F} \Delta_{N-1}^s(\mu)$; $S_N^{\mathrm{opt}} = S_{N-1}^{\mathrm{opt}} \cup \mu_N^*$; END]. Since the marginal cost to evaluate the error bound $\Delta_N^s(\mu)$ is small (online), our input sample Ξ^F can be large, and the maximization problem for μ_N^* may be solved directly—by calculating $\Delta_{N-1}^s(\mu)$ for all $\mu \in \Xi^F$. (Multi-start gradient-based search techniques can also be exploited to effectively determine the true maximum of the (clearly oscillatory) error bound $\Delta_N^s(\mu)$ over $\mathcal{D}$.) Note that in contrast to

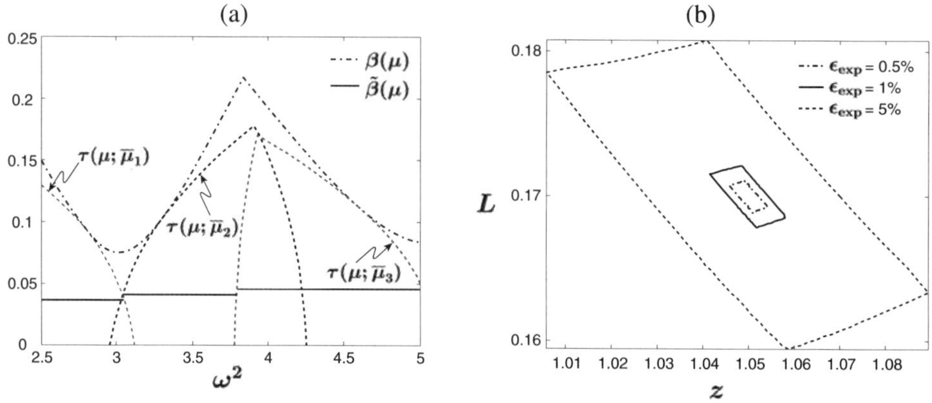

Figure 10.1. *Helmholtz-elasticity example:* (a) *Plots of* $\beta(\mu)$*;* $\tau(\mu;\overline{\mu}_j)$ *for* $\mu \in \mathcal{D}^{\overline{\mu}_j}$, $1 \le j \le J$*; and* $\tilde{\beta}(\mu)$. (b) *Crack parameter uncertainty region* $\mathcal{R}$ *for RB Model* I *with* $N^{\mathrm{I}} = 25$.

POD economization procedures [33], *we never form the rejected snapshots*: our inexpensive bound $\Delta_N^s(\mu)$ serves as a (good) surrogate for the actual error.

10.5 Assess-Act Example: Helmholtz Elasticity

We apply the RB method here to a Helmholtz-elasticity equation often encountered in solid mechanics: inverse analyses based on the Helmholtz-elasticity PDE can gainfully serve in nondestructive evaluation (NDE) procedures for crack characterization [11, 16, 18] and damage assessment [14, 17]. The RB method significantly improves the efficiency of these inverse procedures—accelerating the *many* evaluations [15] of the PDE outputs.

We consider a two-dimensional thin plate with a horizontal crack at the (say) interface of two lamina: the (original) domain $\Omega^{\mathrm{o}}(z, L) \subset \mathbb{R}^2$ is defined as $[0, 2] \times [0, 1] \setminus \Gamma_{\mathrm{C}}^{\mathrm{o}}$, where $\Gamma_{\mathrm{C}}^{\mathrm{o}} \equiv \{x_1 \in [z - L/2, z + L/2], x_2 = 1/2\}$ defines the idealized crack. The left surface of the plate is secured; the top and bottom boundaries are stress-free; and the right boundary is subject to a vertical oscillatory uniform force of frequency ω. We model the plate as plane-stress linear isotropic elastic with (scaled) density unity, Young's modulus unity, and Poisson ratio 0.25; the latter determine the (parameter-independent) constitutive tensor $E_{ijk\ell}$. Our input is $\mu \equiv (\mu_{(1)}, \mu_{(2)}, \mu_{(3)}) \equiv (\omega^2, z, L)$; our output is the (oscillatory) amplitude of the average vertical displacement on the right edge of the plate.

The governing equation for the displacement $u^{\mathrm{o}}(x^{\mathrm{o}}; \mu) \in X^{\mathrm{o}}(\mu)$ is $a^{\mathrm{o}}(u^{\mathrm{o}}(\mu), v; \mu) = f^{\mathrm{o}}(v)$, for all $v \in X^{\mathrm{o}}(\mu)$, where $X^{\mathrm{o}}(\mu)$ is a quadratic finite element truth approximation subspace (of dimension $\mathcal{N} = 14{,}662$) of $X^{\mathrm{e}}(\mu) = \{v \in (H^1(\Omega^{\mathrm{o}}(z, L)))^2 \mid v|_{x_1^{\mathrm{o}}=0} = 0\}$; here $a^{\mathrm{o}}(w, v; \mu) \equiv \int_{\Omega^{\mathrm{o}}(z,L)} w_{i,j} E_{ijk\ell} v_{k,\ell} - \omega^2 w_i v_i$ ($v_{i,j}$ denotes $\partial v_i/\partial x_j$ and repeated physical indices imply summation), and $f^{\mathrm{o}}(v) \equiv \int_{x_1^{\mathrm{o}}=2} v_2$. The crack surface is hence modeled extremely simplistically—as a stress-free boundary; note that no crack-tip element is needed as the output of interest is far from the crack. The output $s^{\mathrm{o}}(\mu)$ is given by $s^{\mathrm{o}}(\mu) = \ell(u^{\mathrm{o}}(\mu))$, where $\ell^{\mathrm{o}}(v) = f^{\mathrm{o}}(v)$; we are thus "in compliance." (For the damped

Table 10.1. *Numerical results for Model* I.

N	$\Delta_{N,\max,\mathrm{rel}}$	$\eta_{N,\mathrm{ave}}$	$\Delta^s_{N,\max,\mathrm{rel}}$	$\eta^s_{N,\mathrm{ave}}$
10	6.19E-01	13.11	8.40E-01	22.50
15	5.76E-02	13.44	4.74E-03	17.22
20	1.58E-02	13.22	4.50E-04	15.44
25	5.69E-03	12.57	4.47E-05	14.50
30	1.32E-03	12.47	2.95E-06	14.27

example of Section 10.4 we suitably complexify our field variable and space and replace $E_{ijk\ell}$ with a very simple "hysteretic" Kelvin model [4] $E_{ijk\ell}(1 + \sqrt{-1}d_m)$; here d_m is a material damping constant.)

We now map $\Omega^o(z, L)$ via a continuous piecewise affine transformation to a fixed domain Ω. This new problem can now be cast precisely in the desired abstract form of Section 10.2, in which Ω, X, and (w, v) are independent of the parameter μ: as required, all parameter dependence now enters through the bilinear and linear forms. Furthermore, it is readily demonstrated that our affine assumption applies for $Q = 10$; the $\Theta^q(\mu)$ are of the form $\mu_{(1)}^{y_1}\mu_{(2)}^{y_2}\mu_{(3)}^{y_3}$ for exponents $y_1 = 0$ or 1, $y_2 = -1, 0,$ or 1, and $y_3 = -1, 0,$ or 1. See [22] for a detailed description of the $\Theta^q(\mu)$, $a^q(w, v)$, $1 \le q \le Q$, and the "bound conditioner" $(\cdot, \cdot)$.

We shall consider two different models. In Model I, relevant to the Assess stage, we consider the parameter domain $\mathcal{D}^{\mathrm{I}} \equiv [3.2, 4.8] \times \mathcal{D}^{z,L}$, where $\mathcal{D}^{z,L} \equiv [0.9, 1.1] \times [0.15, 0.25]$. Note that $\mathcal{D}^{\mathrm{I}}$ does not contain any resonances, and hence $\beta(\mu)$ is bounded away from zero; however, $\omega^2 = 3.2$ and $\omega^2 = 4.8$—the two frequency extremes of our parameter domain—are quite close to corresponding natural frequencies, and hence the problem is distinctly noncoercive. In Model II, relevant to the Act stage, we consider the parameter domain $\mathcal{D}^{\mathrm{II}} \equiv [\omega^2 = 0] \times \mathcal{D}^{z,L}$, where—as in Model I—$\mathcal{D}^{z,L} \equiv [0.9, 1.1] \times [0.15, 0.25]$. Note that Model II is essentially steady linear elasticity, and thus the problem is coercive and relatively easy; we shall hence focus our attention on Model I.

We first present basic numerical results. For our RB spaces we pursue the optimal sampling strategy described in Section 10.4 for $N^{\mathrm{I}}_{\max} = 32$ (Model I) and $N^{\mathrm{II}}_{\max} = 6$ (Model II); for our inf-sup lower bound samples we choose $\overline{\epsilon} = 1/5$ which yields $J^{\mathrm{I}} = 84$ and $J^{\mathrm{II}} = 1$. We present in Table 10.1 $\Delta_{N,\max,\mathrm{rel}}$, $\eta_{N,\mathrm{ave}}$, $\Delta^s_{N,\max,\mathrm{rel}}$, and $\eta^s_{N,\mathrm{ave}}$ as functions of $N = N^{\mathrm{I}}$ for Model I. Here $\Delta_{N,\max,\mathrm{rel}}$ is the maximum over Ξ_{Test} of $\Delta_N(\mu)/\|u_{N_{\max}}(\mu)\|_{\max}$, $\eta_{N,\mathrm{ave}}$ is the average over Ξ_{Test} of $\Delta_N(\mu)/\|u(\mu) - u_N(\mu)\|$, $\Delta^s_{N,\max,\mathrm{rel}}$ is the maximum over Ξ_{Test} of $\Delta^s_N(\mu)/|s_{N_{\max}}(\mu)|_{\max}$, and $\eta^s_{N,\mathrm{ave}}$ is the average over Ξ_{Test} of $\Delta^s_N(\mu)/|s(\mu) - s_N(\mu)|$. Here $\Xi_{\mathrm{Test}} \in (\mathcal{D}^{\mathrm{I}})^{343}$ is a random parameter sample of size 343, $\|u_{N_{\max}}(\mu)\|_{\max} \equiv \max_{\mu\in\Xi_{\mathrm{Test}}} \|u_{N_{\max}}(\mu)\| = 2.0775$, and $|s_{N_{\max}}(\mu)|_{\max} \equiv \max_{\mu\in\Xi_{\mathrm{Test}}} |s_{N_{\max}}(\mu)| = 0.089966$. We observe that the RB approximation converges very rapidly and that our rigorous error bounds are in fact quite sharp. The effectivities are not quite $O(1)$ primarily due to the relatively crude inf-sup lower bound. (Thanks to the rapid convergence of RB approximations, $O(10)$ effectivities do not significantly (adversely) affect efficiency.)

Turning now to computational effort (again for Model I), for $N = N^{\mathrm{I}} = 25$ and any given μ (say, (4.0, 1.0, 0.2))—for which the error in the RB output $s_N(\mu)$ relative to the truth (approximation) $s(\mu)$ is *certifiably* less than $\Delta_N^s(\mu)$ (say, 2.38×10^{-7})—the online time to compute both $s_N(\mu)$ and $\Delta_N^s(\mu)$ is less than 1/330 the time to directly calculate $s(\mu) = \ell(u(\mu))$. Clearly, the savings will be even larger for problems with more complex geometry and solution structure, in particular in three space dimensions. Nevertheless, even for our current very modest example, the computational economies are very significant. (Note, however, that this comparison does not include the RB offline effort and is hence meaningful only in the real-time context or in the limit of many evaluations.)

We now consider an Assess-Act scenario that illustrates the new capabilities enabled by rapid certified input-output evaluation [1]. We first consider the Assess stage (based on Model I): given experimental measurements in the form of intervals $[s(\omega_k^2, z^*, L^*)(1 - \epsilon_{\text{exp}}), s(\omega_k^2, z^*, L^*)(1 + \epsilon_{\text{exp}})]$, $1 \le k \le K$, we wish to determine a region $\mathcal{R} \in \mathcal{D}^{z,L}$ in which the true but unknown crack parameters, (z^*, L^*), must reside. We first introduce $s_N^{\pm}(\mu) \equiv s_N(\mu) \pm \Delta_N^s(\mu)$, and recall that—thanks to our bound theorem (10.2)—$s(\mu) \in [s_N^-(\mu), s_N^+(\mu)]$. We may then define

$$\mathcal{R} \equiv \big\{(z, L) \in \mathcal{D}^{z,L} \mid [s_N^-(\omega_k^2, z, L), s_N^+(\omega_k^2, z, L)] \cap [s(\omega_k^2, z^*, L^*)(1 - \epsilon_{\text{exp}}), s(\omega_k^2, z^*, L^*)(1 + \epsilon_{\text{exp}})] \neq \emptyset, 1 \le k \le K\big\};$$

clearly, we have accommodated both *numerical* and *experimental* error and uncertainty (within our model assumptions), and hence $(z^*, L^*) \in \mathcal{R}$.

In Figure 10.1(b) we present $\mathcal{R}$ for $K = 2$ and $(\omega_1^2 = 3.2, \omega_2^2 = 4.8)$ for $\epsilon_{\text{exp}} = 0.5\%, 1\%, 5\%$. (In actual practice, we first find one point in $\mathcal{R}$; we then conduct a binary chop at different angles to map out the boundary of $\mathcal{R}$.) As expected, as ϵ_{exp} decreases, $\mathcal{R}$ shrinks towards the exact (synthetic) value, $z^* = 1.05$, $L^* = 0.17$. Note that in this example for our RB Model I we choose $N^{\text{I}} = 25$, and it is hence clear from Table 10.1 that the RB error contributes negligibly to the uncertainty region $\mathcal{R}$; we could hence achieve even faster parameter estimation response—at little cost in precision—by decreasing N^{I}.

Most importantly, for any finite ϵ_{exp}, $\mathcal{R}$ *rigorously captures the uncertainty* in our assessment of the crack parameters without a priori regularization hypotheses [7]. The crucial new ingredient is reliable fast evaluations that permit us to conduct a much more extensive search over parameter space; for a given ϵ_{exp}, $\mathcal{R}$ may be generated online in less than 51 seconds (even for $N^{\text{I}} = 25$) on a Pentium 1.6 GHz laptop. Our search over possible crack parameters will certainly never be truly exhaustive, and hence there may be small undiscovered "pockets of possibility" in $\mathcal{D}^{z,L}$; however, we have clearly reduced the uncertainty relative to more conventional approaches. (Of course, our procedure can also only *characterize* cracks within the specified low-dimensional parameterization; however, more general null hypotheses can be constructed to *detect* model deviation.)

Finally, we consider the Act stage (based on Model II). We presume here that the component must withstand an in-service steady force (normalized to unity) such that the deflection $s(0, z^*, L^*)$ in the "next mission" does not exceed a specified value $s_{\max}$. Of course, in practice, we will not be privy to (z^*, L^*). To address this difficulty we first define $s_{\mathcal{R}}^+ \equiv \max_{(z,L)\in\mathcal{R}} s_N^+(0, z, L)$, where $s_N^+(0, z, L) = s_N(0, z, L) + \Delta_N^s(0, z, L)$; our corresponding "go/no-go" criterion is then given by $s_{\mathcal{R}}^+ \le s_{\max}$. It is readily observed that $s_{\mathcal{R}}^+$ rigorously accommodates both experimental (crack) and numerical uncertainty—$s(0, z^*, L^*) \le s_{\mathcal{R}}^+$—and that the associated go/no-go discriminator is hence *fail-safe*. Furthermore, as ϵ_{exp} tends to zero and N^{I} and N^{II} increase, $s^+_{\mathcal{R}}$ will tend to $s(0, z^*, L^*)$; indeed, for $\epsilon_{\text{exp}} = 1\%$ and $N^{\text{I}} = 25$, $N^{\text{II}} = 6$, $[s^+_{\mathcal{R}} - s(0, z^*, L^*)]/|s(0, z^*, L^*)| = 4.73\text{E-}05$. In summary, in real time, we can both Assess the current state of the crack and subsequently Act to ensure the safety (or optimality) of the next "sortie."

10.6 Incompressible Navier–Stokes Equations

To illustrate the difficulties that arise in the treatment of nonlinear problems we consider a particular example [36]: the steady incompressible Navier–Stokes equations—Pr(andtl) = 0 natural convection in an enclosure [31, 34].

Our formulation of Section 10.2 is still applicable (except a is no longer bilinear): $\mu \equiv$ Gr $\equiv$ Grashof number; $\mathcal{D} \equiv [1.0, 1.0\text{E}5]$; $u^{\text{e}}(\mu) = (u^{\text{e}}_1(\mu), u^{\text{e}}_2(\mu))$ is the velocity field; $\Omega = [0, 4] \times [0, 1]$; $X^{\text{e}} = \{(H^1_0(\Omega))^2 \mid \nabla \cdot v = 0\}$; $(w, v) = \int_\Omega w_{i,j} v_{i,j}$; $a(w, v) = a_0(w, v) + \frac{1}{2} a_1(w, w, v)$, where $a_0(w, v) \equiv \int_\Omega w_{i,j}\, v_{i,j}$ and $a_1(w, z, v) \equiv -\int_\Omega (w_i z_j + w_j z_i)\, v_{i,j}$ are the viscous and convective terms, respectively; $f(v; \mu) = \mu f_0(v) = \mu \int_\Omega \left(1 - \frac{1}{4} x_1\right) v_2$ is the buoyancy contribution; and $\ell(v) \equiv 2 \int_{\Gamma_0} v_1(x) dx_2$ (for $\Gamma_0 = \{x_1 = 2, x_2 \in [0.5; 1]\}$) measures the flow rate. (Note that the pressure does not appear explicitly, since we pose the problem over divergence-free velocity fields.)

We next introduce X, a reference finite element approximation space. Our reference (or "truth") finite element approximation $u(\mu) \in X$ is then defined by $a(u(\mu), v) = f(v; \mu)$, for all $v \in X$. As before, $u(\mu)$ is a surrogate for $u^{\text{e}}(\mu)$ upon which we build our RB approximation, and relative to which we measure our RB error. Here X is the space (of dimension $\mathcal{N} = 2{,}786$) of discretely divergence-free functions associated with a classical Taylor–Hood $\mathbb{P}_2 - \mathbb{P}_1$ finite element approximation [10]. (For future reference, we also define $\widetilde{X}$, the full Taylor–Hood velocity space.)

The derivative of a plays a central role: here $da(w, v; z) \equiv a_0(w, v) + a_1(w, z, v)$ satisfies $a(z + w, v) = a(z, v) + da(w, v; z) + \frac{1}{2} a_1(w, w, v)$. It is readily shown [36] that $da(w, v; z) \leq \gamma(z) \|w\| \|v\|$ for $\gamma(z) = 1 + \rho^2 \|z\|$; here $\rho \equiv \sqrt{2} \sup_{v \in \widetilde{X}} \|v\|_{L^4(\Omega)} / \|v\|$ is a Sobolev embedding constant [35], and $\|v\|_{L^p(\Omega)} \equiv (\int_\Omega (w_i w_i)^{p/2})^{1/p}$. We shall further assume [36]—and verify a posteriori—that $\{u(\mu) \mid \mu \in \mathcal{D}\}$ is a nonsingular (isolated) solution branch: $\beta(u(\mu)) \geq \beta_0 > 0$, for all $\mu \in \mathcal{D}$, where $\beta(z) \equiv \inf_{w \in X} \sup_{v \in X} da(w, v; z)/\|w\| \|v\|$ is the inf-sup parameter relevant to our nonlinear problem. Numerical simulations [31, 34] demonstrate that the flow *smoothly evolves* from a single-cell structure for the lower Gr in $\mathcal{D}$ to an inertia-dominated three-cell structure for the higher Gr in $\mathcal{D}$.

We may directly apply the RB formulation of Section 10.3 to the incompressible Navier–Stokes equations [13, 25, 36]. The most significant new issue is (efficient) calculation of the nonlinear terms. We consider the inner Newton update: given a current iterate $\overline{u}_N(\mu) = \sum_{n=1}^N \overline{u}_{N\,n}(\mu)\, \zeta_n$, we must find an increment $\delta u_N \in W_N$ such that $da(\delta u_N, v; \overline{u}_N) = R(v; \mu)$, for all $v \in W_N$; here $R(v; \mu) \equiv f(v; \mu) - a(u_N(\mu), v)$, for all $v \in X$ is the residual. The associated algebraic equations are thus

$$\sum_{j=1}^N \left\{ a_0(\zeta_j, \zeta_i) + \sum_{n=1}^N \overline{u}_{N\,n}(\mu) a_1(\zeta_j, \zeta_n, \zeta_i) \right\} \delta u_{N\,j}$$

$$= \mu f_0(\zeta_i) - \sum_{j=1}^{N} \left\{ a_0(\zeta_j, \zeta_i) + \frac{1}{2} \sum_{n=1}^{N} \overline{u}_{N\,n}(\mu) a_1(\zeta_j, \zeta_n, \zeta_i) \right\} \overline{u}_{N\,j}(\mu)\ \forall\, i \in \mathbb{N},$$

where we recall that $f(v; \mu) = \mu f_0(v)$ and $\mu \equiv \mathrm{Gr}$.

We can directly apply the offline-online computations procedure described in Section 10.3 for linear problems, except now we must perform summations both over the affine parameter dependence (rather trivial here) *and* over the RB coefficients (of the current Newton iterate, $\overline{u}_N(\mu)$). In the online stage—for given new μ—at each Newton iteration $\overline{u}_N(\mu) \to \delta u_N$ we first assemble the right-hand side (residual)—at cost $O(N^3)$; we then form and invert the left-hand side (Jacobian)—at cost $O(N^3)$. The complexity of the online stage is independent of $\mathcal{N}$; furthermore, for our quadratic nonlinearity, there is little increased cost relative to the linear case. Unfortunately, for a pth-order nonlinearity, the online cost for the residual assembly and Jacobian formation will scale as $O(N^{p+1})$, and thus standard Galerkin projections are viable only for $p = 2$ or at most $p = 3$ [38]. Fortunately, for larger p and nonpolynomial nonlinearities—and for nonaffine parameter dependence [5]—quite effective collocation-like alternatives are available.

Turning now to a posteriori error estimation,we first "presume" $\tilde{\beta}_N(\mu)$, a (to-be-constructed) positive lower bound for the inf-sup parameter $\beta_N(\mu) \equiv \beta(u_N(\mu))$: $\beta_N(\mu) \geq \tilde{\beta}_N(\mu) \geq 0$, for all $\mu \in \mathcal{D}$. We next recall the dual norm of the residual, $\varepsilon_N(\mu) \equiv \sup_{v \in X} R(v; \mu)/\|v\|$, and introduce $\tau_N(\mu) \equiv 2\rho^2 \varepsilon_N(\mu)/\tilde{\beta}_N^2(\mu)$, where ρ is our $L^4(\Omega)$-$\widetilde{X}$ embedding constant. Finally, we define $N^*(\mu)$ such that $\tau_N(\mu) < 1$ for $N \geq N^*(\mu)$; we require $N^*(\mu) \leq N_{\max}$, for all $\mu \in \mathcal{D}$. (The latter is a condition on $N_{\max}$ that reflects both the convergence rate of the RB approximation and the quality of our inf-sup lower bound.)

We may now define our error estimator: for $N \geq N^*(\mu)$, $\Delta_N(\mu) \equiv (\tilde{\beta}_N(\mu)/\rho^2)(1-\sqrt{1 - \tau_N(\mu)})$; note that, as $\varepsilon_N(\mu) \to 0$, $\Delta_N(\mu)$ tends to the "linear case" $\varepsilon_N(\mu)/\tilde{\beta}_N(\mu)$. Our main result is then the following: Given any $\mu \in \mathcal{D}$, for all $N \geq N^*(\mu)$, there exists a unique (truth approximation) solution $u(\mu) \in X$ in the ball $\mathcal{B}(u_N(\mu), \tilde{\beta}_N(\mu)/\rho^2) \equiv \{z \in X \mid \|z - u_N(\mu)\| < \tilde{\beta}_N(\mu)/\rho^2\}$; furthermore, $\|u(\mu) - u_N(\mu)\| \leq \Delta_N(\mu)$. The proof [36, 37] is a slight specialization of the abstract "Brezzi–Rappaz–Raviart" result [6, 12]; we can further provide several corollaries related to (i) the well-posedness of the truth approximation, and (ii) the effectivity of our error bound [36]. (We may also develop bounds for the output of interest [36].)

The real challenge is computational: how can we compute $\varepsilon_N(\mu)$, ρ, and $\tilde{\beta}_N(\mu)$? (Note that, armed with these quantities, we can evaluate $\tau_N(\mu)$ and hence verify $N \geq N^*(\mu)$.) The RB context is in fact a rare opportunity to render the Brezzi–Rappaz–Raviart theory *completely quantitative*. To begin, we consider $\varepsilon_N(\mu)$: as for the linear case, we invoke duality, our RB expansion, the affine parameter dependence of a (and f), and linear superposition to express

$$\varepsilon_N^2(\mu) = \mu^2(\mathcal{C}, \mathcal{C}) + \sum_{n=1}^{N} u_{N\,n}(\mu) \left\{ 2\mu(\mathcal{C}, \mathcal{L}_n) + \sum_{n'=1}^{N} u_{N\,n'}(\mu) \left\{ 2\mu(\mathcal{C}, \mathcal{Q}_{n\,n'}) + (\mathcal{L}_n, \mathcal{L}_{n'}) \right.\right.$$

$$\left.\left. + \sum_{n''=1}^{N} u_{N\,n''}(\mu) \left\{ 2(\mathcal{L}_n, \mathcal{Q}_{n'n''}) + \sum_{n'''=1}^{N} u_{N\,n'''}(\mu) \left(\mathcal{Q}_{n\,n'}, \mathcal{Q}_{n''n'''}\right) \right\} \right\} \right\}, \quad (10.4)$$

where $(\mathcal{C}, v) = f(v)$, for all $v \in X$, $(\mathcal{L}_n, v) = -a_0(\zeta_n, v)$, for all $v \in X$, for all $n \in \mathbb{N}$, and $(\mathcal{Q}_{n\,n'}, v) = -\frac{1}{2}a_1(\zeta_n, \zeta_{n'}, v)$, for all $v \in X$, for all $(n, n') \in \mathbb{N}^2$; the latter are again simple (vector) Poisson problems.

We can now readily adapt the offline-online procedure developed in the linear case [36, 37]. In the online stage—for each new μ—we perform the sum (10.4) in terms of the *pre*-formed and *stored* inner products (for example, $(\mathcal{Q}_{nn'}, \mathcal{Q}_{n''n'''})$, $1 \le n, n', n'', n''' \le N$) and the RB coefficients $u_{N\,n}(\mu)$, $1 \le n \le N$—at cost $O(N^4)$. Although the N^4 scaling, which arises due to the trilinear term in the residual, is certainly unpleasant, the error bound is calculated only once: in actual practice, the additional online cost attributable to the dual norm of the residual is not too large. Unfortunately, for a pth-order nonlinearity, the online evaluation of $\varepsilon_N(\mu)$ scales as $O(N^{2p})$, and our approach is thus viable only for $p = 2$. Fortunately, for larger p and nonpolynomial nonlinearities—and for nonaffine parameter dependence [5]—collocation-like alternatives are available; however, in general, there will be some *loss of rigor* in our error estimation.

We next turn to the calculation of ρ. The critical observation is that ρ is the supremum of a "Rayleigh quotient." Thus ρ is related to the smallest multiplier of an associated Euler–Lagrange nonlinear eigenproblem [35]: $(\hat{\lambda}, \hat{\psi}) \in (\mathbb{R}_+, \widetilde{X})$ satisfies $(\hat{\psi}, v) = 2\hat{\lambda}^2 \int_\Omega \hat{\psi}_j \hat{\psi}_j \hat{\psi}_i v_i$, for all $v \in \widetilde{X}$, for $\|\hat{\psi}\|^4_{L^4(\Omega)} = 1$; the ground state is denoted $(\hat{\lambda}_{\min}, \hat{\psi}_{\min})$, and $\rho = \hat{\lambda}^{-1}_{\min}$. In practice, it may be difficult to isolate the ground state, and we thus consider a homotopy procedure.

Towards that end, we first introduce a parameterized generalization of the Euler–Lagrange equation: given $\alpha \in [0, 1]$, $(\lambda(\alpha), \psi(\alpha)) \in (\mathbb{R}_+, \widetilde{X})$ satisfies $(\psi(\alpha), v)$ $= 2\lambda^2(\alpha)[\alpha \int_\Omega \psi_j(\alpha)\psi_j(\alpha)\psi_i(\alpha)v_i + (1-\alpha)\int_\Omega \psi_i(\alpha)v_i]$, for all $v \in \widetilde{X}$, for normalization $\alpha\|\psi(\alpha)\|^4_{L^4(\Omega)} + (1-\alpha)\|\psi(\alpha)\|^2_{L^2(\Omega)} = 1$; the ground state is denoted $(\lambda_{\min}(\alpha), \psi_{\min}(\alpha))$, and $\rho = \lambda^{-1}_{\min}(1)$. We may now apply standard Newton continuation methods to proceed from the known ground state at $\alpha = 0$—$(\lambda_{\min}(0), \psi_{\min}(0))$ is the lowest eigenpair of a simple (vector) "Laplacian" linear eigenproblem—to the ground state of interest at $\alpha = 1$; for sufficiently small increments in α, we will remain on the desired (lowest-energy) branch. For our particular domain, we find (offline) $\rho = 0.4416$; since ρ is μ-independent, no online computation is required.

Finally, as regards the inf-sup lower bound, $\tilde{\beta}_N(\mu)$, we may directly apply appropriate extensions [22, 36] of the procedure developed in Section 10.5. The nonlinear case does present a new difficulty: the parameter dependence of the (linearized) operator is now induced by the RB solution $u_N(\mu)$—in our case, through the $a_1(w, u_N(\mu), v)$ term—and hence is *not known* a priori. Fortunately, since $u_N(\mu) \to u(\mu)$ we may develop a "universal" lower bound for sufficiently large N; the complications are thus largely practical in nature. (For our particular problem, $J = 34$—the sample is relatively small despite the rather large range in Grashof.)

We conclude with a brief discussion of the adaptive sampling procedure introduced in Section 10.4. In the nonlinear case a similar procedure may be pursued but with two important differences. First, as already indicated, $\beta_N(\mu)$ and hence $\tilde{\beta}_N(\mu)$ will now depend on the RB solution $u_N(\mu)$; furthermore, $\beta_N(\mu)$ will only be meaningful for larger N. Thus in the sample construction stage we must replace $\tilde{\beta}_N(\mu)$ in $\Delta_N(\mu)$ with a simple but relevant surrogate—for example, a piecewise-constant (over $\mathcal{D}$) approximation to $\beta(u(\mu))$. Second, in the nonlinear context our error bound is conditional—a small solution to the error equation is only assured if $\tau_N(\mu) < 1$. Thus the greedy procedure must first select

Table 10.2. *Error bounds and effectivities for* Gr = 1.0E1 *and* Gr = 8.5E4.

	Gr = 1.0E1			Gr = 8.5E4		
N	τ_N	$\Delta_{N,\,\mathrm{rel}}$	η_N	τ_N	$\Delta_{N,\,\mathrm{rel}}$	η_N
3	2.0 E−2	5.0 E−2	1.0	∞	*	*
6	1.2 E−2	3.0 E−2	1.0	2.8 E+1	*	*
9	4.4 E−3	1.1 E−2	1.0	5.2 E−1	1.5 E−4	14.1
12	2.7 E−6	6.8 E−6	1.0	5.8 E−1	1.7 E−4	20.5
15	3.0 E−7	7.6 E−7	1.0	1.9 E−2	4.6 E−6	17.6

on $\arg\max_{\mu \in \Xi^F} \tau_N(\mu)$—until $\tau_N(\mu) < 1$, for all $\mu \in \Xi^F$—and only subsequently select on $\arg\max_{\mu \in \Xi^F} \Delta_N(\mu)$; the resulting sample will ensure rapid convergence to a *certifiably* accurate solution.

In Table 10.2 we present $\Delta_{N,\,\mathrm{rel}}(\mu) \equiv \Delta_N(\mu)/\|u_{N_{\max}}(\mu)\|$ and $\eta_N(\mu) \equiv \Delta_N(\mu)/\|u(\mu) - u_N(\mu)\|$ as a function of N for $\mu = \mathrm{Gr} = 1.0\mathrm{E}1$ (single-roll) and $\mu = \mathrm{Gr} = 8.5\mathrm{E}4$ (three-roll). The "*" indicates that $N < N^*(\mu)$ — $\tau_N(\mu) \geq 1$: no error bound is available. For Gr = 1.0E1, we find $N^*(\mu) = 1$, and hence we obtain error bounds for all N; the error bound tends to zero very rapidly; and the effectivity is $O(1)$ [22, 36]. For Gr = 8.5E4, we find $N^*(\mu) = 9$, and hence we obtain error bounds only for rather accurate approximations; however, the error bound still tends to zero rapidly with N—our samples S_N^{opt} are constructed to provide uniform convergence; and the effectivity is still quite good. It is perhaps surprising that the Brezzi–Rappaz–Raviart theory, which is not really designed for quantitative service, indeed yields such sharp results; in fact, as $\varepsilon_N(\mu) \to 0$, the cruder bounds—in particular, ρ—no longer play a role.

Finally, we note that the online cost (on a Pentium M 1.6GHz processor) to predict $s_N(\mu)$ and $\Delta_N(\mu)$ (and a bound for the error in the output, $\Delta_N^s(\mu)$ [36]) is typically 10ms and 90ms, respectively—compared to order minutes for direct finite element calculation of $s(\mu) = \ell(u(\mu))$.

10.7 Parabolic Equations

We consider here the extension of the RB methods and associated a posteriori error estimators described in Sections 10.1–10.4 to *parabolic* PDEs—in particular, the heat equation; we shall "simply" treat time as an additional, albeit special, parameter [32]. For further details, we refer the reader to [9]. (There are many approaches to model reduction for initial value problems: POD methods [33]; balanced-truncation techniques [21]; and even RB approaches [28]. However, in general, these frameworks do not accommodate parametric variation (or, typically, rigorous a posteriori error estimation).) For simplicity, we directly consider a K-level time-discrete framework (corresponding to Euler backward discretization, although we can also readily treat higher-order schemes such as Crank–Nicolson) associated with the time interval $[0, t_f]$: we define $\mathbb{T} \equiv \{t^0, \ldots, t^K\}$, where $t^k = k\Delta t$, $0 \leq k \leq K$, and $\Delta t = t_f/K$; for notational convenience, we also introduce $\mathbb{K} = \{1, \ldots, K\}$. (Clearly, our results must be stable as $\Delta t \to 0$, $K \to \infty$.)

Given $\mu \in \mathcal{D} \subset \mathbb{R}^P$, we evaluate the (here, single) output $s(\mu, t^k) = \ell(u(\mu, t^k))$, for all $k \in \mathbb{K}$, where $u(\mu, t^k) \in X$, for all $k \in \mathbb{K}$, satisfies

$$\Delta t^{-1} m(u(\mu, t^k) - u(\mu, t^{k-1}), v) + a(u(\mu, t^k), v; \mu) = b(t^k) f(v) \quad \forall\, v \in X, \tag{10.5}$$

with initial condition (say) $u(\mu, t^0) = 0$. Here μ and $\mathcal{D}$ are the input and input domain; $u(\mu, t^k)$, for all $k \in \mathbb{K}$, is our field variable; $X \subset X^{\mathrm{e}}$ is our truth approximation subspace for X^{e} (and $(\cdot,\cdot)$, $\|\cdot\|$) defined in Section 10.2; $a(\cdot,\cdot;\mu)$ and $m(\cdot,\cdot)$ are X^{e}-continuous and $L^2(\Omega)$-continuous symmetric bilinear forms, respectively; $f(\cdot)$, $\ell(\cdot)$ are $L^2(\Omega)$-continuous linear forms; and $b(t^k)$ is the (here, single) "control" input at time t^k.

We shall make the following assumptions. First, we require that a and m are independent of time—the system is thus linear time-invariant (LTI). Second, we assume that a and m are coercive: $0 < \alpha_0 \le \alpha(\mu) \equiv \inf_{v \in X}[a(v, v; \mu)/\|v\|^2]$ and $0 < \sigma_0 \le \inf_{v \in L^2(\Omega)}[m(v, v)/\|v\|^2_{L^2(\Omega)}]$. Third, we assume that a depends affinely on μ: $a(w, v; \mu) = \sum_{q=1}^{Q} \Theta^q(\mu)\, a^q(w, v)$ for $q = 1, \ldots, Q$ parameter-*dependent* functions $\Theta^q(\mu) : \mathcal{D} \to \mathbb{R}$ and parameter-*independent* continuous bilinear forms $a^q(w, v)$. (For simplicity, we also assume that m, f, and ℓ are parameter-independent.)

To ensure rapid convergence of the RB output approximation we shall need a dual (or adjoint) problem which shall evolve backward in time. Invoking the LTI property, we can express the adjoint for the output at time t^L, $1 \le L \le K$, as $\psi^L(\mu, t^k) \equiv \Psi(\mu, t^{K-L+k})$, $1 \le k \le L$; here $\Psi(\mu, t^k) \in X$, for all $k \in \mathbb{K}$, satisfies $\Delta t^{-1} m(v, \Psi(\mu, t^k) - \Psi(\mu, t^{k+1})) + a(v, \Psi(\mu, t^k); \mu) = 0$, for all $v \in X$, with final condition $m(v, \Psi(\mu, t^{K+1})) \equiv \ell(v)$, for all $v \in X$. In essence, thanks to the primal LTI property and the linearity of the output functional, the dual system is invariant to a shift in time of the final condition; thus, to obtain $\psi^L(\mu, t^k)$, $1 \le k \le L$, for all $L \in \mathbb{K}$, we need only solve once for $\Psi(\mu, t^k)$, for all $k \in \mathbb{K}$, and then appropriately translate the result—we do not need to solve K separate dual problems [9].

We now introduce the nested samples $S^{\mathrm{pr}}_{N_{\mathrm{pr}}} = \{\tilde{\mu}^{\mathrm{pr}}_1, \ldots, \tilde{\mu}^{\mathrm{pr}}_{N_{\mathrm{pr}}}\}$, $1 \le N_{\mathrm{pr}} \le N_{\mathrm{pr,max}}$, and $S^{\mathrm{du}}_{N_{\mathrm{du}}} = \{\tilde{\mu}^{\mathrm{du}}_1, \ldots, \tilde{\mu}^{\mathrm{du}}_{N_{\mathrm{du}}}\}$, $1 \le N_{\mathrm{du}} \le N_{\mathrm{du,max}}$, where $\tilde{\mu} \equiv (\mu, t^k) \subset \widetilde{\mathcal{D}} \equiv \mathcal{D} \times \mathbb{T}$. Note the samples must now reside in the *parameter-time* space $\widetilde{\mathcal{D}}$; we also introduce separate (and different) samples for the primal and dual problems. We then define the associated nested RB spaces $W^{\mathrm{pr}}_{N_{\mathrm{pr}}} = \mathrm{span}\{\zeta^{\mathrm{pr}}_n \equiv u(\tilde{\mu}^{\mathrm{pr}}_n = (\mu_n, t^{k_n})^{\mathrm{pr}}),\ 1 \le n \le N_{\mathrm{pr}}\}$, $1 \le N_{\mathrm{pr}} \le N_{\mathrm{pr,max}}$, and $W^{\mathrm{du}}_{N_{\mathrm{du}}} = \mathrm{span}\{\zeta^{\mathrm{du}}_n \equiv \Psi(\tilde{\mu}^{\mathrm{du}}_n = (\mu_n, t^{k_n})^{\mathrm{du}}),\ 1 \le n \le N_{\mathrm{du}}\}$, $1 \le N_{\mathrm{du}} \le N_{\mathrm{du,max}}$. Note that for the primal basis we choose—as justified by the LTI hypothesis—an impulse input: $b(t^k) = 1$ for $k = 1$, and $b(t^k) = 0$ for $2 \le k \le K$.

Our RB approximation is then the following: given $\mu \in \mathcal{D}$, evaluate $s_N(\mu, t^k) = \ell(u_N(\mu, t^k)) + \sum_{k'=1}^{k} R^{\mathrm{pr}}(\Psi_N(\mu, t^{K-k+k'}); \mu, t^{k'})\, \Delta t$, for all $k \in \mathbb{K}$, where (pr) $u_N(\mu, t^k) \in W^{\mathrm{pr}}_{N_{\mathrm{pr}}}$, for all $k \in \mathbb{K}$, satisfies $\Delta t^{-1} m(u_N(\mu, t^k) - u_N(\mu, t^{k-1}), v) + a(u_N(\mu, t^k), v; \mu) = b(t^k)\, f(v)$, for all $v \in W^{\mathrm{pr}}_{N_{\mathrm{pr}}}$, with initial condition $u_N(\mu, t^0) = 0$, and (du) $\Psi_N(\mu, t^k) \in W^{\mathrm{du}}_{N_{\mathrm{du}}}$, for all $k \in \mathbb{K}$, satisfies $\Delta t^{-1} m(v, \Psi_N(\mu, t^k) - \Psi_N(\mu, t^{k+1})) + a(v, \Psi_N(\mu, t^k); \mu) = 0$, for all $v \in W^{\mathrm{du}}_{N_{\mathrm{du}}}$, with final condition $m(v, \Psi_N(\mu, t^{K+1})) \equiv \ell(v)$, for all $v \in W^{\mathrm{du}}_{N_{\mathrm{du}}}$. Here, for all $k \in \mathbb{K}$, $R^{\mathrm{pr}}(v; \mu, t^k) \equiv b(t^k)\, f(v) - (\Delta t^{-1} m(u_N(\mu, t^k) - u_N(\mu, t^{k-1}), v) + a(u_N(\mu, t^k), v; \mu))$, for all $v \in X$, is the primal residual. Note that we include a residual correction term (the inner product of the primal residual with the dual RB solution) in $s_N(\mu, t^k)$ to improve the accuracy of our output prediction [26] and to obtain the "square" effect in the convergence of the output bound. We could, of course, also increase the accuracy of $s_N(\mu, t^k)$ by improving the primal RB solution $u_N(\mu, t^k)$ (i.e., by increasing N_{pr}); however, in the case of a single or relatively few outputs, the dual formulation is computationally advantageous.

The offline-online computational procedure is similar to the elliptic case of Section 10.3 but with the added complexity of the dual problem and the time dependence [9]. In the *online* stage, we first assemble the requisite RB "stiffness" matrices—at cost $O((N_{\text{pr}}^2 + N_{\text{du}}^2 + N_{\text{pr}}N_{\text{du}})Q)$; we then solve the primal and dual problems—at cost $O(N_{\text{pr}}^3 + N_{\text{du}}^3 + K(N_{\text{pr}}^2 + N_{\text{du}}^2))$; and finally we evaluate the RB output approximation $s_N(\mu; t^k)$, for all $k \in \mathbb{K}$—at cost $O(K(K+1)N_{\text{pr}}N_{\text{du}})$. The online complexity is thus *independent* of $\mathcal{N}$, and in fact not too sensitive (for our LTI system) to K.

We now turn to a posteriori error estimation. We stress that the development of the error bounds is in no way limited to the RB approximation described here: we may consider "any" stable ODE or PDE system and any reduced-order model. To begin, we assume that we are given $\tilde{\alpha}(\mu) : \mathcal{D} \to \mathbb{R}_+$, a positive lower bound for the coercivity constant, $\alpha(\mu) : \alpha(\mu) \geq \tilde{\alpha}(\mu) \geq \tilde{\alpha}_0 > 0$, for all $\mu \in \mathcal{D}$. In our symmetric case $\alpha(\mu) = \beta(\mu)$, and thus $\tilde{\alpha}(\mu)$ can be constructed according to Section 10.4; in fact, thanks to coercivity, much simpler procedures typically suffice [29]. We next recall the dual norm of the primal and dual residuals: for all $k \in \mathbb{K}$, $\varepsilon_{N_{\text{pr}}}^{\text{pr}}(\mu, t^k) \equiv \sup_{v\in X}[R^{\text{pr}}(v; \mu, t^k)/\|v\|]$ and $\varepsilon_{N_{\text{du}}}^{\text{du}}(\mu, t^k) \equiv \sup_{v\in X}[R^{\text{du}}(v; \mu, t^k)/\|v\|]$, where for all $k \in \mathbb{K}$, $R^{\text{du}}(v; \mu, t^k) \equiv -(\Delta t^{-1} m(v, \Psi_N(\mu, t^k) - \Psi_N(\mu, t^{k+1})) + a(v, \Psi_N(\mu, t^k); \mu))$, for all $v \in X$. Finally, we introduce the "spatio-temporal" energy norm, $|||v(\mu, t^k)|||^2 \equiv m(v(\mu, t^k), v(\mu, t^k)) + \sum_{k'=1}^{k} \Delta t\; a(v(\mu, t^{k'}), v(\mu, t^{k'}); \mu)$, for all $v \in X$.

We may now define our error estimators: for all $\mu \in \mathcal{D}$, for all $k \in \mathbb{K}$, $\Delta_{N_{\text{pr}}}^{\text{pr}}(\mu, t^k) \equiv \tilde{\alpha}^{-\frac{1}{2}}(\mu)(\Delta t \sum_{k'=1}^{k} \varepsilon_{N_{\text{pr}}}^{\text{pr}}(\mu, t^{k'})^2)^{\frac{1}{2}}$; $\Delta_{N_{\text{du}}}^{\text{du}}(\mu, t^k) \equiv \tilde{\alpha}^{-\frac{1}{2}}(\mu)(\Delta t \sum_{k'=k}^{K} \varepsilon_{N_{\text{du}}}^{\text{du}}(\mu, t^{k'})^2)^{\frac{1}{2}}$; and

$$\Delta^s(\mu, t^k) \equiv \Delta_{N_{\text{pr}}}^{\text{pr}}(\mu, t^k)\, \Delta_{N_{\text{du}}}^{\text{du}}(\mu, t^{K-k+1}). \tag{10.6}$$

We can then readily demonstrate that $|||u(\mu, t^k) - u_N(\mu, t^k)||| \leq \Delta_{N_{\text{pr}}}^{\text{pr}}(\mu, t^k)$ and $|s(\mu, t^k) - s_N(\mu, t^k)| \leq \Delta^s(\mu, t^k)$, for all $k \in \mathbb{K}$, for all $\mu \in \mathcal{D}$, $1 \leq N_{\text{pr}} \leq N_{\text{pr,max}}$, $1 \leq N_{\text{du}} \leq N_{\text{du,max}}$ [9]—we obtain rigorous (and, as we shall see, rather sharp) upper bounds for the primal error, dual error, *and* output error. (Note that our particular form (10.6) assumes that $\Psi(\mu, t^{K+1})$—here, μ-independent—is a member of $W_{N_{\text{du}}}^{\text{du}}$; this requirement is readily relaxed.)

The offline-online procedure for the computation of $\Delta^s(\mu, t^k)$, for all $k \in \mathbb{K}$—in particular, for the calculation of the requisite primal and dual residual norms—is similar to the elliptic case of Section 10.4 but with the added complexity of the dual problem and the time dependence [9]. In particular, in the *online* stage—for any given new μ—we evaluate the $\varepsilon_{N_{\text{pr}}}^{\text{pr}}(\mu, t^k)^2$ and $\varepsilon_{N_{\text{du}}}^{\text{du}}(\mu, t^k)^2$ sums in terms of $\Theta^q(\mu)$, $u_{N\,n}(\mu, t^k)$, $\Psi_{N\,n'}(\mu, t^{k'})$ and the *precomputed* inner products—at cost $O(K(N_{\text{pr}}^2 + N_{\text{du}}^2)Q^2)$. Thus, all online calculations are indeed *independent* of $\mathcal{N}$.

We now turn to a particular numerical example. We consider the design of a heat shield (one cell of which is shown in Figure 10.2): the left boundary $\partial\Omega_{\text{out}}$ is exposed to a temperature unity and Biot number Bi_{out} "source" for $t \in [0, t_f]$; the right boundary as well as the top and bottom (symmetry) boundaries are insulated; and the internal boundaries $\partial\Omega_{\text{in}}$—corresponding to three square cooling channels—are exposed to a temperature zero and Biot number Bi_{in} "sink." Our input parameter is hence

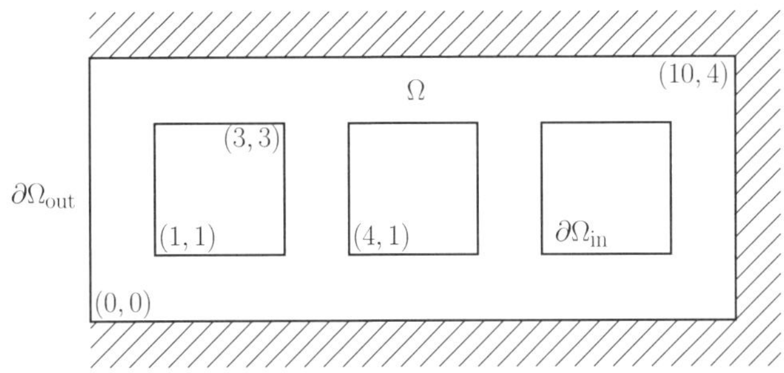

Figure 10.2. *One "cell" of the heat shield.*

Table 10.3. *Convergence results for the heat equation.*

N_{pr}	$\Delta^{\text{pr}}_{\text{max,rel}}$	$\overline{\eta}^{\text{pr}}$	$\Delta^{s}_{\text{max,rel}}$	$\overline{\eta}^{s}$
4	1.6 E−00	5.44	1.6 E−00	95.63
8	6.3 E−02	1.55	6.7 E−03	30.92
12	1.0 E−02	1.03	2.6 E−04	8.43
16	3.2 E−03	1.02	1.5 E−05	11.45
20	8.8 E−04	1.01	1.1 E−06	17.43

$\mu \equiv (\mu_{(1)}, \mu_{(2)}) \equiv (\text{Bi}_{\text{out}}, \text{Bi}_{\text{in}}) \in \mathcal{D} \equiv [0.01, 0.5] \times [0.001, 0.1]$; our output is the average temperature of the structure—a surrogate for the maximum temperature of the (to-be-protected) right boundary for $t \in [0, \infty[$.

The underlying PDE is the heat equation. The (appropriately nondimensionalized) governing equation for the temperature $u(\mu, t^k) \in X$ is thus (10.5), where X is a linear finite element truth approximation subspace (of dimension (exploiting symmetry) $\mathcal{N} = 1{,}396$) of $X^{\text{e}} \equiv H^1(\Omega)$; $a(w, v; \mu) \equiv \int_\Omega \nabla w \cdot \nabla v + \mu_{(1)} \int_{\partial\Omega_{\text{out}}} w\, v + \mu_{(2)} \int_{\partial\Omega_{\text{in}}} w\, v$; $m(w, v) \equiv \int_\Omega w\, v$; $f(v; \mu) \equiv \mu_{(1)} \int_{\partial\Omega_{\text{out}}} v$, which is now (affinely) parameter-dependent; $b(t^k) = 1$, for all $k \in \mathbb{K}$; and $(w, v) \equiv \int_\Omega \nabla w \cdot \nabla v + 0.01 \int_{\partial\Omega_{\text{out}}} w\, v + 0.001 \int_{\partial\Omega_{\text{in}}} w\, v$—hence we may choose $\tilde{\alpha}(\mu) = 1$. The output is given by $s(\mu, t^k) = \ell(u(\mu, t^k))$, where $\ell(v) \equiv |\Omega|^{-1} \int_\Omega v$.

We now present numerical results. Our "optimal" primal and dual samples are constructed (separately) by procedures similar to the greedy approach described for the elliptic case in Section 10.4 [9]: at each step (say, for the primal) we select the parameter value μ^* for which $\Delta^{\text{pr}}_{N_{\text{pr}}}(\mu, t^K)$ is maximized; we then select the time t^{k^*} for which $\varepsilon^{\text{pr}}_{N_{\text{pr}}}(\mu^*, t^k)$ is maximized. In Table 10.3 we present, as a function of N_{pr} ($= N_{\text{du}}$), $\Delta^{\text{pr}}_{\text{max,rel}}$, $\overline{\eta}^{\text{pr}}$, $\Delta^{s}_{\text{max,rel}}$, and $\overline{\eta}^{s}$: $\Delta^{\text{pr}}_{\text{max,rel}}$ is the maximum over Ξ_{Test} of $\Delta^{\text{pr}}_{N_{\text{pr}}}(\mu, t^K)/|||u_N(\mu_u, t^K)|||$, $\overline{\eta}^{\text{pr}}$ is the average over $\Xi_{\text{Test}} \times \mathbb{T}$ of $\Delta^{\text{pr}}_{N_{\text{pr}}}(\mu, t^k)/|||u(\mu, t^k) - u_N(\mu, t^k)|||$, $\Delta^{s}_{\text{max,rel}}$ is the maximum over Ξ_{Test} of $\Delta^{s}_{N}(\mu, t^K)/|s_N(\mu_s, t^K)|$, and $\overline{\eta}^{s}$ is the average over Ξ_{Test} of $\Delta^{s}_{N}(\mu, t_\eta(\mu))/|s(\mu, t_\eta(\mu)) - s_N(\mu, t_\eta(\mu))|$. Here $\Xi_{\text{Test}} \in (\mathcal{D})^{400}$ is a random input sample of size 400; $\mu_u \equiv \arg\max_{\mu \in \Xi_{\text{Test}}} |||u_{N_{\max}}(\mu, t^K)|||$, $\mu_s \equiv \arg\max_{\mu \in \Xi_{\text{Test}}} |s_{N_{\max}}(\mu, t^K)|$ (note the output grows with time), and $t_\eta(\mu) \equiv \arg\max_{t^k \in \mathbb{T}} |s(\mu, t^k) - s_N(\mu, t^k)|$. The output converges rapidly, and the effectivities are reasonably good.

Finally, we note that the calculation of $s_N(\mu, t^k)$ and $\Delta_N^s(\mu, t^k)$, for all $k \in \mathbb{K}$, is (say, for $N_{\text{pr}} = N_{\text{du}} = 12$) roughly 120 times faster than direct calculation of the truth approximation output $s(\mu, t^k) = \ell(u(\mu, t^k))$, for all $k \in \mathbb{K}$. We may thus work with $s_N(\mu, t^k) + \Delta_N^s(\mu, t^k)$ as a *certifiably* conservative (upper bound) and accurate surrogate for the average temperature $s(\mu, t^k)$ in truly interactive design exercises.

Acknowledgments

This work was supported by DARPA and AFOSR under grants F49620-03-1-0356 and FA9550-05-1-0114, by DARPA and GEAE under grant F49620-03-1-0439, and by the Singapore-MIT Alliance.

Bibliography

[1] S. Ali, *Real-Time Optimal Parametric Design using the Assess-Predict-Optimize Strategy*, Ph.D. thesis, Singapore-MIT Alliance, Nanyang Technological University, Singapore, 2003.

[2] B. O. Almroth, P. Stern, and F. A. Brogan, *Automatic choice of global shape functions in structural analysis*, AIAA J., 16 (1978), pp. 525–528.

[3] E. Balmes, *Parametric families of reduced finite element models: Theory and applications*, Mechanical Systems and Signal Processing, 10 (1996), pp. 381–394.

[4] E. Barkanov, *Transient response analysis of structures made from viscoelastic materials*, Internat. J. Numer. Methods Engrg., 44 (1999), pp. 393–403.

[5] M. Barrault, N. C. Nguyen, Y. Maday, and A. T. Patera, *An "empirical interpolation" method: Application to efficient reduced-basis discretization of partial differential equations*, C. R. Math. Acad. Sci. Paris, 339 (2004), pp. 667–672.

[6] G. Caloz and J. Rappaz, *Numerical analysis for nonlinear and bifurcation problems*, in Handbook of Numerical Analysis, Vol. V, P. Ciarlet and J. Lions, eds., Techniques of Scientific Computing (Part 2), North–Holland, Amsterdam, 1997, pp. 487–637.

[7] H. W. Engl, M. Hanke, and A. Neubauer, *Regularization of Inverse Problems*, Kluwer Academic Publishers, Dordrecht, The Netherlands, 1996.

[8] J. P. Fink and W. C. Rheinboldt, *On the error behavior of the reduced basis technique for nonlinear finite element approximations*, Z. Angew. Math. Mech., 63 (1983), pp. 21–28.

[9] M. A. Grepl and A. T. Patera, *A posteriori error bounds for reduced-basis approximations of parametrized parabolic partial differential equations*, M2AN Math. Model. Numer. Anal., 39 (2005), pp. 157–181.

[10] M. D. Gunzburger, *Finite Element Methods for Viscous Incompressible Flows: A Guide to Theory, Practice, and Algorithms*, Academic Press, Boston, 1989.

[11] S. I. Ishak, G. R. Liu, S. P. Lim, and H. M. Shang, *Locating and sizing of delamination in composite laminates using computational and experimental methods*, Composite Part B, 32 (2001), pp. 287–298.

[12] K. Ito and S. S. Ravindran, *A reduced basis method for control problems governed by PDEs*, in Control and Estimation of Distributed Parameter Systems, W. Desch, F. Kappel, and K. Kunisch, eds., Birkhäuser, Basel, 1998, pp. 153–168.

[13] K. Ito and S. S. Ravindran, *A reduced-order method for simulation and control of fluid flows*, J. Comput. Phys., 143 (1998), pp. 403–425.

[14] G. R. Liu and S. C. Chen, *Flaw detection in sandwich plates based on time-harmonic response using genetic algorithm*, Comput. Methods Appl. Mech. Engrg., 190 (2001), pp. 5505–5514.

[15] G. R. Liu, X. Han, and K. Y. Lam, *A combined genetic algorithm and nonlinear least squares method for material characterization using elastic waves*, Comput. Methods Appl. Mech. Engrg., 191 (2002), pp. 1909–1921.

[16] G. R. Liu and K. Y. Lam, *Characterization of a horizontal crack in anisotropic laminated plates*, Internat. J. Solids Structures, 31 (1994), pp. 2965–2977.

[17] G. R. Liu, K. Y. Lam, and J. Tani, *Characterization of flaws in sandwich plates: Numerical experiment*, JSME Int. J. Ser. A Japan, 38 (1995), pp. 554–562.

[18] G. R. Liu, Z. C. Xi, K. Y. Lam, and H. M. Shang, *A strip element method for analyzing wave scattering by a crack in an immersed composite laminate*, J. Appl. Mech., 66 (1999), pp. 898–903.

[19] L. Machiels, Y. Maday, I. B. Oliveira, A. T. Patera, and D. V. Rovas, *Output bounds for reduced-basis approximations of symmetric positive definite eigenvalue problems*, C. R. Acad. Sci. Paris Sér. I Math., 331 (2000), pp. 153–158.

[20] Y. Maday, A. T. Patera, and G. Turinici, *Global a priori convergence theory for reduced-basis approximation of single-parameter symmetric coercive elliptic partial differential equations*, C. R. Math. Acad. Sci. Paris, 335 (2002), pp. 289–294.

[21] B. Moore, *Principal component analysis in linear systems: Controllability, observability, and model reduction*, IEEE Trans. Automat. Control, 26 (1981), pp. 17–32.

[22] N. C. Nguyen, K. Veroy, and A. T. Patera, *Certified real-time solution of parametrized partial differential equations*, in Handbook of Materials Modeling, S. Yip, ed., Springer-Verlag, New York, 2005, pp. 1523–1558.

[23] A. K. Noor and J. M. Peters, *Reduced basis technique for nonlinear analysis of structures*, AIAA J., 18 (1980), pp. 455–462.

[24] I. B. OLIVEIRA AND A. T. PATERA, *Reduced-basis techniques for rapid reliable optimization of systems described by affinely parametrized coercive elliptic partial differential equations*, Optim. Engrg., to appear.

[25] J. S. PETERSON, *The reduced basis method for incompressible viscous flow calculations*, SIAM J. Sci. Stat. Comput., 10 (1989), pp. 777–786.

[26] N. A. PIERCE AND M. B. GILES, *Adjoint recovery of superconvergent functionals from PDE approximations*, SIAM Rev., 42 (2000), pp. 247–264.

[27] T. A. PORSCHING, *Estimation of the error in the reduced basis method solution of nonlinear equations*, Math. Comp., 45 (1985), pp. 487–496.

[28] T. A. PORSCHING AND M. L. LEE, *The reduced basis method for initial value problems*, SIAM J. Numer. Anal., 24 (1987), pp. 1277–1287.

[29] C. PRUD'HOMME, D. ROVAS, K. VEROY, Y. MADAY, A. T. PATERA, AND G. TURINICI, *Reliable real-time solution of parametrized partial differential equations: Reduced-basis output bound methods*, J. Fluids Engineering, 124 (2002), pp. 70–80.

[30] A. QUARTERONI AND A. VALLI, *Numerical Approximation of Partial Differential Equations*, 2nd ed., Springer-Verlag, New York, 1997.

[31] B. ROUX, ED., *Numerical Simulation of Oscillatory Convection in Low-Pr Fluids: A GAMM Workshop*, Notes Numer. Fluid Mech. 27, Vieweg, Braunschweig, 1990.

[32] D. ROVAS, *Reduced-Basis Output Bound Methods for Parametrized Partial Differential Equations*, Ph.D. thesis, Massachusetts Institute of Technology, Cambridge, MA, 2002.

[33] L. SIROVICH, *Turbulence and the dynamics of coherent structures, Part* 1: *Coherent structures*, Quart. Appl. Math., 45 (1987), pp. 561–571.

[34] A. C. SKELDON, D. S. RILEY, AND K. A. CLIFFE, *Convection in a low Prandtl number fluid*, J. Crystal Growth, 162 (1996), pp. 95–106.

[35] G. TALENTI, *Best constant in Sobolev inequality*, Ann. Mat. Pura Appl. (4), 110 (1976), pp. 353–372.

[36] K. VEROY AND A. T. PATERA, *Certified real-time solution of the parametrized steady incompressible Navier-Stokes equations; Rigorous reduced-basis a posteriori error bounds*, Internat. J. Numer. Methods Fluids, 47 (2005), pp. 773–788.

[37] K. VEROY, C. PRUD'HOMME, AND A. T. PATERA, *Reduced-basis approximation of the viscous Burgers equation: Rigorous a posteriori error bounds*, C. R. Math. Acad. Sci. Paris, 337 (2003), pp. 619–624.

[38] K. VEROY, C. PRUD'HOMME, D. V. ROVAS, AND A. T. PATERA, *A posteriori error bounds for reduced-basis approximation of parametrized noncoercive and nonlinear elliptic partial differential equations*, in Proceedings of the 16th Annual AIAA Computational Fluid Dynamics Conference, 2003, AIAA Paper 2003-3847.

Chapter 11

Model Reduction for Large-Scale Applications in Computational Fluid Dynamics

*K. Willcox**

11.1 Introduction

Recent years have seen considerable progress in solution and optimization methods for partial differential equations (PDEs), leading to advances across a broad range of engineering applications. Improvements in methodology, together with a substantial increase in computing power, are such that real-time simulation and optimization of systems governed by PDEs is now an attainable goal; however, a number of challenges remains for applications such as real-time control of dynamic processes. In many cases, computational models for such applications yield very large systems that are computationally intensive to solve. A critical element towards achieving a real-time simulation capability is the development of accurate, efficient models that can be solved sufficiently rapidly to permit control decisions in real time.

Model reduction is a powerful tool that allows the systematic generation of cost-efficient representations of large-scale systems resulting from discretization of PDEs. Reduction methodology has been developed and applied for many different disciplines, including controls, fluid dynamics, structural dynamics, and circuit design. Considerable advances in the field of model reduction for large-scale systems have been made and many different applications have been demonstrated with success; however, a number of open issues remain, including the reliability of reduction techniques, guarantees associated with the quality of the reduced models, and validity of the model over a range of operating conditions. The cost of performing the reduction may also be an issue if there is a need to adapt the reduced-order model in real time.

*Department of Aeronautics and Astronautics, Massachusetts Institute of Technology, Cambridge, MA 02139.

In this chapter, model reduction of computational fluid dynamic (CFD) systems will be discussed, although the techniques presented are applicable to general systems of PDEs. Two reduction methods will be discussed: proper orthogonal decomposition (POD) and Fourier model reduction (FMR) . These methods will be compared in the context of an active flow control application. Their relative advantages and disadvantages will be presented, with particular focus on the issues relevant to real-time implementation. Finally, the chapter concludes with a discussion of outstanding issues and the open question of model reduction for nonlinear systems.

11.1.1 Problem Statement

The system of PDEs governing a general fluid flow can be discretized in the spatial domain using a CFD formulation to yield a set of nonlinear ordinary differential equations (ODEs). For unsteady flows, these ODEs can be linearized about the steady-state solution to obtain a linear CFD model that is valid for small deviations of the flow from steady state conditions. A general linearized CFD model can be written as

$$G: \quad \frac{d}{dt}x(t) = Ax(t) + Bu(t), \qquad y(t) = Cx(t) + Du(t), \tag{11.1}$$

where $x(t) \in \mathbb{R}^n$ contains the n unknown perturbation flow quantities at each point in the computational grid. For example, for two-dimensional, compressible, inviscid flow, which is governed by the Euler equations, the unknowns at each grid point are the perturbations in flow density, Cartesian momentum components, and flow energy. The vectors $u(t)$ and $y(t)$ in (11.1) contain the system inputs and outputs, respectively. The definition of inputs and outputs will depend upon the problem at hand. For active control applications, the output might monitor a flow condition at a particular location which varies in response to a disturbance in the incoming flow, while inputs describe both actuation mechanisms and flow disturbances.

The linearization matrices A, B, C, and D in (11.1) are evaluated at steady state conditions. In comparison with the nonlinear equations, the system (11.1) is relatively efficient, since a time discretization, such as backward Euler, can be applied and the resulting large $n \times n$ system matrix factored just once for a time-dependent calculation. However, the order of the system is still prohibitively high for many applications, and the cost to solve the system is too large for implementation in real time. We therefore consider the task of finding a low-order, stable, linear time invariant (LTI), state-space model

$$\hat{G}: \quad \frac{d}{dt}\hat{x}(t) = \hat{A}\hat{x}(t) + \hat{B}u(t), \qquad \hat{y}(t) = \hat{C}\hat{x}(t) + \hat{D}u(t), \tag{11.2}$$

which approximates well the given stable model (11.1). Typically, A in (11.1) is a sparse, square matrix of very large dimension $n > 10^4$, and the desired order k of $\hat{G}$ is less than 50.

The quality of $\hat{G}$ as an approximation of G is defined as the H-infinity norm of the difference between their transfer functions:

$$\|\hat{G} - G\|_\infty = \sup_{\omega \in \mathbb{R}} |\hat{G}(j\omega) - G(j\omega)|, \tag{11.3}$$

which in turn equals the square root of the maximal energy of the difference $e = \hat{y} - y$, given by

$$\|\hat{y} - y\|_2^2 = \int_t |\hat{y}(t) - y(t)|^2 dt. \tag{11.4}$$

11.1.2 Projection Framework

No efficient (polynomial time) solution is known for the problem of minimizing $\|\hat{G} - G\|_\infty$ subject to the order and stability constraints imposed on $\hat{G}$. Optimal Hankel model reduction and balanced truncation are polynomial-time algorithms that produce suboptimal reduced models with strong guarantees of quality. However, the computational burden of these methods is such that their direct application is impractical for $n > 10^3$. Instead, most model reduction techniques for large-scale systems use a projection framework. The full state vector x is represented in a reduced-space basis

$$x = V\hat{x}, \tag{11.5}$$

where the columns of the matrix $V \in \mathbb{R}^{n \times k}$ contain k basis vectors. Defining a left projection space W so that $W^T V = I$, the governing equations (11.1) can be projected onto the reduced space to yield an mth-order model of the form

$$\hat{G}: \quad \frac{d}{dt}\hat{x}(t) = W^T A V \hat{x}(t) + W^T B u(t), \qquad \hat{y}(t) = C V \hat{x}(t) + \hat{D} u(t). \tag{11.6}$$

The reduction task is then to find a suitable basis W, V so that $k \ll n$ and the reduction error, defined by (11.4), is small. Several methods for computing the basis exist, including Krylov subspace methods and POD.

11.2 Proper Orthogonal Decomposition

POD has been widely used to determine efficient bases for dynamic systems [1]. It was introduced for the analysis of turbulence by Lumley [2], and is also known as singular value decomposition, Karhunen–Loève decomposition [3], and principal component analysis [4]. POD basis vectors are computed empirically using a set of data that samples the range of relevant system dynamics.

11.2.1 Time-Domain POD Basis

The basis vectors Ψ are chosen so as to maximize the following cost [5]:

$$\Psi = \arg\max \frac{\langle |(x, \Psi)|^2 \rangle}{(\Psi, \Psi)}, \tag{11.7}$$

where (x, Ψ) denotes the scalar product of the basis vector with the field $x(\theta, t)$ which depends on problem geometry θ and time t, and $\langle \, \rangle$ represents a time-averaging operation. It can be shown that a necessary condition for (11.7) to hold is that Ψ is an eigenfunction of the kernel K defined by

$$K(\theta, \theta') = \langle x(\theta, t) \; x^*(\theta', t) \rangle, \tag{11.8}$$

where x^* denotes the complex conjugate transpose of x.

Sirovich introduced the method of snapshots as a way of determining the eigenfunctions Ψ without explicitly calculating the kernel K [6]. The kernel can be approximated as

$$K(\theta, \theta') = \frac{1}{m} \sum_{i=1}^{m} x_i(\theta) x_i^*(\theta'), \tag{11.9}$$

where $x_i(\theta)$ is the instantaneous system state or "snapshot" at a time t_i and the number of snapshots m is sufficiently large. The eigenvectors of K are of the form

$$\Psi = \sum_{i=1}^{m} \beta_i x_i, \tag{11.10}$$

where the constants β_i satisfy the eigenvector equation

$$R\beta = \Lambda\beta \tag{11.11}$$

and R is now the correlation matrix

$$R_{ij} = \frac{1}{m}(x_i, x_j), \tag{11.12}$$

which contains the inner product between every pair of snapshots. The magnitude of the jth eigenvalue, λ_j, describes the relative importance of the jth POD basis vector for reconstruction of the data contained in the snapshot ensemble. For a basis containing the first p POD modes, the least-squares error of data reconstruction is given by the eigenvalues corresponding to the neglected modes:

$$\sum_{i=1}^{m} \|x_i - \tilde{x}_i^p\|_2^2 = \sum_{j=p+1}^{m} \lambda_j, \tag{11.13}$$

where $\tilde{x}_i^p$ is the representation of the ith snapshot, x_i, in the pth-order POD basis.

11.2.2 Frequency-Domain POD Basis

Often, the POD snapshots are obtained from a time simulation of the CFD model. One issue with this approach is an appropriate choice of input to the simulation. This input choice is critical, since the resulting basis will capture only those dynamics present in the snapshot ensemble. This can be a problematic issue for many applications, such as flow control design, where the dynamics of the controlled and uncontrolled systems might differ significantly. An alternative approach is to apply the POD in the frequency domain [7]. Rather than selecting a time-dependent input function, one selects a set of sample frequencies. The corresponding flow snapshots can then be obtained by solving the frequency domain CFD equations

$$X(\omega) = [j\omega_i I - A]^{-1} B, \tag{11.14}$$

where $u(t) = e^{j\omega_i t}$, $x(t) = Xe^{j\omega_i t}$, and ω_i is the ith sample frequency.

Frequency domain POD approaches typically yield better results; however, the computational cost of the method is high. The nth-order system given by (11.14) must be solved for each frequency selected. In a typical CFD application, a large number of frequency points is required to obtain satisfactory models. In particular, for three-dimensional applications, the cost of this approach can be prohibitive. While not discussed in this chapter, Krylov-based reduction techniques are computationally more efficient than the POD. These techniques are widely used in integrated circuit applications, and have also been applied to CFD systems. It can be shown that the approach of using the Arnoldi method with multiple interpolation points has strong connections to the frequency-domain POD approach [8].

11.2.3 POD Reduced-Order Model

Once the set of POD basis vectors has been computed, using time- or frequency-domain snapshots, the reduced-order model is obtained using the projection (11.5). It is important to note that, while the POD basis is optimal in the sense that it provides the most efficient representation of the data contained in the snapshot ensemble, one can make no statement regarding the quality of the resulting reduced-order model. No estimate or bound for the reduction error is available. Since the POD basis is orthonormal, that is, $W = V$, the projection $\hat{A} = V^T A V$ preserves definiteness; however, in general, this is not sufficient to preserve stability. In practice, unstable models are often generated, and some trial and error is required to determine an appropriate snapshot ensemble and basis size.

The POD is a useful reduction technique and has been shown to yield satisfactory results for a wide range of applications (see, for example, [9]); however, its lack of rigorous guarantees should not be overlooked.

11.2.4 Balanced POD

The concept of a balanced realization of a system was first introduced by Moore [10]. The underlying idea is to take account of both the inputs and the outputs of the system when determining which states to retain in the reduced-state representation but to do so with appropriate internal scaling. The controllability and observability gramians of the linear system (11.1) are defined, respectively, as

$$W_c = \int_0^\infty e^{At} B B^* e^{A^* t} dt \tag{11.15}$$

and

$$W_o = \int_0^\infty e^{A^* t} C^* C e^{At} dt. \tag{11.16}$$

The above matrices describe the controllable and observable subspaces of the linear system (11.1). These two subspaces can be interpreted geometrically as described in [11]. The controllable subspace is that set of states which can be obtained with zero initial state and a given input $u(t)$, while the observable subspace comprises those states which as initial conditions could produce a certain output $y(t)$ with no external input.

To obtain a balanced realization of the system (11.1), a state transformation is chosen so that the controllability and observability gramians are diagonal and equal. This transformation can be computed by first calculating the matrix $W_{co} = W_c W_o$ and then determining its eigenmodes:

$$W_{co} = T^{-1}\Lambda T. \tag{11.17}$$

The columns of T then contain the basis vectors which describe the balancing transformation. The eigenvalues λ_i contained in the diagonal matrix Λ are positive, real numbers, and $\sigma_i = \sqrt{\lambda_i}$ are known as the Hankel singular values of the system. These values are independent of the particular realization of the system and describe the importance of the corresponding state for transmission of input to output. In a balanced truncation, only those states are retained which correspond to large Hankel singular values. A bound on the error of the kth-order balanced reduced model, $\hat{G}_k$, is given by the Hankel singular values of the truncated modes

$$||G - \hat{G}_k||_\infty \leq 2 \sum_{i=k+1}^{m} \sigma_i. \tag{11.18}$$

Approximate Balanced Realization via the Method of Snapshots

For large systems, it is not practical to explicitly compute the gramians using (11.15) and (11.16); however, it can be shown that there is a strong connection between the controllability gramian and the POD kernel function [12, 13]. By noting that for a SISO system the quantity $x_\delta(t) = e^{At}B$ is simply the impulse response of the system (set $u(t) = \delta(t)$ in (11.1)), the controllability gramian can also be written

$$W_c = \int_0^\infty x_\delta(t) x_\delta^*(t) dt \tag{11.19}$$

and compared with the POD kernel function K defined by (11.9). Similarly, the observability gramian can be written as

$$W_o = \int_0^\infty \overline{x}_\delta(t) \overline{x}_\delta^*(t) dt, \tag{11.20}$$

where $\overline{x}_\delta(t) = e^{A^*t}C^*$ is the impulse response of the dual SISO system which is given by

$$\overline{G}: \quad \frac{d}{dt}\overline{x}(t) = A^*\overline{x}(t) + C^*\overline{u}(t), \qquad \overline{y}(t) = B^*\overline{x}(t). \tag{11.21}$$

By considering general, nonimpulsive inputs, the POD computes the most controllable modes of the system within a certain restricted range of dynamics. It is a natural extension to consider a POD analysis which determines the most observable modes. Furthermore, to obtain a balanced representation of the system, concepts from a traditional control balanced realization can then be used. The direct POD method can be used to obtain approximations to the system gramians for small systems [12]. For large-scale systems, the POD method of snapshots can be used to approximate the gramians in a computationally efficient manner that does not require computation of large $n \times n$ matrices [13].

By obtaining snapshots of the primal and dual systems (11.1) and (11.21), and performing the POD method of snapshots analysis to determine the POD basis vectors, p eigenmodes of the controllability and observability kernel functions are obtained, respectively. Let the eigenvectors of the controllability kernel K be contained in the columns of the matrix V, with corresponding eigenvalues on the diagonal entries of the matrix Λ_c.

Similarly, let the eigenvectors of the observability kernel be contained in the columns of the matrix X, with corresponding eigenvalues on the diagonal entries of the matrix Λ_o. Low-rank approximations to the controllability and observability gramians can then be made as follows:

$$W_c^p = V \Lambda_c V^*, \tag{11.22}$$

$$W_o^p = X \Lambda_o X^*, \tag{11.23}$$

where the superscript p denotes a pth-order approximation.

Through use of an efficient eigenvalue solver, the eigenmodes of the product $W_c^p W_o^p$ can then be calculated using only matrix-vector multiplications, and hence the large matrices W_c^p and W_o^p need never be explicitly formed.

The balancing algorithm can therefore be summarized as follows:

1. Use method of snapshots to obtain p POD eigenmodes (V, Λ_c) for the primal system.
2. Use method of snapshots to obtain p POD eigenmodes (X, Λ_o) for the dual system.
3. Calculate the low-rank approximations $W_c^p = V \Lambda_c V^*$ and $W_o^p = X \Lambda_o X^*$.
4. Obtain the eigenvectors of the product $W_c^p W_o^p$ to determine the balancing transformation T.

Once again, it is important to note the limitations of this reduction approach. Balanced truncation is a rigorous method for small systems that yields reduced-order models with strong guarantees of quality and a computable error bound given by (11.18). The approximate method using the POD draws upon a strong analogy with balanced truncation and has been shown to work effectively for many cases; however, as for the conventional POD, this method does not offer rigorous error or stability guarantees.

In the next section, a different approach to model reduction of large-scale systems that does not use a projection framework will be described. This method has associated with it a rigorous, although noncomputable, error bound.

11.3 Fourier Model Reduction

The FMR method is described in [14] and uses discrete-time Fourier coefficients of the CFD system to form a reduced-order model with a rigorous error bound. Many coefficients can be calculated, which results in a very accurate representation of the system dynamics, but only a single factorization of the large system is required. As shown in Figure 11.1, FMR can be combined with an efficient second reduction step using explicit formulae for balanced truncation, which allows use of the Hankel singular values to rigorously select the number of reduced states. FMR yields very accurate, low-order models when the original transfer function is smooth and thus is an attractive method for reduction of CFD systems.

11.3.1 Fourier Series of Discrete-Time Systems

Consider the discrete-time (DT) model g corresponding to the system (11.1), which is defined by the difference equations

$$g: \quad x(t+1) = ax(t) + bu(t), \quad y(t) = cx(t) + du(t), \tag{11.24}$$

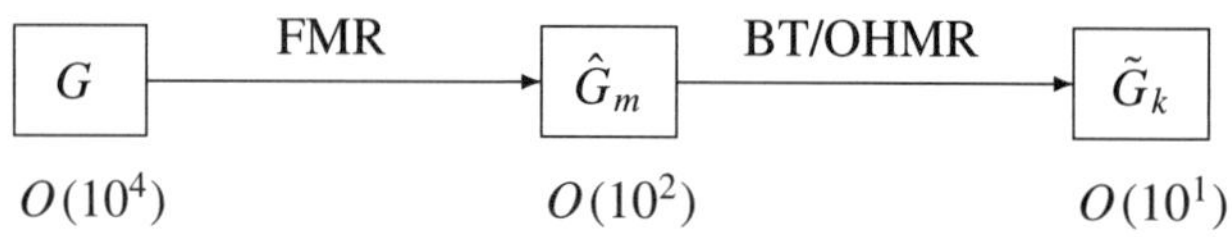

Figure 11.1. *Two-step model reduction process: FMR with rigorous but non-computable error bound followed by balanced truncation (BT) or optimal Hankel model reduction (OHMR).*

where a, b, c, d are the DT matrices. The transfer function

$$g(z) = d + c(zI - a)^{-1}b \tag{11.25}$$

has the Fourier decomposition

$$g(z) = \sum_{k=0}^{\infty} g_k z^{-k}, \tag{11.26}$$

where

$$g_0 = d, \quad g_k = ca^{k-1}b \; (k = 1, 2, \ldots). \tag{11.27}$$

The Fourier expansion converges exponentially for $|z| > \rho(a)$, where $\rho(a)$ denotes the spectral radius of a, defined as the maximal absolute value of its eigenvalues. Note that the first m Fourier coefficients g_k are easy to calculate using the "cheap" iterative process

$$g_k = ch_{k-1}, \quad h_k = ah_{k-1} \; (k = 1, \ldots, m), \quad \text{where } h_0 = b. \tag{11.28}$$

Let $\hat{g}_m$ denote the mth-order approximation of g based on the Fourier series expansion:

$$\hat{g}_m(z) = \sum_{k=0}^{m} g_k z^{-k}. \tag{11.29}$$

The following simple result provides an estimate of the approximation error:

$$\|g - \hat{g}_m\|_\infty = \max_{|z|=1} |g(z) - \hat{g}_m(z)|, \tag{11.30}$$

which ties it to the smoothness of G as follows.

Theorem 11.1. *For* $q = 1, 2, \ldots,$

$$\|g - \hat{g}_m\|_\infty^2 \leq \frac{m^{1-2q}}{2\pi(2q-1)} \int_{-\pi}^{\pi} |g^{(q)}(e^{j\tau})|^2 d\tau, \tag{11.31}$$

where $g^{(q)}$ *is the* qth *derivative of* g *with respect to* τ.

Proof. See [14].

11.3.2 Fourier Series of Continuous-Time Systems

Consider the full continuous-time LTI system model G defined by the system (11.1), where $u(t)$, $y(t)$ are scalar input and output. It will be assumed that G is stable, i.e., that all roots of the characteristic equation $\det(sI - A) = 0$ have negative real part, and that $C(sI - A)^{-1}B$ remains bounded as $s \to \infty$.

Let $\omega_0 > 0$ be a fixed positive real number. The transfer function

$$G(s) = D + C(sI - A)^{-1}B \tag{11.32}$$

has the Fourier decomposition

$$G(s) = \sum_{k=0}^{\infty} G_k \left(\frac{s - \omega_0}{s + \omega_0}\right)^k. \tag{11.33}$$

This decomposition is suggested in [15] and an approximate FFT algorithm is used to calculate the Fourier coefficients G_k. In [14], an efficient iterative procedure is proposed to directly calculate the Fourier coefficients as follows.

Consider the identity

$$G(s) = g(z) = d + c(zI - a)^{-1}b \quad \text{for } z = \frac{s + \omega_0}{s - \omega_0},$$

which allows one to apply the observations and theorem from the previous subsection to this case. Note that by comparing (11.26) and (11.33), it can be seen that $G_k = g_k$. The Fourier coefficients are therefore given by the following formulae:

$$G_0 = d, \quad G_k = ca^{k-1}b \; (k = 1, 2, \ldots), \tag{11.34}$$

$$d = D + C(\omega_0 I - A)^{-1}B, \tag{11.35}$$

$$a = (\omega_0 I + A)(\omega_0 I - A)^{-1}, \tag{11.36}$$

$$c = 2\omega_0 C(\omega_0 I - A)^{-1}, \tag{11.37}$$

$$b = -(\omega_0 I - A)^{-1}B, \tag{11.38}$$

which are relatively straightforward to obtain using algebraic manipulations as described in [14].

11.3.3 Reduced Model Construction

To construct an mth-order reduced model, one first calculates the Fourier coefficients, $g_0, g_1, \ldots, g_m$. The DT reduced model is then given by

$$\hat{g}: \quad \begin{aligned} \hat{x}[t+1] &= \hat{a}\hat{x}[t] + \hat{b}u[t], \\ \hat{y}[t] &= \hat{c}\hat{x}[t] + \hat{d}u[t], \end{aligned} \tag{11.39}$$

where

$$\hat{a} = \begin{bmatrix} 0 & 0 & 0 & 0 & \cdots \\ 1 & 0 & 0 & 0 & \cdots \\ 0 & 1 & 0 & 0 & \cdots \\ 0 & 0 & 1 & 0 & \cdots \\ 0 & 0 & 0 & \ddots & \end{bmatrix}, \quad \hat{b} = \begin{bmatrix} 1 \\ 0 \\ 0 \\ \vdots \end{bmatrix}, \tag{11.40}$$
$$\hat{c} = \begin{bmatrix} g_1 & g_2 & \cdots & g_m \end{bmatrix}, \quad \hat{d} = g_0.$$

It should be noted that if the original system g is stable, then the reduced system $\hat{g}$ is also guaranteed to be stable. This is due to the orthogonality properties of the Fourier expansion; taking a truncated number of Fourier terms, as in (11.29), automatically results in a stable approximation of a stable system.

Different alternatives could be chosen for the DT system representation. The controller canonical form above was selected, as in [15], for the purpose of an efficient second step of reduction. As shown in Figure 11.1, an effective approach is to use the efficient iterative procedure to calculate several hundred coefficients, resulting in an intermediate reduced model of the form (11.39). A second reduction step using balanced truncation can now be performed easily, since the expressions for the gramians are known explicitly. For the DT reduced model (11.39), the controllability matrix is the identity matrix and the observability matrix is the Hankel matrix that has $\hat{c}$ as its first row. The balancing vectors can therefore be obtained by computing the singular vectors of the mth-order Hankel matrix

$$\Gamma = \begin{bmatrix} g_1 & g_2 & g_3 & \cdots & g_{m-1} & g_m \\ g_2 & g_3 & g_4 & \cdots & g_m & 0 \\ g_3 & g_4 & g_5 & \cdots & 0 & 0 \\ \vdots & \vdots & \vdots & & \vdots & \vdots \\ g_{m-1} & g_m & 0 & \cdots & 0 & 0 \\ g_m & 0 & 0 & \cdots & 0 & 0 \end{bmatrix}. \tag{11.41}$$

The Hankel singular values, σ_i, $i = 1, 2, \ldots, m$, of the intermediate reduced system are given by the singular values of Γ and can be used to quantitatively guide the second reduction step via the error bound given in (11.18).

11.3.4 FMR Algorithm

The FMR algorithm is summarized in the following steps.

1. Choose a value of ω_0. The value of ω_0 should reflect the frequency range of interest. The nominal value is unity; however, if the response at high frequencies is of interest, a higher value of ω_0 should be chosen. One can visualize the transformation from continuous to discrete time as a mapping of the imaginary axis in the s-plane to the unit circle in the z-plane. The value of ω_0 then describes the compression of frequencies around the unit circle.

2. Calculate $m + 1$ Fourier coefficients using (11.34)–(11.38). Using the iterative procedure, any number of coefficients can be calculated with a single nth-order matrix factorization.

3. Using (11.41), calculate the mth-order Hankel matrix. Calculate its singular values and singular vectors. Note that in the general case of p inputs and q outputs, each entry g_k will be a block of size $q \times p$.

4. Using balanced truncation, construct a kth-order DT system, $\tilde{g}$. The value of k is chosen according to the distribution of Hankel singular values of the intermediate system.

5. Convert the kth-order DT reduced model to a continuous-time model using the relationships

$$\hat{A} = \omega_0 \left(\hat{a} - I\right)^{-1} \left(\hat{a} + I\right), \tag{11.42}$$

$$\hat{B} = 2\omega_0 \left(\hat{a} - I\right)^{-1} \hat{b}, \tag{11.43}$$

$$\hat{C} = -\hat{c} \left(\hat{a} - I\right)^{-1}, \tag{11.44}$$

$$\hat{D} = \hat{d} - \hat{c} \left(\hat{a} - I\right)^{-1} \hat{b}. \tag{11.45}$$

The error bounds corresponding to the above algorithm are given in (11.31) and (11.18) for the first and second stages of the reduction, respectively. The Hankel singular values of the intermediate system provide a straightforward, quantitative means to choose k, the size of the final reduced model; however, the error bound given in (11.31) is not readily computable. One way to determine the appropriate number of Fourier coefficients, m, is to monitor the magnitudes of the coefficients, g_k, which tend to decrease quickly with increasing index k. A tolerance on the coefficient magnitude can thus be used to set a stopping criterion for the iterative procedure. Note that the exact distribution of coefficients will depend upon the transfer function under consideration. For near resonant systems, the magnitudes of the Fourier coefficients will not decrease quickly and many coefficients will be needed to gain an accurate representation. For such systems, other algorithms, such as POD or Krylov subspace methods, may yield better results.

11.4 Active Flow Control of a Supersonic Diffuser

Figure 11.2 shows the contours of Mach number at steady state conditions inside the fixed geometry of a supersonic diffuser that operates at a freestream Mach number of 2.2. In steady state operation, a shock forms downstream of the throat. In practice, the incoming supersonic flow is subject to perturbations, such as atmospheric density disturbances. Such perturbations in the flow may cause the shock to move upstream of the throat, and eventually to be expelled from the diffuser. This phenomenon, known as inlet unstart, causes huge losses in engine performance and thus is highly undesirable. In order to prevent inlet unstart for this diffuser, an active control mechanism of the shock is required.

Figure 11.3 presents a schematic of the actuation mechanism. Incoming flow with possible disturbances enters the inlet and is sensed using pressure sensors. The controller then adjusts the bleed upstream of the throat in order to control the position of the shock and to prevent it from moving upstream. In simulations, it is difficult to automatically determine the shock location. The average Mach number at the diffuser throat provides an appropriate

Figure 11.2. *Contours of Mach number for steady flow through supersonic diffuser. Steady state inflow Mach number is* 2.2.

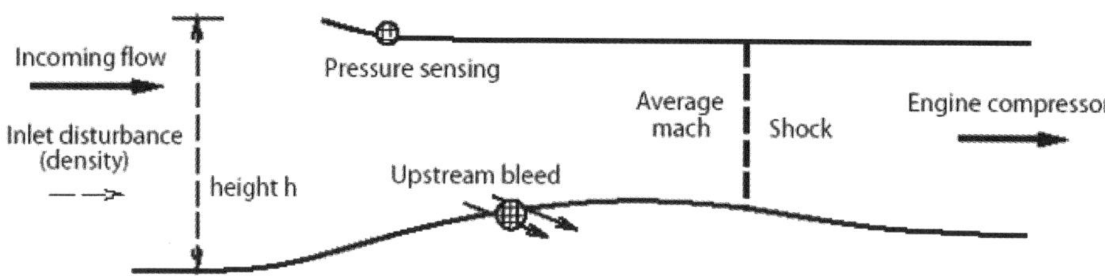

Figure 11.3. *Supersonic diffuser active flow control problem setup.*

surrogate that can be easily computed. A CFD model provides an accurate representation of the complicated flow dynamics but is not suitable for controller design purposes.

The governing equations considered are the two-dimensional unsteady Euler equations, linearized about steady state conditions. The CFD model considered here has 3078 grid points and 11,730 unknowns and is described in [16]. The first transfer function of interest is that between bleed actuation and average Mach number at the throat. Bleed occurs through small slots located on the lower wall between 46% and 49% of the inlet overall length. Frequencies of practical interest lie in the range $f/f_0 = 0$ to $f/f_0 = 2$, where $f_0 = a_0/h$, a_0 is the freestream speed of sound and h is the height of the diffuser.

Figure 11.4 shows the magnitude and phase of this transfer function as calculated by the CFD model and three reduced-order models each of size $k = 10$. The FMR model was calculated by using 201 Fourier coefficients (calculated at the cost of a single CFD matrix inversion) with $\omega_0 = 5$ to construct the Hankel matrix in (11.41). This 200th-order system was then further reduced to 10 states using explicit balanced truncation. The POD model was obtained by computing 41 snapshots at 21 equally spaced frequencies from $f/f_0 = 0$ to $f/f_0 = 2$. For comparison, a tenth-order Arnoldi-based model, derived using the methodology in [16], is also shown in Figure 11.4.

It can be seen from Figure 11.4 that the FMR model matches the CFD results well over the entire frequency range plotted, with a small discrepancy at higher frequencies. The Arnoldi model matches well for low frequencies but shows considerable error for $f/f_0 > 1.3$. The POD model has some undesirable oscillations at low frequencies, and

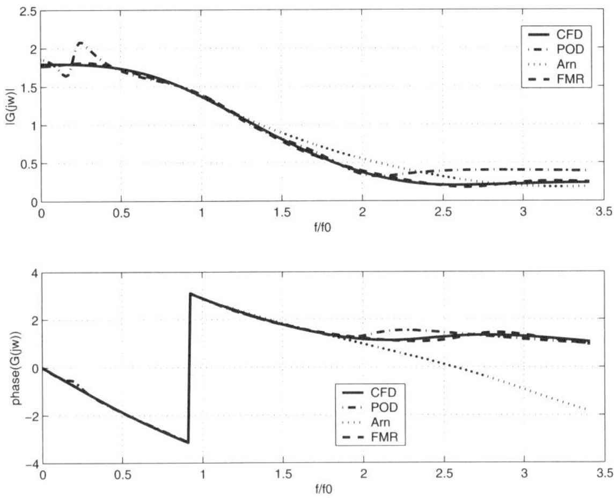

Figure 11.4. *Transfer function from bleed actuation to average throat Mach number for supersonic diffuser. Results from CFD model ($n = 11,730$) are compared to FMR, POD, and Arnoldi models with $k = 10$ states.*

strictly is only valid over the frequency range sampled in the snapshot ensemble ($f/f_0 < 2$).

The performance of the POD and Arnoldi models can be improved by increasing the size of the reduced-order models. Figure 11.5 shows the results using 30 Arnoldi vectors and 15 POD basis vectors. The agreement at low frequencies is now very good for all models, but the POD and Arnoldi models still show discrepancy at higher frequencies. The POD model could be further improved by including more snapshots in the ensemble; however, each additional frequency considered requires an nth-order complex matrix inversion. The Arnoldi model could be further improved by increasing the size of the basis; however, this was found to result in unstable reduced-order models. This result highlights one of the major problems with the commonly used POD and Krylov-based reduction techniques. Because no rigorous statement about model quality can be made, the reduction becomes an ad hoc process that requires trial and error to obtain accurate, stable reduced-order models. In particular, for the POD, the choice of snapshot ensemble is critical.

FMR is also applied to the transfer function between an incoming density perturbation and the average Mach number at the diffuser throat. This transfer function represents the dynamics of the disturbance to be controlled and is shown in Figure 11.6. As the figure shows, the dynamics contain a delay and are thus more difficult for the reduced-order model to approximate. Results are shown for FMR with $m = 200$ and $\omega_0 = 5, 10$. With $\omega_0 = 5$, the model has significant error for frequencies above $f/f_0 = 2$. Choosing a higher value of ω_0 improves the fit, although some discrepancy remains. These higher frequencies are unlikely to occur in typical atmospheric disturbances; however, if they are thought to be important, the model could be further improved by either evaluating more Fourier coefficients, or by choosing a higher value of ω_0. The $\omega_0 = 10$ model is further reduced via balanced truncation to a system with 30 states without a noticeable loss in accuracy.

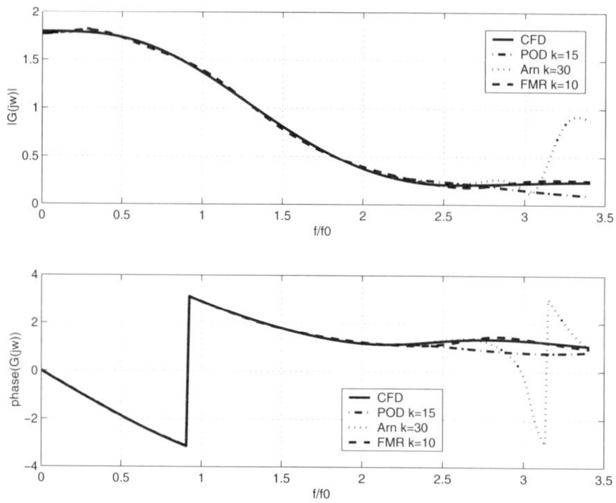

Figure 11.5. *Transfer function from bleed actuation to average throat Mach number for supersonic diffuser. Results from CFD model ($n = 11,730$) are compared to FMR with $k = 10$ states, POD with $k = 15$ states, and Arnoldi with $k = 30$ states.*

11.5 Conclusion

Model reduction is an essential component to achieving real-time simulation and control of PDEs. Several model reduction techniques are available and in use across a broad range of applications. For large-scale systems, the POD and Krylov-based methods have been used with considerable success. The three orders of magnitude reduction from $O(10^4)$ states to $O(10^1)$ states demonstrated in the supersonic diffuser example is representative of what can be achieved in many applications. However, it is important to note that many open questions and unresolved issues remain. The popular reduction approaches for large-scale systems do not offer rigorous guarantees regarding the quality of the reduced-order model and remain ultimately ad hoc. FMR goes some way to addressing this issue by providing a rigorous, although noncomputable, error bound. Using this error bound to combine FMR with a more rigorous technique, such as balanced truncation, yields an efficient, systematic, two-step reduction process.

In addition, reduction of large-scale nonlinear systems remains an open question. Direct projection of the nonlinear governing equations onto a reduced subspace yields a model that has low order but that cannot be implemented efficiently. A trajectory piecewise linear approach that can be used efficiently in conjunction with large-scale reduction methods has been proposed and demonstrated for integrated circuit applications [17]. This approach has also been shown to combine effectively with the POD for nonlinear CFD applications [18]. Many open questions remain regarding the robustness and rigor of this approach; however, it represents a significant step towards achieving efficient, low-order, nonlinear models.

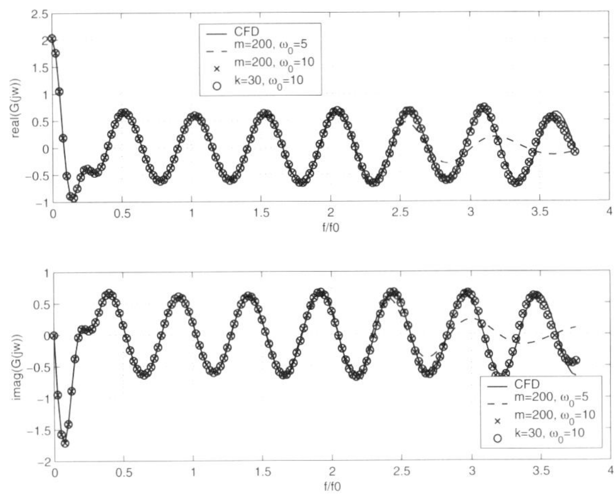

Figure 11.6. *Transfer function from incoming density perturbation to average throat Mach number for supersonic diffuser. Results from CFD model ($n = 11,730$) are compared to 200th-order FMR models with $\omega_0 = 5, 10$. The $\omega_0 = 10$ model is further reduced to $k = 30$ via balanced truncation.*

Bibliography

[1] P. HOLMES, J. L. LUMLEY, AND G. BERKOOZ, *Turbulence, Coherent Structures, Dynamical Systems and Symmetry*, Cambridge University Press, Cambridge, UK, 1996.

[2] J. L. LUMLEY, *The structures of inhomogeneous turbulent flow*, in Atmospheric Turbulence and Radio Wave Propagation, Nauka, Moscow, 1967, pp. 166–178.

[3] M. LOÈVE, *Probability Theory*, Van Nostrand, New York, 1955.

[4] H. HOTELLING, *Analysis of a complex of statistical variables with principal components*, J. Educational Psychology, 24 (1933), pp. 417–441, 498–520.

[5] G. BERKOOZ, P. HOLMES, AND J. L. LUMLEY, *The proper orthogonal decomposition in the analysis of turbulent flows*, Annu. Rev. Fluid Mech., 25 (1993), pp. 539–575.

[6] L. SIROVICH, *Turbulence and the dynamics of coherent structures. Part 1: Coherent structures*, Quart. Appl. Math., 45 (1987), pp. 561–571.

[7] T. KIM, *Frequency-domain Karhunen-Loeve method and its application to linear dynamic systems*, AIAA J., 36 (1998), pp. 2117–2123.

[8] K. WILLCOX, J. PERAIRE, AND J. WHITE, *An Arnoldi approach for generation of reduced-order models for turbomachinery*, Comput. & Fluids, 31 (2002), pp. 369–89.

[9] E. H. DOWELL AND K. C. HALL, *Modeling of fluid-structure interaction*, Annu. Rev. Fluid Mech., 33 (2001), pp. 445–90.

[10] B. C. Moore, *Principal component analysis in linear systems: Controllability, observability, and model reduction*, IEEE Trans. Automat. Control, 26 (1981), pp. 17–31.

[11] D. F. Enns, *Model Reduction for Control System Design*, Ph.D. thesis, Department of Aeronautics and Astronautics, Stanford University, Stanford, CA, 1984.

[12] S. Lall, J. E. Marsden, and S. Glavaski, *A subspace approach to balanced truncation for model reduction of nonlinear control systems*, Internat. J. Robust Nonlinear Control, 12 (2002), pp. 519–535.

[13] K. Willcox and J. Peraire, *Balanced model reduction via the proper orthogonal decomposition*, AIAA J., 40 (2002), pp. 2323–2330.

[14] K. Willcox and A. Megretski, *Fourier series for accurate, stable, reduced-order models in large-scale linear applications*, SIAM J. Sci. Comput., 26 (2005), pp. 944–962.

[15] G. Gu, P. P. Khargonekar, and E. B. Lee, *Approximation of infinite-dimensional systems*, IEEE Trans. Automat. Control, 34 (1989), pp. 610–18.

[16] G. Lassaux, *High-Fidelity Reduced-Order Aerodynamic Models: Application to Active Control of Engine Inlets*, SM thesis, Department of Aeronautics and Astronautics, MIT, Cambridge, MA, 2002.

[17] M. Rewienski, *A Trajectory Piecewise-Linear Approach to Model Order Reduction of Nonlinear Dynamical Systems*, Ph.D. thesis, Department of Electrical Engineering and Computer Science, MIT, Cambridge, MA, 2003.

[18] D. Gratton and K. Willcox, *Reduced-order, trajectory piecewise-linear models for nonlinear computational fluid dynamics*, AIAA 2004-2329, 2004.

Chapter 12

Suboptimal Feedback Control of Flow Separation by POD Model Reduction

Karl Kunisch and Lei Xie**

12.1 Introduction

Optimal design for fluid dynamical systems, which is a useful computational tool in aerodynamic drag reduction, turbulence delay, and combustion control, etc., has been an interesting research area for many years. This research stands at the crossroads of optimization (to carry out optimal design) and computational fluid dynamics. Many computational approaches and theoretical investigations were carried out in the past two decades, such as [6, 8, 12, 21], to name a few.

Generally speaking it is infeasible to carry out feedback design on infinite-dimensional fluid dynamical systems by numerically solving the dynamic programming equation (DPE) or the Hamilton–Jacobi–Bellman (HJB) equation. The development of *proper orthogonal decomposition* (POD) techniques in recent years provides a promising approach to perform suboptimal feedback design on reduced-order models. In this technique, a set of basis functions is constructed from preselected snapshots. Subsequently the fluid dynamic system is projected onto the space spanned by the basis functions, and the infinite-dimensional nonlinear partial differential equations (PDEs) of the fluid system are reduced to finite-dimensional ordinary differential equations (ODEs). Then the optimal design is performed on this reduced system at reasonable computational cost. This suboptimal approach has been applied for open-loop design on the one-dimensional Burgers equation in [13] and on the two-dimensional Navier–Stokes (NS) equations with different geometrical shapes in [9, 19, 20]. POD-based optimal control for distributed-parameter systems was also investigated in [16], [1, 2], for example. Ito and Schröter [11] and Kunisch and Volkwein [13]

*Department of Mathematics and Scientific Computing, University of Graz, A-8010 Graz, Austria.

implemented an approach to carry out feedback control for distributed-parameter systems on the basis of model reduction combined with solving the optimality systems for an array of open-loop problems and interpolating the results thus obtained to define the closed-loop synthesis. In [7] it was proved for the two-dimensional NS equations that the value function is the unique viscosity solution to the HJB equations. In [10] a suboptimal solution strategy for conservative systems (such as the NS equations) based on an ansatz for a solution to the HJB equation is proposed, and in the authors' previous work [14, 15] solving the HJB equations for a POD-based reduced system of the Burgers equation is investigated.

The prohibitive computational cost of feedback design for nonlinear systems strongly suggests the use of parallel computations. In this work several computers are exploited concurrently; furthermore, on different computers, a different number of MATLAB sessions is invoked, depending on the overall performance and the available computational power of each computer. The message passing interface based on MATLAB software is utilized to exchange data between MATLAB sessions on different computers. This parallel implementation reduces the computing time to an acceptable level.

The chapter is arranged as follows. The distributed volume control problem and the numerical NS solver are briefly described in Section 12.2. Then the POD technique is applied in Section 12.3 to reduce the NS equations to a set of nonlinear ODEs, on which feedback design will be carried out. Numerical methods to solve the discretized DPE are discussed in Section 12.4. The computational results from the volume control are presented in Section 12.5. In these sections, the methodology is described for the distributed volume control problem. But it can also be applied to the boundary control problem, which is addressed in Section 12.6.

12.2 Distributed Volume Control Problem

The dynamical system to be considered is incompressible viscous flow over a forward-facing step, with a uniform injection on the upper boundary of the channel. Let $\Omega \in R^2$ denote the flow domain with boundary Γ and let $y = (u, v)^T$ stand for the velocity. Then we have the NS equations in nondimensional form,

$$\begin{aligned} y_t - \tfrac{1}{Re}\Delta y + (y \cdot \nabla)y + \nabla p &= f, \\ \nabla \cdot y &= 0 \end{aligned} \tag{12.1}$$

in the domain $\Omega \times [0, \infty)$, where p denotes pressure, Re is the Reynold's number, and f is the body force. Associated with the above governing equation we have the following boundary conditions:

$$\begin{aligned} y = y_i \quad &\text{in} \quad \Gamma_i \times [0, \infty), \\ pn - \tfrac{1}{Re}\tfrac{\partial y}{\partial n} = 0 \quad &\text{in} \quad \Gamma_o \times [0, \infty), \\ y = c(t) \quad &\text{in} \quad \Gamma_c \times [0, \infty), \\ y = 0 \quad &\text{in} \quad \Gamma \backslash (\Gamma_i \cup \Gamma_o \cup \Gamma_c) \times [0, \infty), \end{aligned} \tag{12.2}$$

where Γ_i, Γ_o are the inflow and the outflow surfaces, respectively, Γ_c is the injection surface, $c(t)$ is the injection velocity, and n denotes the unit normal to the boundary. The boundary condition on the outflow boundary is the stress-free condition. The initial condition is $y(0) = y_0$.

In this distributed volume control problem, the right-hand side of the first component of the NS equations is set as a linear combination of a set of shape functions, $f = \Psi v(t)^T$, see Section 12.3, and the coefficients $v(t)$ are taken as the design variables. The cost functional to be minimized is defined as

$$\mathcal{J} = \frac{1}{2}\left(\int_0^\infty \int_{\Omega_o} e^{-\mu t}\|y - y_{Stokes}\|^2 d\Omega dt + \int_0^\infty \beta e^{-\mu t}\|v\|^2 dt\right), \tag{12.3}$$

where the positive constants β and μ are the weight coefficient and the discount rate, respectively. The first term in (12.3) is of tracking type. Here y_{Stokes} denotes the solution to the Stokes equation associated with (12.1) with otherwise the same geometric and data settings. Since the Stokes solution is separation-free, a reduction of the recirculation regions should be observed for the successfully controlled NS solution. By Ω_o we denote the observation region. The set of admissible controls is

$$U_{ad} = \{v \in L^2(0, T) : v_a \leq v \leq v_b\}.$$

The weak form of the NS equations is given by

$$\begin{aligned} \langle y_t + (y \cdot \nabla)y, \psi\rangle_\Omega + \tfrac{1}{Re}\langle \nabla y, \nabla \psi\rangle_\Omega - \langle \nabla \cdot \psi, p\rangle_\Omega &= \langle f, \psi\rangle_\Omega, \\ \langle \nabla \cdot y, q\rangle_\Omega &= 0 \end{aligned} \tag{12.4}$$

with boundary conditions

$$\begin{aligned} y &= y_i && \text{in } \Gamma_i \times [0, \infty), \\ y &= c(t) && \text{in } \Gamma_c \times [0, \infty), \\ y &= 0 && \text{in } \Gamma \backslash (\Gamma_i \cup \Gamma_o \cup \Gamma_c) \times [0, \infty) \end{aligned} \tag{12.5}$$

and initial condition

$$y(0) = y_0, \tag{12.6}$$

for all test function pairs $(\psi, q) \in V \times L^2(\Omega)$, where

$$V = \{v \in H^1(\Omega) : v = 0 \quad \text{on} \quad \Gamma \backslash \Gamma_o\}$$

and $\langle \cdot, \cdot \rangle_\Omega$ denotes the L_2 inner product over the flow domain Ω. This weak formulation will be utilized to obtain the reduced-order model for the dynamic system.

In the numerical NS solver, a finite difference method is applied to approximate the spatial differentiation over a nonuniform grid system, with an upwind scheme for the nonlinear convection term. A staggered grid system is applied; i.e., the horizontal velocities are located at the midpoints of the vertical cell edges, the vertical velocities are located at the midpoints of the horizontal cell edges, and the pressure is defined at each cell center. At the inflow boundary, only the horizontal velocity is specified as a parabolic distribution with maximum equal to 1. The flow is initialized with the stationary Stokes flow. For time integration, we used the fractional-step-θ-scheme as applied in [22]. A macrotime step from

t^n to $t^{n+1} = t^n + \Delta t$ is split into three substeps, $t^n \to t^{n+\theta} \to t^{n+1-\theta} \to t^{n+1}$:

$$\begin{aligned}
(I + \alpha\theta\Delta t D(y^{n+\theta}))y^{n+\theta} + \theta\Delta t\nabla p^{n+\theta} &= \\
(I - \beta\theta\Delta t D(y^n))y^n &+ \theta\Delta t f^n \\
\nabla \cdot y^{n+\theta} &= 0, \\
(I + \beta\theta'\Delta t D(y^{n+1-\theta}))y^{n+1-\theta} + \theta'\Delta t\nabla p^{n+1-\theta} &= \\
(I - \alpha\theta'\Delta t D(y^{n+\theta}))y^{n+\theta} &+ \theta'\Delta t f^{n+1-\theta} \\
\nabla \cdot y^{n+1-\theta} &= 0, \\
(I + \alpha\theta\Delta t D(y^{n+1}))y^{n+1} + \theta'\Delta t\nabla p^{n+1} &= \\
(I - \beta\theta\Delta t D(y^{n+1-\theta}))y^{n+1-\theta} &+ \theta\Delta t f^{n+1} \\
\nabla \cdot y^{n+1} &= 0,
\end{aligned} \tag{12.7}$$

where D denotes the diffusive and advective terms, $\theta = 1 - \frac{\sqrt{2}}{2}, \theta' = 1 - 2\theta$, and $\alpha = \frac{1-2\theta}{1-\theta}, \beta = 1 - \alpha$. This fractional-step scheme combines advantages from the classical Crank–Nicolson method (second-order accuracy) and the backward Euler method (strongly A-stable).

To avoid solving nonlinear equations and to ensure second-order accuracy in time, the nonlinear convective term $y^{n+1} \cdot \nabla y^{n+1}$ in the left-hand side of the first component of (12.7) is replaced by $(2y^n - y^{n-1}) \cdot \nabla y^{n+1}$, which is of second-order accuracy.

12.3 POD-Based Reduced-Order Model

It is not yet possible to carry out feedback optimal control directly on the full NS equations. A suboptimal alternative is to utilize a reduced-order model. In this section, the principles of proper orthogonal decomposition will be introduced to construct the reduced-order model for incompressible NS equations. In Section 12.3.1 POD will be briefly introduced. Then this principle is applied to the NS equation to get the reduced-order model, which is presented in Section 12.3.2. Section 12.3 describes the method to obtain the shape functions for the distributed volume control.

12.3.1 Proper Orthogonal Decomposition

POD emerges from choosing a "best" set of orthonormal basis functions from a given collection frequently referred to as the snapshots. It can be readily verified that this problem is equivalent to an eigenvalue problem. Readers are referred to [1, 2, 4, 16], and the references given there, for more details. We will only briefly sketch the methodology of POD and recapitulate its properties.

Let the preselected collection of snapshots be denoted by

$$S = \{S_j, j = 1, 2, \ldots, N | S_j \in H^1(\Omega)\}$$

and suppose that S has rank l, that is, $l = \dim\{span\{S\}\}$. If $\{\varphi_i\}_{i=1}^l$ is an orthonormal basis for S, each element in S can be expressed as $S_j = \sum_{i=1}^l \langle S_j, \varphi_i\rangle\varphi_i$ for $j = 1, 2, \ldots, N$. The orthonormal basis can be constructed in many ways. However, in POD we require the orthonormal basis to have the property of minimizing

$$\sum_{j=1}^{N} \left\| S_j - \sum_{i=1}^{k} \langle S_j, \varphi_i \rangle \varphi_i \right\|^2$$

for every $k \in \{1, 2, \ldots, l\}$. Introduce the bounded linear operator $\mathcal{S} : R^n \mapsto L_2$,

$$\mathcal{S}v = \sum_{j=1}^{N} v_j S_j.$$

Its adjoint $\mathcal{S}^* : L_2 \mapsto R^n$ is given by

$$\mathcal{S}^* z = (\langle z, S_1 \rangle, \ldots, \langle z, S_N \rangle)^T.$$

The necessary optimality condition for this constrained minimization problem requires that basis functions satisfy

$$\mathcal{S}\mathcal{S}^* \varphi = \lambda \varphi.$$

Since $\mathcal{S}\mathcal{S}^*$ is nonnegative and symmetric, its eigenvalues must be real and nonnegative.

In numerical implementation, we do not need to calculate the matrix $\mathcal{S}\mathcal{S}^*$; instead, the *singular value decomposition* (SVD) will be applied to $\mathcal{S}$ such that $\mathcal{S} = U\Sigma V^T$, and the POD basis function is given by [13]

$$\varphi_i = \frac{1}{\Sigma_i} \mathcal{S} v_i, \quad 1 \le i \le l, \tag{12.8}$$

where Σ_i is the ith singular value in Σ and v_i is the associated right eigenvector in V.

Let us reiterate that the POD basis is the optimal orthonormal basis approximating in the mean all snapshots at arbitrary truncation level. In view of our application to the NS equations let us note that the basis elements inherit the boundary and incompressibility condition of the snapshots.

Our next task is to properly construct the snapshots from which the basis functions will be deduced. There are several requirements for the snapshots: first, they must capture as much dynamics of the flow system as possible, since the basis functions inherit the information already contained in the snapshots; second, they should satisfy homogeneous boundary conditions; and third, the snapshots must be solenoidal for incompressible flow.

The immediate idea is to take the snapshots as the solutions of the NS equations at specified time instances. To satisfy the second requirement, a term accounting for the boundary conditions is subtracted from the solutions at each time instant. With these considerations in mind, we can propose the snapshots

$$\tilde{S}_k = \{y(t^k) - \bar{y}\}, \quad k = 1, 2, \ldots, N,$$

where $y(t^k)$ is the NS solution at $t = t^k$ and the mean flow $\bar{y}$ is given by

$$\bar{y} = \frac{1}{N} \sum_{k=1}^{N} y(t^k).$$

These snapshots satisfy homogeneous boundary conditions on the Dirichlet boundaries, and the stress-free boundary condition is also satisfied by the snapshots on the outflow boundary. Furthermore, to account for the effect of the nonuniform grid system, a diagonal matrix $\mathcal{M} \in R^{m\times m}$ is introduced, whose entries are the measures of the grid cells; hence the snapshot matrix now takes the following form:

$$S = \{\mathcal{M}^{\frac{1}{2}}\tilde{S}_1, \mathcal{M}^{\frac{1}{2}}\tilde{S}_2, \dots, \mathcal{M}^{\frac{1}{2}}\tilde{S}_N\}.$$

Then the SVD is applied to the snapshot matrix using the MATLAB subroutine *svds* to find the largest $\ell < N$ eigenvalues and the corresponding eigenvectors,

$$S^\ell = P_\ell \Sigma_\ell Q_\ell^T, \tag{12.9}$$

where $P_\ell \in R^{m\times\ell}$, $\Sigma_\ell \in R^{\ell\times\ell}$, and $Q_\ell \in R^{N\times\ell}$. Actually S^ℓ is the best rank-ℓ approximation of the snapshot matrix S; that is, we perform truncated SVD.

Let the column of Q_ℓ be $w_i, i = 1, 2, \dots, \ell$, and let the singular value in Σ_ℓ be denoted by $\sigma_i, i = 1, 2, \dots, \ell$. To determine the number of basis elements for the actual computations we choose $M \leq \ell$ such that $r = \Sigma_{k=1}^{M}\sigma_k^2 / \Sigma_{k=1}^{\ell}\sigma_k^2 > 99.5\%$. Then the basis functions can be expressed as

$$\varphi_i = \frac{\tilde{S} w_i}{\sigma_i}, \qquad i = 1, 2, \dots, M.$$

The ratio r is the percentage of the total energy captured in the first M POD basis functions and therefore is an index to measure how closely the M-dimensional subspace $\{\varphi_1, \varphi_2, \dots, \varphi_M\}$ approximates the snapshot set. In this way the basis functions are orthonormal with respect to the L_2-norm. Furthermore, they are solenoidal and satisfy homogeneous boundary conditions on the Dirichlet boundaries and the stress-free boundary condition on the outflow boundary. These properties are inherited from the snapshots. The solution to NS equations is therefore modeled by

$$\hat{y} = \bar{y} + \sum_{i=1}^{M} \alpha_i(t)\varphi_i, \tag{12.10}$$

with $\alpha_i(t)$ to be determined from a Galerkin procedure.

12.3.2 Reduced-Order Model

With the basis functions in hand, the reduced-order model can be derived by restricting the modeled solution (12.10) to the weak form (12.4) of the NS equations. Furthermore, recalling that the basis functions are solenoidal and satisfy homogeneous boundary conditions on the Dirichlet boundaries, we can arrive at the ODE governing α,

$$\dot{\alpha} = -\mathcal{A}\alpha - (\alpha^T H \alpha + P) + G, \tag{12.11}$$

where the components of matrices are determined as follows:

$$\mathcal{A}_{i,j} = \frac{1}{Re}\langle \nabla\varphi_j, \nabla\varphi_i\rangle + \langle \varphi_j \cdot \nabla\bar{y}, \varphi_i\rangle + \langle \bar{y} \cdot \nabla\varphi_j, \varphi_i\rangle,$$

$$H_{i,j,k} = \langle \varphi_j \cdot \nabla \varphi_k, \varphi_i \rangle,$$
$$P_i = \langle \bar{y} \cdot \nabla \bar{y}, \nabla \varphi_i \rangle + \frac{1}{Re} \langle \nabla \bar{y}, \nabla \varphi_i \rangle,$$
$$G_i = \langle \Psi v^T, \varphi_i \rangle.$$

The boundary integral term on the outflow boundary vanishes, since we chose stress-free outflow boundary conditions. The initial conditions for the reduced system can be derived from evaluating (12.10) at the initial time $t = 0$ by

$$y(0) = \bar{y} + \sum_{i=1}^{M} \alpha_i(0) \varphi_i,$$

which can be put in the following form:

$$\alpha_i(0) = \langle y_0 - \bar{y}, \varphi_i \rangle, \quad i = 1, 2, \ldots, M. \tag{12.12}$$

Classical Galerkin approximations are typically obtained via sets of test functions that are not related to the dynamic system. Hence the resulting finite element model is a system with a large number of unknowns. Contrary to that approach, the POD technique takes basis functions that reflect the system dynamics. In such a way, the reduced-order model can be of much smaller scale. The stiffness matrix is full in this case, while for finite element or finite difference approximations it is typically sparse. This reduced-order model is a nonlinear initial value problem, on which the feedback design will be carried out to minimize the cost functional,

$$\hat{\mathcal{J}} = \frac{1}{2} \int_0^T \int_\Omega e^{-\mu t} \| \hat{y} - y_{Stokes} \|^2 d\Omega dt + \frac{\beta}{2} \int_0^T e^{-\mu t} \| v \|^2 dt, \tag{12.13}$$

where $\hat{y}$ is given by (12.10), and the α_i's are the solutions to (12.11) with the initial condition (12.12).

12.3.3 Construction of Shape Functions

In the control problem, the distributed volume control is achieved by the linear combination of a set of shape functions, which are constructed in a similar way as the construction of the basis functions for the states. First, a set of snapshots is defined as

$$\mathcal{S} = y^0(t^k) - y_{Stokes},$$

where $y^0(t^k)$ is the time-dependent solution to the NS equations without any control, and y_{Stokes} is the Stokes solution; both these terms are under the action of the injection. The same method as stated in Section 12.3.1 is then followed to find the basis functions from these snapshots; and the first two of them are taken as the shape functions. With the snapshot constructed in this way, the regions are filtered out for the separation bubbles. Therefore, this method will be very efficient, as can be seen from the numerical tests to be presented.

12.4 Feedback Design for the Reduced Model

Feedback control can now be performed on the reduced ODE system. In this section, we first discretize the dynamic programming equation in the time and space domains. Then a multilevel acceleration method and a parallel computation method are briefly described. Finally, the method of retrieving the synthesis or the feedback law is introduced.

12.4.1 Discretization of DPE

For the optimal control problem

$$\min_{\nu \in U_{ad}} \quad J = \int_0^\infty e^{-\mu t} L(\alpha(t), \nu(t)) dt$$
$$\text{subject to} \quad \dot{\alpha} = F(\alpha(t), \nu(t)), \quad \alpha(0) = \alpha_0 \in R^M,$$

where $L : R^M \times R^K \mapsto R$ and $F : R^M \times R^K \mapsto R^M$, the value function can be defined as

$$\Im(\alpha_0) = \inf_{\nu \in U_{ad}} \int_0^\infty e^{-\mu t} L(\alpha, \nu) dt : R^M \mapsto R$$

for any initial value α_0. The principle of optimality by Bellman and a simple argument based on the semigroup property lead to the *dynamic programming equation* (DPE) (or *dynamic programming principle*) [3]

$$\Im(\alpha_0) = \inf_{\nu \in U_{ad}} \left\{ \int_0^{\Delta T} e^{-\mu t} L(\alpha(t), \nu(t)) dt + \Im(\alpha(\Delta T)) e^{-\mu \Delta T} \right\} \tag{12.14}$$

for any $\Delta T > 0$ and any $\alpha_0 \in R^M$.

To derive the discrete DPE, we assume that the control ν is constant in the time interval $[0, \Delta T]$. Thus we have to first-order accuracy

$$\begin{aligned} \Im(\alpha_0) = \inf_{\nu \in U_{ad}} \Big\{ & \big(L(\alpha(0), \nu) + e^{-\mu \Delta T} L(\alpha(\Delta T), \nu)\big) \frac{\Delta T}{2} \\ & + \Im(\alpha(\Delta T)) e^{-\mu \Delta T} \Big\}, \end{aligned} \tag{12.15}$$

where the trapezoid rule is applied for the integration of the cost functional on the interval $[0, \Delta T]$. Next, the problem arises of calculating $\alpha(\Delta T)$ from the initial condition $\alpha(0) = \alpha_0$. There are several candidates for this, such as the explicit Euler method, the modified Euler method (Heun method), and semi-implicit methods. In the numerical tests, an explicit scheme is applied based on a Runge–Kutta (4, 5) formula (the Dormand–Prince pair).

Then we introduce a bounded polyhedron $\Upsilon = [a_1, b_1] \times [a_2, b_2] \times \cdots \times [a_M, b_M]$. The intervals $[a_i, b_i]$ are chosen in such a way that we projected the solution of the uncontrolled NS equations onto the POD basis. This allows us to anticipate reasonable bounds for the controlled trajectories in POD space. Along each of its dimensions we distribute equidistant grid points to obtain a discretized polyhedron Υ_h. Let us denote these grid points by α_{0j}, $j = 1, 2, \ldots, \hbar$. Rewriting the semidiscretized DPE (12.15) at each grid point gives

$$\begin{aligned} \Im(\alpha_{0j}) = \inf_{\nu \in U_{ad}} \Big\{ & \big(L(\alpha_{0j}, \nu) + e^{-\mu \Delta T} L(\alpha_j(\Delta T), \nu)\big) \frac{\Delta T}{2} \\ & + \Im(\alpha_j(\Delta T)) e^{-\mu \Delta T} \Big\}, \end{aligned} \tag{12.16}$$

where $\alpha_j(\Delta T)$ is the trajectory at $t = \Delta T$ from the initial position α_{0j}.

The viability assumption is made for the polyhedron, i.e., for sufficiently small δt

$$x + \delta t F(x, u) \in \Upsilon \quad \forall y \in \Upsilon, \text{and} \quad \forall u \in U_{ad}.$$

Since for our control problem, where no a priori constraint is imposed on the system state, the value function is continuous over the whole domain [3, Chapter 3], we can find a piecewise linear approximate solution to (12.16). For example the value function can be approximated at $\alpha_j(\Delta T)$ by the convex combination

$$\Im(\alpha_j(\Delta T)) = \sum_{i=1}^{\hbar} \Lambda_{ij} \Im(\alpha_{0i}),$$

where the coefficients Λ_{ij} satisfy

$$0 \leq \Lambda_{ij} \leq 1 \quad \text{and} \quad \sum_{i=1}^{\hbar} \Lambda_{ij} = 1.$$

The construction of these coefficients is straightforward. In this way we arrive at the fully discretized DPE

$$\Im(\alpha_{0j}) = \inf_{v \in U_{ad}} \left\{ \big(L(\alpha_{0j}, v) + e^{-\mu \Delta T} L(\alpha_j(\Delta T), v)\big) \frac{\Delta T}{2} + e^{-\mu \Delta T} \sum_{i=1}^{\hbar} \Lambda_{ij} \Im(\alpha_{0i}) \right\} \tag{12.17}$$

for each $j = 1, 2, \ldots, \hbar$. It can be proved [3, Appendix A] that the above equation is equivalent to the fixed point problem

$$\Im(\alpha_0) = \mathcal{T}(\Im(\alpha_0))$$

with the map $\mathcal{T} : R^{\hbar} \mapsto R^{\hbar}$ defined by

$$\mathcal{T}(\alpha_{0_j}) = \inf_{v \in U_{ad}} \left\{ \big(L(\alpha_{0j}, v) + e^{-\mu \Delta T} L(\alpha_j(\Delta T), v)\big) \frac{\Delta T}{2} + e^{-\mu \Delta T} \sum_{i=1}^{\hbar} \Lambda_{ij} \Im(\alpha_{0i}) \right\}.$$

Close examination reveals that the map $\mathcal{T}$ is a contraction map with respect to the maximum norm in $R^{\hbar}$ with contraction coefficient given by $e^{-\mu \Delta T}$. Consequently, the discretized equation (12.17) has a unique solution, which can be found by the common technique of fixed point iteration.

12.4.2 Numerical Methods

The fixed point iteration begins at an initial guess to the value function, which is zero in our tests. After the constrained minimization problem on the right-hand side of (12.17) is solved, the minimum is set as the new guess to the value function. This iteration is repeated until the stopping criterion is satisfied. The convergence rate of the fixed point iteration depends on $\mu\Delta T$. When $\mu\Delta T$ approaches 0 the contraction coefficient $e^{-\mu\Delta T}$ tends to 1, and therefore only very slow convergence behavior will be observed. Falcone in [3, Appendix A] suggested several techniques to speed up convergence for the fixed point iteration. But they are too expensive to carry out in our numerical tests.

However, in this study by using a multilevel method, a different method for speed-up is carried out. In this strategy the discretized HJB equation (12.17) is first solved for $\Delta T = 0.2$ until the relative residual is decreased to less than a prescribed tolerance. These results are taken as the initial guess to solve the HJB equations for $\Delta T = 0.05$.

In solving the DPE, a constrained minimization problem must be solved independently at each point of the polyhedron. It is possible to perform these tasks in a data parallel fashion, running the same code simultaneously on different pieces of data on different processors. The message passing interface (MPI) [17] is used to send and receive data among processors. Interested readers are referred to [18] for additional information. The Gauss–Seidel scheme can also be used to accelerate convergence; i.e., the newest available unknowns are utilized to calculate the remaining unknowns at each iteration. But this technique does not lend itself to parallel computations.

In practical implementations, the slaves are first started remotely from the master computer via host-based authentication. Then the MATLAB sessions are started on the slaves. (On each slave, more than one MATLAB session can be initialized.) For the parallel computations to be performed, we need to transfer the required data to each MATLAB session via MPI. On receiving data, each session on different slaves can execute computations concurrently. The portion of the computational work, to be performed on each session, can be adjusted according to the performance of the slaves. After all computations are finished on all slaves, the data are collected from the MATLAB sessions.

12.4.3 Retrieving Optimal Control

After solving the discretized DPEs, we will have the value function $\Im(\alpha_{0j})$ and the corresponding control $\nu(\alpha_{0j})$ at each point of the polyhedron. The feedback control law or synthesis can be established through these data. Assuming the reduced-order state at a time instant t^k is known as $\alpha(t^k)$, this state can be located in the polyhedron to find out the coefficients Λ_j for the convex combination,

$$\alpha(t^k) = \sum_{j=1}^{\hbar} \Lambda_j \alpha_{0j}.$$

The optimal control corresponding to the state at time t^k is then set as

$$\nu^*(t^k) = \sum_{j=1}^{\hbar} \Lambda_j \nu(\alpha_{0j}).$$

Under the assumption that the optimal control is piecewise constant, the NS equations are integrated in the time interval from t^k to t^{k+1} to find out the state of the flow at t^{k+1} and hence the reduced-order state $\alpha(t^{k+1})$ at that time instance. This procedure is repeated until

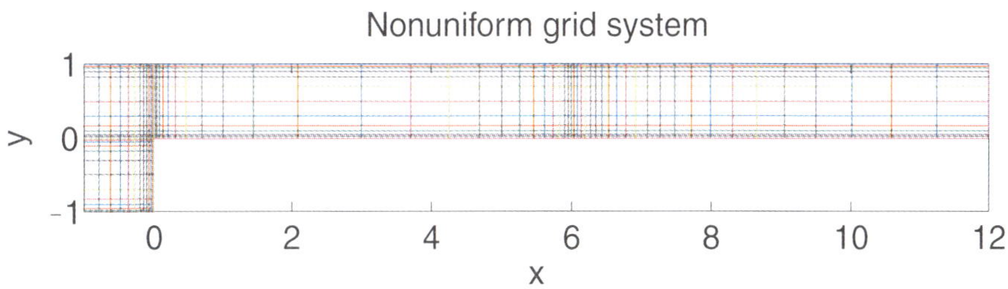

Figure 12.1. *Shape of the forward-facing step and the nonuniform grid system (shown in every other grid line).*

reaching the final time. In this way the optimal trajectory and the associated piecewise constant optimal control are determined.

12.5 Numerical Results from Volume Control

In the numerical experiments, a nonuniform 24×48 grid is chosen in the region $[-1, 0] \times [-1, 1]$ and a 96×24 grid for the region $[0, 12] \times [0, 1]$. The injection is located at $(6, 1)$ with width 0.125 and its uniform velocity is $(-1, -1)^T$. Figure 12.1 depicts the shape of the step and the nonuniform grid system used to solve the NS equations. The grid lines are concentrated toward the wall and the injection to better capture the separation bubbles. The Reynolds number is 1000 and the observation region Ω_o is $[-1, 12] \times [0, 1]$.

Figure 12.2 shows the streamlines of the desired flow, which is the Stokes flow obtained with otherwise the same data. The streamlines for the uncontrolled flow at time instant $t = 10$ are shown in Figure 12.3. There are separation bubbles around the step corner and behind the injection flow. (The separation in the corner around the point $(0, -1)$ is not shown here or in what follows, since it is not of concern from the point of view of design target.) The design target is to eliminate these two bubbles by tracking the desired Stokes flow. In the POD treatment, 6 basis functions are taken to capture more than 99.5% of the total energy. The distributed volume control is modeled by two shape functions, so that there are two design variables. The discount rate is $\mu = 2$. In the feedback design, there are 18480 grid points distributed in the six-dimensional polyhedron. These data are summarized in Table 12.1.

Six computers are utilized for the numerical experiments, on each of which a different number of MATLAB sessions is invoked. The finest time discretization is $\Delta T = 0.05$. The first computational case corresponds to $\beta = 0.1$. As shown in the first row of Table 12.2, the cost functional $\mathcal{J}$ is 5.1374×10^{-2} under no control; and is decreased to 2.9264×10^{-2} with optimal control. The value function, corresponding to the specified initial state, is 2.8089×10^{-2}. The agreement between them is satisfactory. (Theoretically, the value function should be equal to the cost functional with the optimal control.) The streamlines are depicted in Figure 12.4 for the optimally controlled flow at $t = 10$. Comparisons of Figure 12.3 and 12.4 demonstrate that the separation bubbles are substantially attenuated in the sense that the separation lengths are significantly decreased.

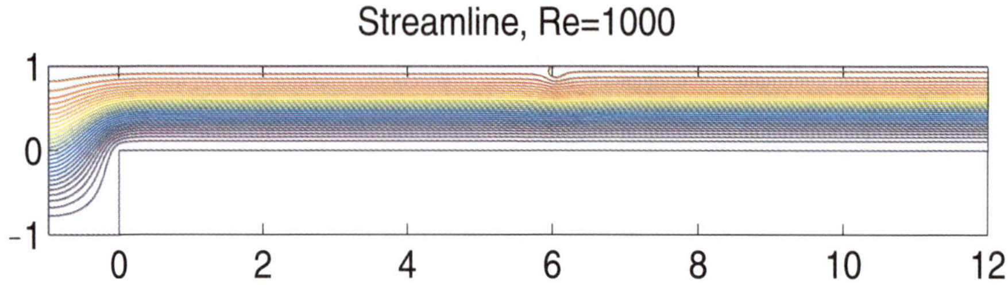

Figure 12.2. *Streamlines for the desired Stokes flow.*

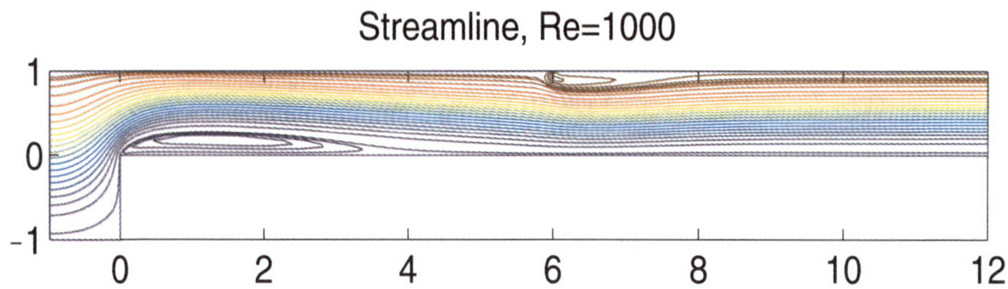

Figure 12.3. *Streamlines for the uncontrolled flow at $t = 10$.*

Table 12.1. *Parameter settings.*

Reynolds number, Re	1000
Time horizon, T	10
No. of basis functions	6
No. of controls	2
Bounds on controls	$[-0.5, 0.5]$
Grid system	$12 \times 6 \times 6 \times 4 \times 3 \times 3$
Discount rate, μ	2
Observation region, Ω_o	$[-1, 12] \times [0, 1]$

Table 12.2. *Computational results.*

	Cost w/o control	Cost w. opt. control	Value function
$\beta = 0.1$	5.1374e-2	2.9264e-2	2.8089e-2
$\beta = 0.05$	5.1374e-2	2.7547e-2	2.5762e-2

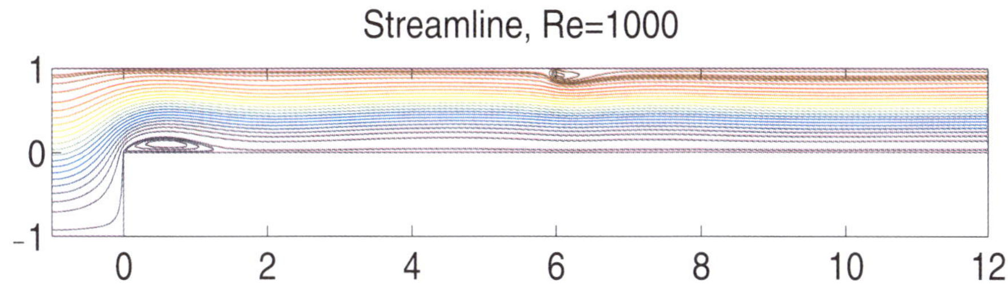

Figure 12.4. *Streamlines for the controlled flow at* $t = 10$, $\beta = 0.1$.

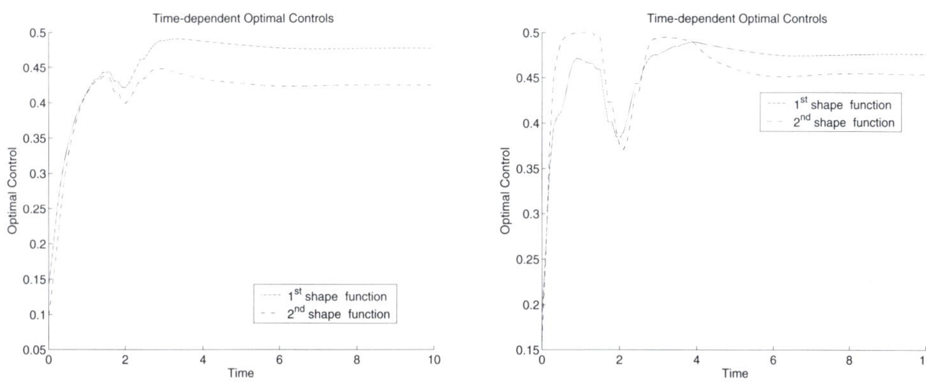

Figure 12.5. *Optimal control for the distributed volume control problem;* $\beta = 0.1$ *(left) and* $\beta = 0.05$ *(right).*

We also solve the DPE for $\beta = 0.05$. The computational results are listed in the second row of Table 12.2. Again we observe good agreement between the value function and the cost functional with optimal control. The streamlines for this case are not shown here, since they are similar to the streamlines for the case $\beta = 0.1$. As expected, the cost functional and the value function are smaller than the corresponding values for the case $\beta = 0.1$ where the controls are more expensive. Figure 12.5 presents the optimal control for two cases, showing that all controls approach steady state as time evolves. It is worthwhile to note the difference between two cases. In the case $\beta = 0.05$, the second control hits the upper bound in some time interval. However, the bounds are not active throughout the whole time horizon in the case $\beta = 0.1$. This is consistent, since we expect larger controls when the weighting coefficient β becomes smaller.

Let us briefly comment on the computational efforts that are involved in solving the discrete dynamic programming principle (12.14) over directly solving the optimization problem defined just above (12.14) for each grid point α_{0_i}. Obviously the infinite time horizon problems would have to be replaced by finite time horizons. For the grid system described in Table 12.1, this would require solving 18480 open-loop optimal control problems on "large" time horizons. Apparently this is numerically unfeasible.

The main motivation for the construction of feedback design is its use in the context

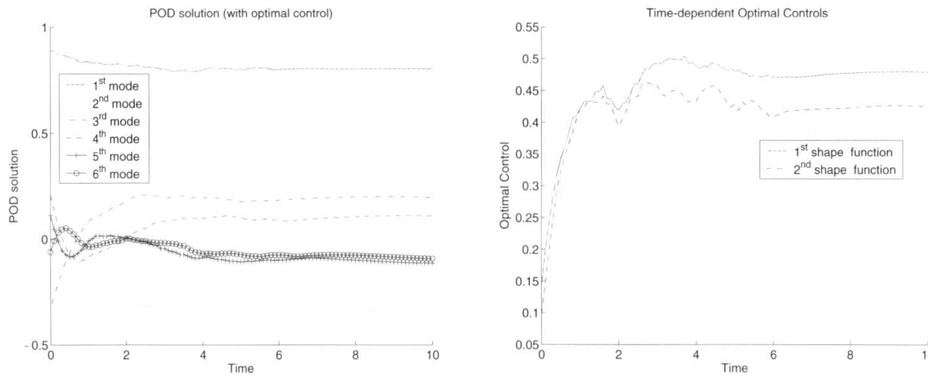

Figure 12.6. *POD solution and the optimal controls: with uniform noise in* $[-0.5, 0.5]$ *imposed on the right-hand side during the time interval* $[0, 6]$, $\beta = 0.1$.

of noise to the system or to data. We performed a test in which uniform noise with values in the interval $[-0.5, 0.5]$ is added to the right-hand side of the NS equations during the time interval $[0, 6]$. The optimal control subject to the noise can be easily retrieved from the feedback law which is already obtained from solving the DPE. At time level t^n the numerical solution of the NS equation is projected onto the POD basis to obtain the reduced state $\alpha(t^n)$. An optimal control is then determined by the interpolation procedure described in Section 12.4.3 and applied over the interval $[t^n, t^{n+1}]$. The POD solution and the optimal control are shown in Figure 12.6. The effect of the noise on the controls becomes apparent by comparing Figures 12.5 and 12.6. The controls in the latter are oscillatory up until $t = 6$ when the noise in the forcing function is deactivated. The controlled streamlines are indistinguishable from those in the noise-free case and are therefore not shown.

12.6 Feedback Design for Boundary Control

In this section the proposed methodology is extended to a Dirichlet boundary control problem, in which a vertical blowing/suction actuator is located at the surface $([0, 3], 0)$ to attenuate the separation bubble (as shown in Figure 12.7) in the flow over the forward-facing step of length 8. In this case no injection is set on the upper channel surface. Consequently we concentrate the grid lines only toward the wall surfaces. The actuator takes the following form:

$$v_c = \nu(t) \times \left(-\frac{0.05}{3}(x - 3) \right),$$

with $\nu(t)$ representing the time-dependent control. When ν is positive, the control is blowing, and otherwise it is suction. The cost functional to be considered in this design problem is

$$\mathcal{J} = \frac{1}{2} \left(\int_0^{\infty} \int_{\Omega_o} e^{-\mu t} \|u - u_{Stokes}\|^2 d\Omega dt + \beta \int_0^{\infty} e^{-\mu t} |\nu|^2 dt \right), \tag{12.18}$$

where u_{Stokes} is the u-component of the Stokes solution, and the observation window is $\Omega_o = [0, 3] \times [0, 0.5]$. The reason for using the u-component in the cost functional is that

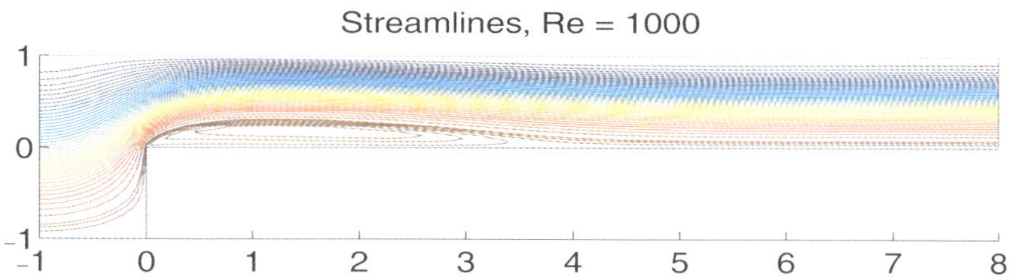

Figure 12.7. *Streamlines for the uncontrolled flow at* $t = 10$.

the separation is directly related to the negative u-component velocity and the control is vertical blowing/suction. The dynamical system is the NS equations (12.1) where f is set to 0, together with boundary and initial conditions as in (12.2), except for the inject condition. Once snapshots are chosen, we proceed in an analogous way described in the previous sections for the volume control. This concerns the construction of the basis elements, the Galerkin process for obtaining the reduced-order model, the approach of conducting feedback design, and the method of retrieving the optimal control. We just describe in detail the snapshot construction. We first take

$$\hat{S} = \{y(t^k)|_{\nu(\cdot)=0}, \quad y(t^k)|_{\nu(\cdot)=2)} - y_r\}, \quad k = 1, 2, \ldots, N,$$

where $y(t^k)_{\nu(\cdot)=0}$ and $y(t^k)_{\nu(\cdot)=2}$ denote the solution of the NS equation at time instant t^k without control and with constant control $\nu = 2$, respectively. The reference solution y_r is defined as

$$y_r = y_{Stokes}|_{\nu=2} - y_{Stokes}|_{\nu=0}$$

such that all the elements in the snapshot collection satisfy homogeneous boundary condition on the Dirichlet boundaries. Finally, the mean is subtracted from the snapshot set $\hat{S}$ resulting in

$$\tilde{S} = \{\hat{S} - y_m\},$$

with $y_m = \frac{1}{N}\sum_{k=1}^{N}\hat{S}_k$. After applying POD technique to obtain the basis functions φ_i, we can model the controlled NS solution as

$$\hat{y} = y_m + \sum_{i=1}^{M}\alpha_i\varphi_i + \nu(t)y_r.$$

The reduced-order model can now be obtained by a Galerkin process, namely, by restricting the modeled solution to the weak form (12.4). It is worthwhile to point out that the control now enters the modeled solution to account for the boundary conditions, differently from the reduced-order model solution (12.10) in the volume control case. As a result, the reduced-order ODEs have $M + 1$ components. In other words, given the same number of basis functions, the boundary control problem should deal with an ODE system with one more component than the distributed volume control problem. This is partly why the boundary control problem is more challenging.

In the design, the weight coefficient is set to $\beta = 0.001$ and the discount rate is $\mu = 0.5$. The Reynolds number, the time horizon, and the number of basis functions are set

Table 12.3. *Computational results for the boundary control problem.*

	Cost w/o control	Cost w. opt. control	Value function
$\beta = 0.001$	2.1674e-1	1.7096e-1	1.5920e-1

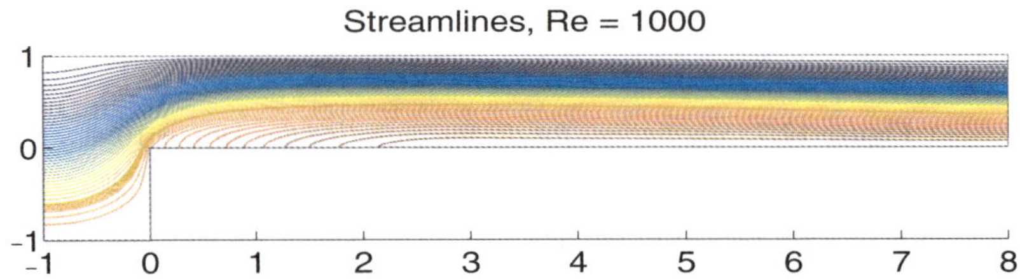

Figure 12.8. *Streamlines for the flow with optimal boundary control, at* $t = 10$.

to the same values as in the previous design case. As shown in Table 12.3, the cost functional $\mathcal{J}$ is decreased from 0.21674 to 0.17096. Meanwhile the value function (0.15920) agrees well with the cost functional under the optimal control. The optimal control turns out to be blowing. Figures 12.7 and 12.8 present the streamlines at $t = 10$ of the uncontrolled and the optimally controlled flow, respectively. The design objective is fulfilled in the sense that the separation bubble disappears in the controlled flow. The local flow pattern experiences large changes, which result in a relative small reduction in the cost functional.

We also tested a parabolic shape function for the spatial part of the actuator. Such a shape function is less effective in the sense that the reduction in the cost functional is smaller and that a small separation bulb still remains around the corner.

Acknowledgment

The support by the Fonds zur Förderung der wissenschaftlichen Forschung under SFB 03 "Optimierung und Kontrolle" is acknowledged.

Bibliography

[1] J. A. Atwell, *Proper Orthogonal Decomposition for Reduced Order Control of Partial Differential Equations*, Ph.D. thesis, Virginia Tech, Blacksburg, VA, 2000.

[2] J. A. Atwell and B. B. King, *Proper orthogonal decomposition for reduced basis feedback controllers for parabolic equations*, Math. Comput. Modelling, 33 (2001), pp. 1–19.

[3] M. Bardi and I. C.-Dolcetta, *Optimal Control and Viscosity Solutions of Hamilton-Jacobi-Bellman Equations*, in Systems & Control: Foundations & Applications, Birkhäuser, Boston, 1997.

[4] M. Fahl, *Computation of POD basis functions for fluid flows with Lanczos methods*, Math. Comput. Modelling, 34 (2001), pp. 91–107.

[5] T. FURLANI, *Introduction to Parallel Computing at CCR*, http://www.ccr.buffalo.edu/documents/CCR-parallel_process_intro.pdf.

[6] A. V. FURSIKOV, M. D. GUNZBURGER, AND L. S. HOU, *Boundary value problems and optimal boundary control for the Navier–Stokes system: The two-dimensional case*, SIAM J. Control Optim., 36 (1998), pp. 852–894.

[7] F. GOZZI, S. S. SRITHARAN, AND A. ŚWIĘCH, *Viscosity solutions of dynamic-programming equations for the optimal control of the two-dimensional Navier-Stokes equations*, Arch. Ration. Mech. Anal., 163 (2002), pp. 295–327.

[8] R. M. HICKS AND P. A. HENNE, *Wing design by numerical optimization*, in Proceedings of the AIAA Aircraft System and Technology Conference, AIAA 77-1247, Seattle, WA, 1977.

[9] M. HINZE AND K. KUNISCH, *Three control methods for time-dependent fluid flow*, Flow, Turbulence and Combustion, 65 (2000), pp. 273–298.

[10] K. ITO AND S. KANG, *A dissipative feedback control synthesis for systems arising in fluid dynamics*, SIAM J. Control Optim., 32 (1994), pp. 831–854.

[11] K. ITO AND J. D. SCHRÖTER, *Reduced order feedback synthesis for viscous incompressible flows*, Math. Comput. Modelling, 33 (2001), pp. 173–192.

[12] A. JAMESON, *Aerodynamic design via control theory*, J. Sci. Comput., 3 (1988), pp. 233–260.

[13] K. KUNISCH AND S. VOLKWEIN, *Control of Burgers' equation by reduced order approach using proper orthogonal decomposition*, J. Optim. Theory Appl., 102 (1999), pp. 345–371.

[14] K. KUNISCH, S. VOLKWEIN, AND L. XIE, *HJB-POD-based feedback design for the optimal control of evolution problems*, SIAM J. Appl. Dyn. Syst., 3 (2004), pp. 701–722.

[15] K. KUNISCH AND L. XIE, *POD-based feedback control of the Burgers equation by solving the evolutionary HJB equation*, Comput. Math. Appl., 49 (2005), pp. 1113–1126.

[16] H. V. LY AND H. T. TRAN, *Modelling and control of physical processes using proper orthogonal decomposition*, Math. Comput. Modelling, 33 (2001), pp. 223–236.

[17] E. HEIBERG, *Matlab Parallization Toolkit*, http://hem.passagen.se/einar_heiberg/

[18] http://mpi-forum.org/

[19] S. S. RAVINDRAN, *Reduced-order adaptive controllers for fluid flows using POD*, J. Sci. Comput., 15 (2002), pp. 457–478.

[20] S. S. RAVINDRAN, *Control of flow separation over a forward-facing step by model reduction*, Comput. Methods Appl. Mech. Engrg., 191 (2002), pp. 4599–4617.

[21] S. S. Sritharan, ed., *Optimal Control of Viscous Flow*, SIAM, Philadelphia, 1998.

[22] S. Turek, *A comparative study of some time-stepping techniques for the incompressible Navier-Stokes equations: From fully implicit nonlinear schemes to semi-implicit projection methods*, Internat. J. Numer. Methods Fluids, 22 (1996), pp. 987–1011.

Part IV

Applications

Chapter 13

A Combined Shape-Newton and Topology Optimization Technique in Real-Time Image Segmentation

*Michael Hintermüller**

13.1 Introduction and Motivation

Recent computational challenges in *image segmentation* are due to image guided surgery, in particular brain surgery. In a preoperative phase high resolution images, e.g., magnetic resonance images (MRIs), of a patient are taken, enhanced (denoised, deblurred, etc.), and then used for a three-dimensional rendering of the region of interest which includes a segmentation phase for giving a precise account of tumor boundaries. Based on this reconstruction the surgeon plans the operative intervention. In brain surgery, where high precision is required, any intervention changes the local situation, e.g., the tumor location. Therefore, in modern regimes [8, 19, 20] a correction of the segmented image is attempted intraoperatively by taking new MRI scans and redoing the image processing part. This should ideally reduce the operation time and accurately guide the surgeon. Besides certain modeling aspects, on the computational side this poses several challenges such as fast and reliable segmentation.

In image segmentation, which is the task of identifying boundary curves (contours) of regions of approximately homogeneous features, there are several paradigm models such as edge detectors, Mumford–Shah model, shape-based approaches, discrete approaches, etc., which address different aspects in the segmentation. In this chapter we consider a variational approach based on edge detectors. In this case, a widely used model problem is given by

$$\min J(\Gamma) = \int_{\Gamma} g \, dS + \nu \int_{\Omega} g \, dx, \tag{13.1}$$

*Department of Mathematics and Scientific Computing, University of Graz, A-8010 Graz, Austria.

where $\Gamma \in \mathcal{D}$ denotes the contour and $\mathcal{D}$ a set of shapes, g is a given edge detector defined on the image domain $D \subset \mathbb{R}^2$, $\Omega \subseteq D$ is the set of homogeneous features with boundary Γ, and $\nu \in \mathbb{R}$ acts as a regularization parameter. In Figure 13.1, for the contrast agent based kidney image (left plot) we provide a standard edge detector based on Gaussian smoothing $G \star \tilde{I}$ of the intensity map $\tilde{I}$ of the image (right plot). In general, at ideal edges we have $g = 0$; otherwise $g > 0$. However, due to rest noise in the image, we can usually only expect $g \approx 0$. Let us mention that more sophisticated edge detectors compared to the one presented here are available.

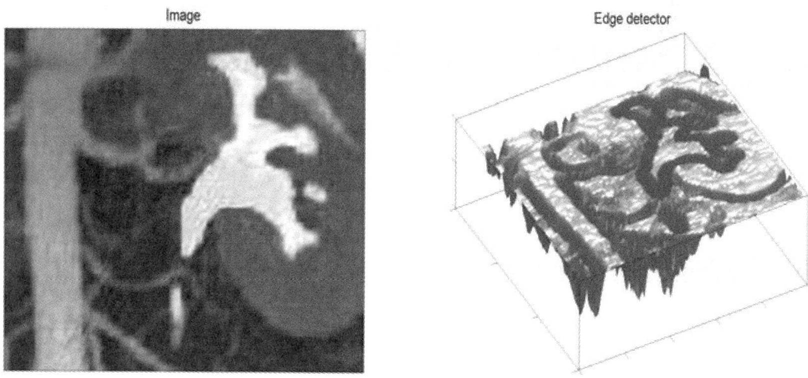

Figure 13.1. *Image intensity map $\tilde{I}$ (left) and corresponding edge detector $g = 1/(1 + \|\nabla(G \star \tilde{I})\|_2^2)$ (right).*

Active contour approaches in image segmentation consider Γ as the optimization variable which is adjusted iteratively. Earlier accounts for solving (13.1) were based on parameterizations of Γ, like in [10]. A particular choice was given by the arc length parameterization; see, e.g., [21]. These techniques, however, have several drawbacks including the need for expensive reparameterizations in case of topological changes. Borrowing a concept from Osher and Sethian [13] and with the aim of devising a geometrically intrinsic, i.e., parameter-independent approach, Caselles et al. [4] proposed the representation of Γ as the zero-level set of a function φ and the approximation of the desired contour by means of an iterative time-marching process. Based on some initial estimate Γ_0, the evolution towards the solution contour is embedded in the propagation of the continuous level-set function $\varphi : D \times [0, T] \to \mathbb{R}$ as

$$\Gamma(t) = \{x \in D : \varphi(x, t) = 0\}.$$

For this reason φ is sometimes called geometrical implicit function, since it implicitly contains the information of interest [12]. Let us assume that the components of $\Gamma(t)$ are closed curves, and let $\Omega(t)$ denote the domain with boundary $\Gamma(t)$. Then, in order to have a well-conditioned representation of $\Gamma(t)$ one usually requires

$$\Omega(t) = \{x \in D : \varphi(x, t) < 0\} \quad \text{and} \quad D \setminus \Omega(t) = \{x \in D : \varphi(x, t) \geq 0\}. \tag{13.2}$$

In this case, the outward unit normal $n(x,t)$ to $\Gamma(t)$ at $x \in \Gamma(t)$ is given by

$$n(x,t) = \frac{\nabla\varphi(x,t)}{\|\nabla\varphi(x,t)\|_2}$$

whenever φ is sufficiently smooth. A popular choice satisfying (13.2) is given by signed distance functions. In this case we have $\|\nabla\varphi(x,t)\|_2 = 1$ almost everywhere and, hence, $n(x,t) = \nabla\varphi(x,t)$. Since φ is usually defined on D, these definitions and properties naturally extend to D.

Now, let us assume that the velocity of the above-mentioned evolution in normal direction to $\Gamma(t)$ is given by $F(x(t),t)$ with F a scalar-valued function defined on $\Gamma(t)$. Then it is well known that the transport of φ is governed by the *level-set equation*

$$\varphi_t + F_{\text{ext}}\|\nabla\varphi\|_2 = 0 \text{ on } D; \tag{13.3}$$

see [12, 15]. Here F_{ext} denotes some suitable extension of F to D.

The *level-set method* is celebrated for its numerical robustness, flexibility with respect to topological changes, wide applicability, and many more attributes. In connection with velocity fields F coming from shape sensitivity analysis [5, 18], as for instance F being the negative shape gradient of J or the corresponding Newton direction [9], it is a versatile tool for shape optimization including image segmentation via (13.1). While the merging or splitting of existing contours can be handled easily by level-set-based shape optimization algorithms [2, 9], the creation of new components is in general hard (if not impossible) to accomplish by using classical shape sensitivity concepts only. As a remedy we propose the blending of shape sensitivity information with topological sensitivity. The latter concept goes back to [7, 17] and allows the creation of holes in a domain, rather than operating with existing boundaries only as it is the case with shape sensitivity analysis. For a given function $\mathcal{J} : \mathcal{D} \to \mathbb{R}$, where $\mathcal{D}$ is an appropriate set of shapes, the *topological derivative* (or *topological gradient*) at $x \in D$ is defined as

$$\mathcal{T}(x) := \lim_{\rho\downarrow 0} \frac{\mathcal{J}(D \setminus \overline{B_\rho(x)}) - \mathcal{J}(D)}{|\overline{B_\rho(x)}|}, \tag{13.4}$$

where $B_\rho(x) := \{y \in \mathbb{R}^n : \|y-x\| < \rho\} \cap D$, $\overline{B_\rho(x)}$ denotes the closure of $B_\rho(x)$, and $|\overline{B_\rho(x)}|$ the volume of $\overline{B_\rho(x)}$. In the case of two-dimensional image segmentation we have $n = 2$.

In this chapter, we study edge-detector-based image segmentation algorithms which utilize both shape as well as topological sensitivity information. Our technique is a phase-I–phase-II-type approach, which means that the shape sensitivity phase and the phase using the topological derivative are employed in a serial fashion; i.e., no synchronous blending of information takes place. This is similar to [3]. Typically, the algorithm is started by the topological phase for computing initial guesses. Then, if necessary, shape sensitivity is used to drive the contour. After this phase is stopped, one may again enter the topology optimization phase and then continue altering the phases until some stopping rule is satisfied.

An additional benefit of the new approach is given by the fact that the segmentation becomes fully automatic as opposed to state-of-the-art techniques which require a (more or less) manual selection of an initial contour. This reduction of required user interaction is an important feature in real-time applications.

The rest of the chapter is organized as follows. In Section 13.2 we derive a model for reconstructing topological information in a given image. The corresponding minimization problem involves a geometrically regularized least-squares objective function subject to a linear elliptic partial differential equation. Further, the topological gradient for the objective in the model problem is computed. The section finishes with the phase-I algorithm. First- and second-order shape sensitivity analysis and the phase-II algorithm, which is a shape-Newton method, are the subjects of Section 13.3. In the final section (Section 13.4) we address aspects of an efficient numerical realization and report on results obtained by the new method. It turns out that the topology optimization phase-I can act as a segmentation algorithm in its own right.

Notation. We use the standard notation $L^p(S)$ and $H^1(S)$, $H_0^1(S)$ for Lebesgue and Sobolev spaces, where $S \subset \mathbb{R}^n$ is some measurable set. By $(\cdot, \cdot)_S$ the standard $L^2(S)$ inner-product is denoted. In a slight misuse of notation we also write $(\cdot, \cdot)_S$ for vector-valued arguments. The norm on $H_0^1(S)$ is given by $\|\nabla v\|_{L^2(S)}$ for $v \in H_0^1(S)$. Moreover, $\|w\|_{L^2(S)}$, $w \in L^2(S)$, denotes the standard $L^2(S)$-norm; analogously for $\|w\|_{H^1(S)}$. Further, the measure of a set S is denoted by $|S|$.

13.2 Model and Its Topological Sensitivity

In this section we study the segmentation of an image by means of the topological derivative. Here we assume that images are given in terms of their intensity map I, which is also at the core of computing suitable edge detectors. We start by establishing an appropriate minimization problem and compute the topological derivative of its objective functional afterwards. We conclude this section by specifying the phase-I algorithm for topology optimization.

13.2.1 A Simple Model

Let us assume that $u_d \in H_0^1(D)$ is given data related to the intensity map $I : D \to [0, 1]$ of a gray-scale image. For simplicity we assume that the image intensities are scaled to [0,1], where features of interest have intensity 1, ideally. Since I might contain some rest noise, one option for computing u_d is given by solving

$$-\Delta u_d = I \text{ in } D, \quad u_d = 0 \text{ on } \partial D, \tag{13.5}$$

which can be interpreted as the convolution of the intensity map I with a Gaussian kernel. Our aim is now to find a function $u^* \in H_0^1(D)$ which is close to u_d. Then $-\Delta u^*$ is an approximation to the intensity map I. A natural choice realizing this objective is given by

$$\min \mathcal{J}_0(u, \Omega) := \frac{\beta_1}{2}\|u - u_d\|^2_{L^2(D)} + \frac{\beta_2}{2}\|u - u_d\|^2_{H_0^1(D)} \tag{13.6a}$$

$$\text{subject to } -\Delta u = f_\Omega \text{ in } D, \quad u = 0 \text{ on } \partial D, \tag{13.6b}$$

where $\mathcal{J}_0 : \mathcal{D} \to \mathbb{R}$, and $\beta_1, \beta_2 \in \mathbb{R}_0^+$, with $\beta_1 + \beta_2 > 0$. The right-hand side of the governing equation is defined as

$$f_\Omega(x) = \begin{cases} 1 & \text{if } x \in \Omega, \\ 0 & \text{else.} \end{cases}$$

Since the presence of noise in I may adversely affect the minimization of (13.6), we propose to add a regularization term in the form of a volume penalty to the objective function. Then the resulting model for recovering u^* is as follows:

$$\min \mathcal{J}_\mu(u,\Omega) := \mathcal{J}_0(u,\Omega) + \mu \int_\Omega 1dx \tag{13.7a}$$

$$\text{subject to } -\Delta u = f_\Omega \text{ in } D, \quad u = 0 \text{ on } \partial D, \tag{13.7b}$$

where $\mu \in \mathbb{R}_0^+$ is fixed.

The state equation (13.6b) admits a unique solution $u = u(\Omega)$. This fact can be used in problem (13.7) to derive a corresponding reduced problem:

$$\min \mathcal{J}(\Omega) = \mathcal{J}_\mu(u(\Omega),\Omega) \quad \text{over} \quad \Omega \in \mathcal{D}. \tag{13.8}$$

We consider (13.8) as our model problem for the topology optimization phase.

Before addressing topological sensitivity issues of $\mathcal{J}(\Omega)$, let us point out that different, more sophisticated approaches and models are possible.

13.2.2 Topological Derivative

Now we compute the topological derivative of $\mathcal{J}$. For this purpose note that we are not only interested in creating holes in Ω but also in adding components in $D \setminus \overline{\Omega}$. This is accomplished by considering for $x \in D \setminus \overline{\Omega}$

$$\mathcal{T}(x) = \lim_{\rho \downarrow 0} \frac{\mathcal{J}(\Omega \cup \overline{B_\rho(x)}) - \mathcal{J}(\Omega)}{|\overline{B_\rho(x)}|}. \tag{13.9}$$

Notice the difference from (13.4). In (13.9) we do not "subtract material"; rather we "add material" with the aim of changing the topology in $D \setminus \overline{\Omega}$. Strictly speaking, we have to consider $\mathcal{J}(D \cap (\Omega \cup \overline{B_\rho(x)}))$ and $|D \cap (\Omega \cup \overline{B_\rho(x)})|$ in (13.9).

Let us first recall a result from [3, Proposition 2.2] which is useful in the proof of our main result, Theorem 13.2.

Proposition 13.1. *Let $\Omega \subset D$ be some measurable domain with positive measure, and let $u_{\rho,x} \in H_0^1(D)$ and $u \in H_0^1(D)$ be the solutions of*

$$-\Delta u_{\rho,x} = f_{\Omega \setminus \overline{B_\rho(x)}} \quad \text{and} \quad -\Delta u = f_\Omega,$$

respectively. Then there exists a constant $C > 0$ such that

$$\|u_{\rho,x} - u\|_{H^1(\Omega)} \le C|B_\rho(x) \cap \Omega|^{1/s} \tag{13.10}$$

for $s \in (\frac{2n}{n+2}, +\infty]$.

Now we are prepared for proving the main result of this section.

Theorem 13.2. *Let $\Omega \subset D$ be some measurable domain with positive measure. Then the topological derivative of $\mathcal{J}$ in (13.8) is given by*

$$\mathcal{T}(x) = \begin{cases} -p(x) - \mu & \text{for a.a. } x \in \overline{\Omega}, \\ p(x) + \mu & \text{for a.a. } x \in D \setminus \overline{\Omega}, \end{cases} \tag{13.11}$$

where $p \in H_0^1(\Omega)$ solves the adjoint problem

$$-\Delta p = \beta_1(u - u_d) - \beta_2\Delta(u - u_d) \text{ in } D, \quad p = 0 \text{ on } \partial D. \tag{13.12}$$

Proof. We consider first $x \in \Omega$ and (13.4). Then

$$\begin{aligned}
\mathcal{J}(\Omega \setminus \overline{B_\rho(x)}) - \mathcal{J}(\Omega) &= \frac{\beta_1}{2}\|u_{\rho,x} - u\|^2_{L^2(D)} + \beta_1(u_{\rho,x} - u, u - u_d)_D \\
&\quad + \frac{\beta_2}{2}\|u_{\rho,x} - u\|^2_{H_0^1(D)} \\
&\quad + \beta_2(\nabla(u_{\rho,x} - u), \nabla(u - u_d))_D - \mu|\overline{B_\rho(x)}| \\
&= \frac{\beta_1}{2}\|u_{\rho,x} - u\|^2_{L^2(D)} + (\nabla p, \nabla(u_{\rho,x} - u))_D - \mu|\overline{B_\rho(x)}| \\
&= \frac{\beta_1}{2}\|u_{\rho,x} - u\|^2_{L^2(D)} + (f_{\Omega\setminus\overline{B_\rho(x)}} - f_\Omega, p)_D - \mu|\overline{B_\rho(x)}|,
\end{aligned}$$

where we used (13.12) in the next to the last inequality and the governing equation for the last identity. Using Proposition 13.1, we obtain

$$\lim_{\rho\downarrow 0}\frac{\mathcal{J}(\Omega \setminus \overline{B_\rho(x)}) - \mathcal{J}(\Omega)}{|\overline{B_\rho(x)}|} = -\lim_{\rho\downarrow 0}\frac{1}{|\overline{B_\rho(x))}|}\int_{\overline{B_\rho(x))}\cap D} p(z)\,dz - \mu.$$

From the localization theorem we infer

$$-\lim_{\rho\downarrow 0}\frac{1}{|\overline{B_\rho(x)}|}\int_{\overline{B_\rho(x)}\cap D} p(z)\,dz = -p(x)$$

and thus

$$\lim_{\rho\downarrow 0}\frac{\mathcal{J}(\Omega \setminus \overline{B_\rho(x)}) - \mathcal{J}(\Omega)}{|\overline{B_\rho(x)}|} = -p(x) - \mu.$$

For $x \in D \setminus \overline{\Omega}$ and (13.9) we obtain in a similar way

$$\mathcal{J}(\Omega \cup \overline{B_\rho(x)}) - \mathcal{J}(\Omega) = \frac{\beta_1}{2}\|u_{\rho,x} - u\|^2_{L^2(D)} + (f_{\Omega\cup\overline{B_\rho(x)}} - f_\Omega, p)_D + \mu|\overline{B_\rho(x))}|$$

and further

$$\begin{aligned}
\lim_{\rho\downarrow 0}\frac{\mathcal{J}(\Omega \cup \overline{B_\rho(x)}) - \mathcal{J}(\Omega)}{|\overline{B_\rho(x))}|} &= \lim_{\rho\downarrow 0}\frac{1}{|\overline{B_\rho(x))}|}\int_{\overline{B_\rho(x))}\cap D} p(z)\,dz + \mu \\
&= p(x) + \mu,
\end{aligned}$$

which completes the proof. □

According to (13.11) the topological gradient has a simple structure, since—up to an additive constant—it coincides with the adjoint. In general, the computational effort is related to the solution of (13.12) which can be obtained efficiently by, e.g., multigrid methods or FFT techniques. In the special case $\beta_1 = 0$ and $\beta_2 > 0$ we have

$$\mathcal{T}(x) = \begin{cases} \beta_2(u_d(x) - u(x)) - \mu & \text{for a.a. } x \in \overline{\Omega}, \\ \beta_2(u(x) - u_d(x)) + \mu & \text{for a.a. } x \in D \setminus \overline{\Omega}, \end{cases}$$

which shows that for computing $\mathcal{T}$ only the state equation has to be solved.

As it is the case for classical gradient techniques, the objective function $\mathcal{J}$ can be decreased if there exists a search (update) direction $\delta^{\mathcal{T}}$ such that for some sufficiently small $\delta_t > 0$

$$\mathcal{J}(\Omega(t+\delta_t)) - \mathcal{J}(\Omega(t)) < 0, \tag{13.13}$$

where the new domain $\Omega(t+\delta_t)$ is given by

$$\Omega(t+\delta_t) = \{x \in D : \varphi(x, t+\delta_t) < 0\}$$

with φ satisfying

$$\varphi_t(x,t) = \delta^{\mathcal{T}}(x,t). \tag{13.14}$$

It remains to address how (13.11)–(13.12) can be used if the geometry Γ is represented by a geometrical implicit function φ and such that (13.13) is realized. Throughout we assume that φ satisfies the sign convention (13.2). We further point out that the evolution of $\Gamma(t)$, respectively, $\Omega(t)$, is related to a time-dependent process via (13.3), respectively, (13.14). As a consequence we write $\mathcal{T}(x,t)$ for the topological derivative of $\mathcal{J}(\Omega(t))$ at $x \in D$. Following a reasoning similar to the one in [3], we find the following:

- For $x \in \Omega(t)$, i.e., for $\varphi(x,t) < 0$, a reduction of $\mathcal{J}(\Omega(t))(x)$ is expected if $\mathcal{T}(x,t) < 0$ which means that the topology should change; i.e., a hole should be created. Since $\varphi(x,t) < 0$, creating a hole can be accomplished by adding a positive quantity to $\varphi(x,t)$. Since $\mathcal{T}(x,t) < 0$, the search (update) direction can be chosen as

$$\delta^{\mathcal{T}}(x,t) = -\mathcal{T}(x,t) > 0 \quad \text{for } x \in \Omega(t). \tag{13.15}$$

- For $x \in D \setminus \overline{\Omega(t)}$, i.e., for $\varphi(x,t) > 0$, we again expect a reduction if $\mathcal{T}(x,t) < 0$. Changing the topology can now be achieved by subtracting a positive quantity from φ. This suggests

$$\delta^{\mathcal{T}}(x,t) = \mathcal{T}(x,t) < 0 \quad \text{for } x \in D \setminus \overline{\Omega(t)}. \tag{13.16}$$

From (13.11), (13.15), and (13.16) we infer

$$\delta^{\mathcal{T}}(x,t) = p(x) + \mu \qquad \text{for a.a. } x \in D,$$

where p solves the adjoint problem (13.12) with $u = (-\Delta)^{-1} f_{\Omega(t)} \in H_0^1(D)$. Note that in general we do not expect that $\|\delta^{\mathcal{T}}\|_{H^1(D)} = 0$ at the solution Ω^*, since the solution set Ω^* is invariant with respect to certain update directions. In fact, Ω^* remains unchanged if $\delta^{\mathcal{T}} \le 0$ on Ω^* and $\delta^{\mathcal{T}} \ge 0$ on $D \setminus \overline{\Omega^*}$.

13.2.3 Phase-I Algorithm for Topology Optimization

In Sections 13.2.1 and 13.2.2 we gathered all the tools needed for devising the algorithm for phase-I, i.e., the optimization with respect to the topology. Before we state the method we briefly discuss a strategy for adjusting the step length δ_t in (13.13). We propose a technique very similar to classical line search methods [11].

As a descent measure we propose with where χ_S denotes the characteristic function of a set S, and $(\cdot)_+ = \max(0, \cdot)$, $(\cdot)_- = \min(0, \cdot)$. Then starting with $\delta_t^0 > 0$ and setting $l = 0$ we check whether

$$\mathcal{J}(\Omega(t + \delta_t^l)) - \mathcal{J}(\Omega(t)) \le \omega \delta_t^l m(t) < 0, \tag{13.17}$$

where $0 < \omega < 1$ is fixed. If (13.17) holds true, then we set $\delta_t = \delta_t^l$; otherwise δ_t^l is reduced, e.g., by $\delta_t^{l+1} = q\delta_t^l$, with $0 < q < 1$ fixed, and (13.17) is checked again. For this procedure to work well, the initial δ_t^0 has to be chosen sufficiently large in order to achieve a topological change.

The algorithm is as follows.

Algorithm 13.3 (Phase-I Algorithm for Topology Optimization).

(i) Choose an initial contour Γ_0, with associated domain Ω_0, and the corresponding $\varphi(x, 0)$, $x \in D$; set $t_0 = 0$ and $k = 0$.

(ii) Solve for $u_k \in H_0^1(\Omega)$: $-\Delta u_k = f_{\Omega(t_k)}$.

(iii) Solve the adjoint problem (13.12) with $u = u_k$, and let p_k denote the solution. Determine $\delta_k^{\mathcal{T}}$ according to (13.15) and (13.16).

(iv) Choose $\delta_t^0 > 0$, and compute the first element of the sequence $\{\delta_t^l\}_{l=0}^{\infty} = \{\delta_t^0, q\delta_t^0, q^2\delta_t^0, \dots\}$ such that (13.17) is satisfied. Denote this stepsize by $\delta_{t,k}$.

(v) Update φ according to (13.14) with time step $\delta_{t,k}$; set $t_{k+1} := t_k + \delta_{t,k}$, $k := k + 1$, and go to (ii).

Several remarks concerning the algorithm are in order. To ease the notation we subsequently use $\mathcal{J}_k = \mathcal{J}(\Omega(t_k))$ and $\Omega_k = \Omega(t_k)$.

- Numerically we stop the algorithm if $\delta_{t,k} \le \epsilon_1$, or if the difference of two successive function values is small: $|\mathcal{J}_{k+1} - \mathcal{J}_k| \le \epsilon_2$. Here ϵ_1, ϵ_2 are user-specified tolerances which typically depend on the mesh size of the discretization.
- In our numerical tests we found that a stopping rule solely based on $m(t)$ is not adequate.
- There is no need for sophisticated initializations. Step (i) merely serves the purpose of computing a reasonably scaled φ which contains geometrical information and can be used in the updating process. One may choose φ to be a signed distance function initially. However, we point out that different choices satisfying (13.2) are possible.
- In contrast to level-set-based shape optimization methods there is no need to maintain φ to be a signed distance function in all iterations.
- In our tests we choose the line search parameters $\delta_t^0 = \max(1, \|\delta^{\mathcal{T}}\|_{H^1(D)}^{-1})$, $q = 0.5$, and $\omega = 0.001$.

13.3 Shape Sensitivity

Now we collect some results concerning the shape sensitivity of the objective functional of (13.1), and devise a level-set-based shape Newton algorithm for solving (13.1). In connection with Algorithm 13.3 this provides the phase-II method for shape optimization.

13.3.1 Shape Gradient and Shape Hessian

Suppose that we are given an open set $\Omega \subset \mathbb{R}^n$ with boundary $\Gamma = \partial\Omega$, and we would like to compute the sensitivity of

$$J(\Gamma) := J_1(\Gamma) + J_2(\Omega) = \int_\Gamma g\, dS + \nu \int_\Omega g\, dx \tag{13.18}$$

with respect to small perturbations of Ω and, hence, Γ. Further we assume that these perturbations are due to a sufficiently smooth vector field $V : \mathbb{R}^2 \to \mathbb{R}^2$ with compact support in $\mathbb{R}^2$. In a Lagrangian frame, the perturbations are given by

$$X'(\tau) = V(X(\tau)), \quad X(0) = x \in \mathbb{R}^2. \tag{13.19}$$

From this we define $P_V : \mathbb{R}^2 \times [0, +\infty) \to \mathbb{R}$ with

$$P_V(x, \tau) = X(\tau), \quad \text{and} \quad \Gamma(\tau) = P_V(\Gamma, \tau), \quad \Omega(\tau) = P_V(\Omega, \tau).$$

We have $\Gamma(0) = \Gamma$ and $\Omega(0) = \Omega$. Then the *Eulerian semiderivative* of J at Γ in direction V is defined by

$$dJ_1(\Gamma; V) = \lim_{\tau \downarrow 0} \frac{J(\Gamma(\tau)) - J(\Gamma)}{\tau}; \tag{13.20}$$

analogously one defines $dJ_2(\Omega; V)$. If $dJ_1(\Gamma; V)$ and $dJ_2(\Omega; V)$ exist for all $V \in B$, with B a Banach space of feasible velocity fields, and is linear and continuous in V, then J is called *shape differentiable*. By the Zolesio–Hadamard structure theorem [18, Theorem 2.27] we have that

$$dJ_2(\Omega; V) = \langle j_2(\Omega)n, V\rangle_{\mathcal{C}'(\Gamma),\mathcal{C}(\Gamma)},$$

where $\langle \cdot, \cdot\rangle_{\mathcal{C}'(\Gamma),\mathcal{C}(\Gamma)}$ denotes an appropriate duality pairing. Then the quantity $j_2(\Omega)n$ is called the *shape gradient* of J_2. Note that the structure theorem asserts that the shape gradient always admits a representation as a distribution on Γ. Since this is generically true for J_1, the shape gradient of J can be written as

$$j(\Gamma) = j_1(\Gamma) + j_2(\Omega)$$

with $j_1(\Gamma)$ the shape gradient of J_1.

In a similar fashion one defines the *second-order Eulerian semiderivative* of J at Γ in direction $W \in B$:

$$d^2J(\Gamma; V; W) = \lim_{\tau \downarrow 0} \frac{dJ(\Gamma(\tau); V) - dJ(\Gamma; V)}{\tau},$$

where $\Gamma(\tau) = P_W(\Gamma; \tau)$.

In [9] the first- and second-order Eulerian semiderivatives of the function in (13.18) were computed. We provide only the formulae and refer the reader to [9] for details. The first-order Eulerian derivative is given by

$$dJ(\Gamma; V) = \int_{\Gamma} \left(\frac{\partial g}{\partial n} + g(\kappa + \nu) \right) n \cdot V \, dS, \tag{13.21}$$

where κ denotes the mean curvature of Γ; see [5]. Thus, we identify

$$j(\Gamma) = \frac{\partial g}{\partial n} + g(\kappa + \nu).$$

The second-order Eulerian derivative, or shape Hessian, is found to be

$$d^2 J(\Gamma; V_F; V_G) = \int_{\Gamma} \left(\frac{\partial^2 g}{\partial n^2} + (2\kappa + \nu) \frac{\partial g}{\partial n} + \nu\kappa g \right) FG + g(\nabla_\Gamma F \cdot \nabla_\Gamma G) \, dS. \tag{13.22}$$

Above we used velocity fields $V_F, V_G \in B$ satisfying

$$V_F(x, t) = F_{\text{ext}}(x, t) \frac{\nabla\varphi(x, t)}{\|\nabla\varphi(x, t)\|_2}, \tag{13.23}$$

with the properties

$$\nabla F_{\text{ext}}(x, t) \cdot \frac{\nabla\varphi(x, t)}{\|\nabla\varphi(x, t)\|_2} = 0, \qquad F_{\text{ext}}|_\Gamma = F; \tag{13.24}$$

analogously for V_G. Above, φ denotes a geometrical implicit function satisfying (13.2). Therefore,

$$(V_F \cdot n)_{|\Gamma} = F \quad \text{and} \quad (V_G \cdot n)_{|\Gamma} = G. \tag{13.25}$$

This choice is motivated by two facts: (i) Analytically it makes nonsymmetric terms in the shape Hessian vanish; for details see [9, p. 454]. (ii) Extensions of velocities originally defined only on Γ to a neighborhood are frequently chosen to be constant in normal direction; see [1, 9].

Further, $\nabla_\Gamma v$ denotes the tangential gradient of v and is defined by

$$\nabla_\Gamma v_{|\Gamma} = (\nabla v)_{|\Gamma} - \frac{\partial v}{\partial n} n.$$

The term $(\nabla_\Gamma F, \nabla_\Gamma G)_\Gamma$ for all G denotes the weak form of the Laplace–Beltrami operator applied to F which is sometimes written as $-\Delta_\Gamma F$.

13.3.2 Newton-Type Flow and Descent Properties

Using the shape gradient and the shape Hessian at $\Gamma = \Gamma(t)$ we can specify the system satisfied by a *Newton-type velocity field* $N(\Gamma(t))$. Observe that, for instance, for computing the shape sensitivity information required for obtaining the Newton direction at $\Gamma(t)$, in (13.20) the set $\Gamma(\tau)$ has to be replaced by $\Gamma(\tau + t)$. The Newton equation is given by

$$d^2 J(\Gamma(t); V_F; V_{N(\Gamma(t))}) = -dJ(\Gamma(t); V_F) \quad \forall V_F \in B. \tag{13.26}$$

In general the bilinear form related to d^2J need not be coercive. In our particular example (13.18) we have that the Laplace–Beltrami operator implies only positive semidefiniteness properties, and the first term in (13.22) may be negative on subsets of $\Gamma(t)$, which is true especially far away from the optimal shape. Therefore, it is of interest to consider coercive modifications $S(\Gamma(t); V_F, V_G)$ of $d^2J(\Gamma(t); V_F; V_G)$. Here we consider

$$S(\Gamma(t); V_F; V_G) = \int_\Gamma \mathfrak{s}(\Gamma(t))FG + g_{\epsilon_g}(\nabla_\Gamma F \cdot \nabla_\Gamma G)\, dS \tag{13.27}$$

with

$$\mathfrak{s}(\Gamma(t))(x) = \max\left(\left(\frac{\partial^2 g}{\partial n^2} + (2\kappa + \nu)\frac{\partial g}{\partial n} + \nu\kappa g\right)(x), \epsilon_{\mathfrak{s}}\right)$$

whenever the first term in the max expression is smooth enough; otherwise we set $\mathfrak{s}(\Gamma(t)) = \epsilon_{\mathfrak{s}}$ for small $\epsilon_{\mathfrak{s}} > 0$. Notice that we replaced g by g_{ϵ_g} with $g_{\epsilon_g} \geq \epsilon_g$ for small $\epsilon_g > 0$.

As a result there exists $\epsilon > 0$ such that

$$S(\Gamma(t); V_F; V_F) \geq \epsilon \|F\|^2_{H^1(\Gamma)} \quad \forall V_F \in B. \tag{13.28}$$

Let $N^S(\Gamma(t))$ denote the solution of

$$S(\Gamma(t); V_F; V_{N^S(\Gamma(t))}) = -dJ(\Gamma(t); V_F) \quad \forall V_F \in B. \tag{13.29}$$

Further assume that $N^S(\Gamma(t))$ is sufficiently smooth and that $N^S_{\text{ext}}(\Gamma(t))$ is an extension satisfying (13.24). Further we assume that $V_{N^S(\Gamma(t))} \in B$ satisfies

$$V_{N^S(\Gamma(t))} \cdot \nabla\varphi = N^S_{\text{ext}}(\Gamma(t))\|\nabla\varphi\|_2.$$

Plugging $V_{N^S(\Gamma(t))}$ into (13.19) (replacing $V(X(t))$), and applying the chain rule to $\varphi(X(t), t) = 0$ (recall that the $\Gamma(t)$ is required to be the zero-level set of the geometrical implicit function φ at all times t), we get

$$\begin{aligned} 0 &= \left(\varphi_t(X(\tau), \tau) + \nabla\varphi(X(\tau), \tau) \cdot X'(\tau)\right)_{\tau=t} \\ &= \varphi_t(x, t) + \|\nabla\varphi(x, t)\|_2 \left(\frac{\nabla\varphi(x, t)}{\|\nabla\varphi(x, t)\|_2}\right) \cdot V_{N^S(\Gamma(t))}(x, t) \\ &= \varphi_t(x, t) + N^S(\Gamma(t))_{\text{ext}}(x, t)\|\nabla\varphi(x, t)\|_2 \end{aligned} \tag{13.30}$$

which is the level-set equation (13.3) with $F_{\text{ext}} = N^S_{\text{ext}}$.

Now suppose that $\Gamma(t + \delta_t)$, with $\delta_t > 0$ sufficiently small, is obtained by evolving φ via (13.30). From our shape sensitivity analysis and (13.29) we infer

$$\begin{aligned} J(\Gamma(t + \delta_t)) &= J(\Gamma(t)) + \delta_t dJ(\Gamma(t); V_{N^S(\Gamma(t))}) + \mathcal{O}(\delta_t) \\ &= J(\Gamma(t)) - \delta_t S(\Gamma(t); V_{N^S(\Gamma(t))}; V_{N^S(\Gamma(t))}) + \mathcal{O}(\delta_t) \\ &\leq J(\Gamma(t)) - \epsilon\delta_t \|N^S(\Gamma(t))\|^2_{H^1(\Gamma)} + \mathcal{O}(\delta_t), \end{aligned}$$

where we used (13.28) for the last inequality. Thus, if $\delta_t > 0$ is sufficiently small we have

$$J(\Gamma(t + \delta_t)) \leq J(\Gamma(t)) - \underline{\epsilon}\delta_t \|N^S(\Gamma(t))\|^2_{H^1(\Gamma)} \tag{13.31}$$

for some fixed $\underline{\epsilon}$ satisfying $0 < \underline{\epsilon} < \epsilon$. This proves that the Newton-type flow is a descent flow; i.e., if $\Gamma(t)$ (respectively, $\Omega(t)$) is not optimal, then there exists a small positive time step in (13.30) such that the objective value is sufficiently reduced. We summarize this result in the following proposition.

Proposition 13.4. *Let $N_S(\Gamma(t))$ be the solution of* (13.29) *with extension $N^S_{ext}(\Gamma(t))$ satisfying* (13.23)–(13.24). *Then there exist $\underline{\epsilon} > 0$ and $\delta_t > 0$ such that* (13.31) *is satisfied.*

Proposition 13.4 and (13.31) are the basis for a strategy for selecting δ_t. In the spirit of classical backtracking line search techniques an initial step length $\delta_t^0 > 0$ is selected and, if necessary reduced, until (13.31) is satisfied. This can be done by determining the first element of the sequence $\{\delta_t^l\}_{l=0}^{\infty} = (\delta_t^0, q\delta_t^0, q^2\delta_t^0, \dots)$, with $0 < q < 1$, which satisfies (13.31).

13.3.3 Phase-II Algorithm for Shape Optimization

Now we specify the shape-Newton algorithm for solving (13.1). Here the method is presented on the continuous level. Details for its discrete counterpart are the subject of Section 13.4. For ease of notation, below subscript k refers to quantities at time t_k.

Algorithm 13.5 (Phase-II Algorithm for Shape Optimization).

(i) Choose an initial contour Γ_0 with associated domain Ω_0. Compute the corresponding geometrical implicit function φ_0 which satisfies the sign convention (13.2); set $t_0 := 0$ and $k := 0$.

(ii) Solve the Newton-type equation

$$S_k(\Gamma_k; V_F; V_{N_k}) = -dJ(\Gamma_k; V_F) \quad \forall V_F \in B.$$

(iii) Extend N_k to a neighborhood of Γ_k to obtain $N_{k,\text{ext}}$.

(iv) Compute φ_{k+1} by using (13.30) with velocity $N_{k,\text{ext}}$ and a time step $\delta_{t,k} > 0$ such that (13.31) is satisfied.

(v) If necessary reinitialize φ_{k+1}; set $t_{k+1} := t_k + \delta_{t,k}$ and $k := k + 1$. Go to (ii).

A few words on details and the realization of the steps of the algorithm are in order.

- Compared to our phase-I algorithm (Algorithm 13.3) the initialization of Γ is more delicate. In fact, depending on the sign of the parameter ν, see (13.1), the initial contour has to be a closed curve which is either completely contained in the region of interest ($\nu < 0$), or the region of interest is a subset of Ω_0 ($\nu > 0$). A poor initial choice results in a segmentation failure.
- Choosing φ_k to be a signed distance function has several implications:
  - Flatness in the implicit function is avoided, which allows a numerically stable identification of $\Omega(t)$.

- Geometric information can be easily computed; e.g., $n(\Gamma(t), t) = \nabla\varphi(\Gamma(t), t)$, $\kappa(\Gamma(t), t) = \mathrm{div}(n(\Gamma(t), t)) = \Delta\varphi(\Gamma(t), t)$.
- If the aforementioned relations are used to simplify geometrical expressions, then it is essential that φ remains a signed distance function for all times t; otherwise the geometrical terms are inaccurate.
- Numerically, the latter aspect requires a reinitialization; i.e., if φ deviates sufficiently from being a signed distance function, then it has to be "reshaped" into a signed distance function.

- There are several options for the reinitialization process; see [12, 15]. We focus on solving the *Eikonal* equation
$$\|\nabla\varphi\| = 1 \text{ in } D \text{ with } \varphi_{|\Gamma_{k+1}} = \varphi_{k+1|\Gamma_{k+1}} = 0. \tag{13.32}$$
The solution of (13.32) is called φ_{k+1} again.
- We propose the following extension:
$$\nabla N_{k,\mathrm{ext}} \cdot n_k = 0 \quad \text{and} \quad (N_{k,\mathrm{ext}})_{|\Gamma_k} = N_k.$$
Then $V_{N_k} = N_{k,\mathrm{ext}} n_k$ satisfies (13.25).

13.4 Numerical Realization and Results

In this section we discuss several aspects of the implementation of our phase-I–phase-II algorithm. Especially phase-II requires some care in its numerical realization. This is necessitated by the discretization of the shape gradient and shape Hessian. We close the section by a report on extensive test runs. All computations were performed on a Dell Latitude C laptop computer with a Pentium 4-M processor with 2.00GHz.

13.4.1 Numerical Realization

The Laplacian operators in the state equation (13.7b) and the adjoint equation (13.12) are discretized by the standard 5-point stencil. The solution of these Poisson problems is achieved by a multigrid technique with a Gauss–Seidel-type smoother.

Phase-I. Compared to the phase-II method, Algorithm 13.3 is rather easy to implement.

- We initialize the method by some easy to generate shape Γ_0 like a circle, ellipse, or $\{x \in D : \|x - x_m\|_q^q = r\}$, $1 \le q < +\infty$, $r > 0$, and $x_m \in \mathbb{R}^2$. In our experience the method behaves robustly with respect to the initialization of the geometry.
- As described earlier, we stop the method if $\delta_{t,k}$ or $|\mathcal{J}_{k+1} - \mathcal{J}_k|$ is smaller than some prescribed tolerances $\epsilon_1, \epsilon_2 > 0$. In our test runs we typically choose $\epsilon_1 = \sqrt{\epsilon_M}$, with ϵ_M equal to the machine precision, and $\epsilon_2 = (\tau^0/(n_{x_1} n_{x_2}))\mathcal{J}_0$, where $\tau^0 > 0$ (typically $\tau^0 = 0.01$ in our tests) and n_{x_1}, n_{x_2} are set to the numbers of pixels in the x_1- and x_2-direction.

Phase-II. The challenges in the discretization of Algorithm 13.5 for shape optimization are primarily related to the shape gradient and the shape Hessian. In level-set-based front

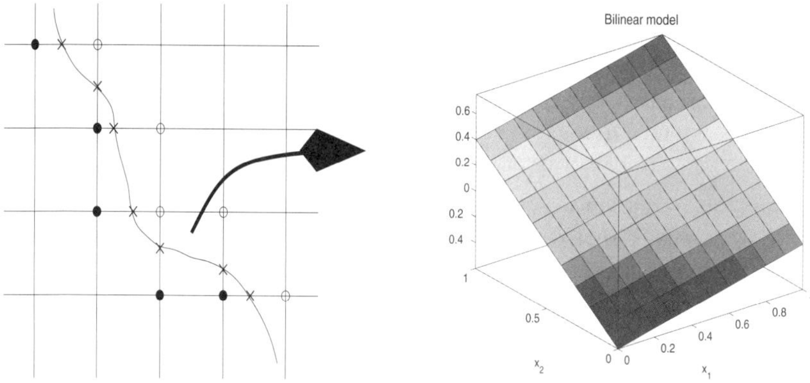

Figure 13.2. *Left plot: Interior (black circles), exterior (white circles) interface neighbors, and intersection points (crosses). Right plot: Bilinear interpolation model.*

propagation one typically avoids the resolution of the interface Γ_k. While in many cases this may lead to satisfactory evolution (see [12] and the references therein), in the case of optimization techniques one encounters the following troubles: An inaccurate shape Hessian approximation, as long as it induces a coercive bilinear form, may yield no improved convergence speed compared to a pure shape gradient flow. This is in contrast to one of the aims for employing higher-order derivatives in optimization. Inaccurate shape gradients typically cause the algorithm to get stuck away from the (discrete) solution. Then the distance to the optimal solution depends on the condition of the problem and the size of the error.

One way to overcome these difficulties is to resolve the interface which allows one to compute more accurate approximations to normals, curvatures, function values, and other geometry-dependent information and, thus, yielding rather accurate shape gradients and shape Hessians. Below we provide some details of the numerical realization of Algorithm 13.5.

- The level-set equation (13.3) is discretized on a fixed Cartesian grid. Since it is a PDE of Hamilton–Jacobi-type, one has to be careful in selecting spatial and time discretizations. With respect to time we use a first-order explicit Euler scheme (in our tests it appears that spatial accuracy is more important than accuracy in time). For the spatial discretization we use a second-order essentially nonoscillatory scheme (ENO scheme); see [12, 15, 16].
- The interface is resolved in the following way. In a first sweep we detect interior and exterior interface neighbors among the grid points. An interior neighbor is considered to be in the discrete analogue of Ω_k, and an exterior neighbor is an element of the discrete version of $D \setminus \overline{\Omega_k}$; see Figure 13.2 where the black nodes correspond to interior and the white nodes to exterior interface neighbors. Based on this information, then for every cell with an interface transition a local bilinear interpolation model is established. The function values of the model are given by the signed distance values at the grid points, i.e., the corners of the cell. Then discrete approximations of the

intersection points (crosses in Figure 13.2) are given by the roots along grid lines of the interpolation polynomial.

- We choose a locally linear approximation of the interface. For instance, in the case of the exemplary cell in Figure 13.2, the portion of an interface in a cell is approximated by connecting the interface points. From this we can easily compute the outward normal direction of the interface model. Since we are interested in the outward unit normal at an intersection point, which is not unique due to the kinky structure of the piecewise linear model, we use a linear combination of the normals of neighboring linear pieces. The weights are given by the lengths of the interface in the respective cell. Normal derivatives and curvature values are computed as follows: Standard finite difference models are used for computing the gradient of a function on the fixed grid points. At the intersection point the gradient of a function is computed by weighted averages. Then the normal derivative is obtained as the inner product of the gradient times the normal. The curvature at regular grid points is computed by discretizing $\Delta\varphi$. At intersection points again weighted averages are used.
- For the discretization of the Laplace–Beltrami operator we refer the reader to [6, 9]. In the case of a one-dimensional interface in the form of a closed curve, it is closely related to a nonuniform discretization of the Laplacian operator on an interval with periodic boundary conditions.
- The (re)initialization of φ is based on (13.32). On the numerical level we employ a fast marching technique for solving the *Eikonal* equation; see [14] for details.
- In order to reduce the computational effort a narrow band approach is used; i.e., φ and all related quantities and equations are only considered on a band around the actual interface (contour).

13.4.2 Numerical Results

In this section we report on numerical results obtained from applying our algorithm. Phase-I is initialized by $\varphi_0 = (x_1 - 0.5)^4 + (x_2 - 0.5)^2 - 0.06$. Our initialization of phase-II uses the contour obtained from applying phase-I and computes the corresponding signed distance function. The intensity map I is obtained by a simple thresholding technique. In fact, let $\tilde{I} \in [0, 1]$ denote the original image intensity map. Then, for a grid point x_i, we set

$$I(x_i) = \begin{cases} 1 & \text{if } \tilde{I}(x_i) \geq \theta, \\ 0 & \text{if } \tilde{I}(x_i) < \theta \end{cases} \tag{13.33}$$

with $\theta > 0$ a user-defined threshold. An alternative is to use $I = \tilde{I}$, but, of course, more sophisticated choices are possible.

Phase-I as an Initialization Procedure

We first focus on results where phase-I is used as an (re)initialization procedure for phase-II. To this end note that the second integral in (13.1), depending on the sign of ν, either minimizes or maximizes the area of Ω. If $\nu > 0$, then Ω is intended to be small; otherwise, if $\nu < 0$, then Ω should be large. Thus, depending on the sign of ν the initialization of phase-II is either a contour within the feature of interest ($\nu < 0$), or Ω_0 contains the feature

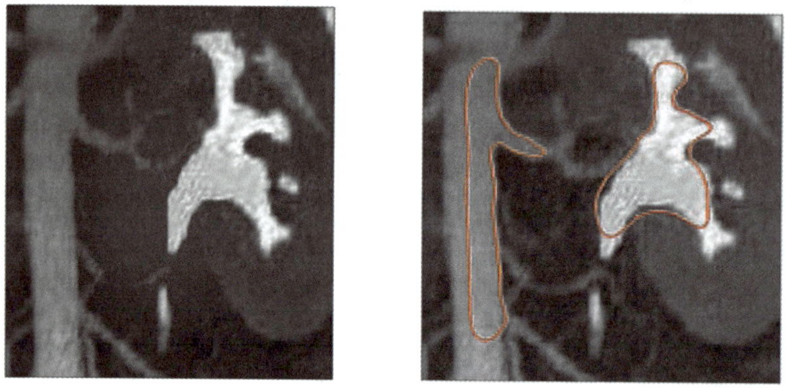

Figure 13.3. *Image (left) and the result obtained from phase-*I *(right);* $\beta_1 = 1$, $\beta_2 = 1.0E4$, $\mu = 50$, $\theta = 0.45$.

of interest ($\nu > 0$). Here we focus on the first case. In this situation phase-I should provide a geometry which is contained in the region of interest. Then phase-II should drive the contour to its optimal location with respect to (13.1). This requires $\nu > 0$, and $\mu > 0$, to be relatively large. Note that if μ is large, then $|\Omega|$ is supposed to be small.

In Figure 13.3 we show the result of phase-I (right plot) when applied to the image in the left plot. For this run the parameters had values $\beta_1 = 1$, $\beta_2 = 1.0E4$, $\mu = 50$, $\theta = 0.45$ and the algorithm terminated after 9 iterations. The initialization of the contour is displayed in the right plot of Figure 13.7. The corresponding solutions u of (13.7b) and p of (13.12) are depicted in the left and right plots of Figure 13.4, respectively. One can see that the contour is far off the desired contours at places and stays within the region of interest. Consequently, the set Ω_*^h, the numerical approximation of the solution to (13.1), upon termination of phase-I is smaller than the optimal set. In Figure 13.5 we present the result for a smaller parameter $\mu = 25$ as compared to the previous run. In this case the algorithm took 14 iterations. Due to the reduction in μ the area of Ω_*^h is larger. Still, Ω_*^h is smaller than the optimal solution. However, the contour in Figure 13.5 is an excellent initial guess for phase-II with $\nu < 0$.

If we use the result of Figure 13.5 as the initial guess for phase-II with $\nu = 1.25$ (φ_0 for phase-II is depicted in the right plot of Figure 13.5), then after 14 phase-II iterations the segmentation result shown in Figure 13.6 is obtained. Observe that there is no change of the initial topology. This reflects the general fact that a level-set-based method using shape sensitivity, except for merging and splitting existing components, is unable to create new topological information. On the other hand, phase-I provides an excellent automatically produced initial guess for phase-II.

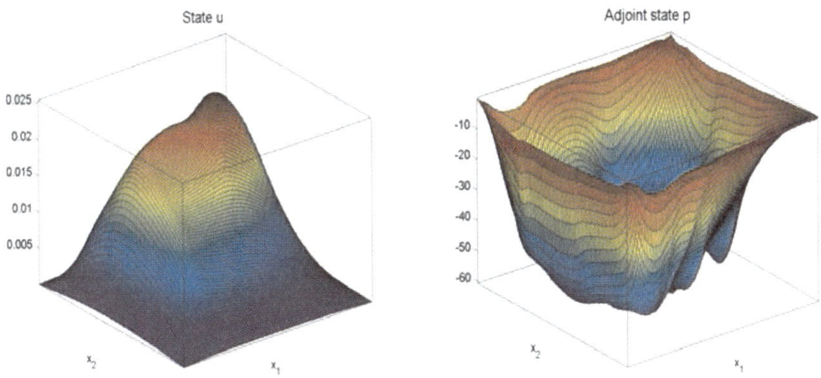

Figure 13.4. *State u (left) and adjoint state (right) for phase-*I*;* $\beta_1 = 1$, $\beta_2 = 1.0E4$, $\mu = 50$, $\theta = 0.45$.

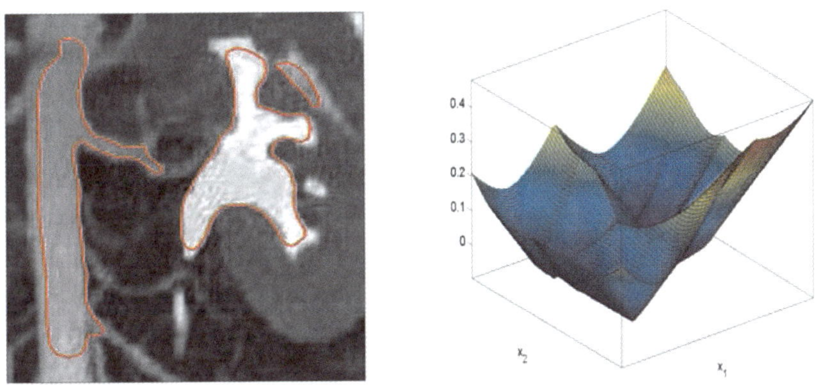

Figure 13.5. *Segmentation result for phase-*I *and the corresponding signed distance function;* $\beta_1 = 1$, $\beta_2 = 1.0E4$, $\mu = 25$, $\theta = 0.45$.

Phase-I as a Segmentation Algorithm

In our numerical tests we found that our new phase-I algorithm for topology optimization has the potential to serve as a segmentation algorithm even without applying phase-II. Then, of course, μ has to be adjusted adequately. But let us mention that in our test runs no "sophisticated" selection strategies appeared to be necessary.

In Figure 13.7 (left plot) we show the segmentation result for the contrast agent based kidney image in Figure 13.3 (left plot) together with the initial contour (right plot in Figure 13.7). We chose the parameters $\beta_1 = 1$, $\beta_2 = 1.0E4$, $\mu = 1.0E - 2$, and $\theta = 0.45$. The algorithm required 20 iterations (23s CPU time). Compared to segmentation runs using the phase-II method with a simple manual initial choice ($>$ 60 iterations) and level-set-based methods using the shape gradient as a descent flow ($>$ 100 iterations), the

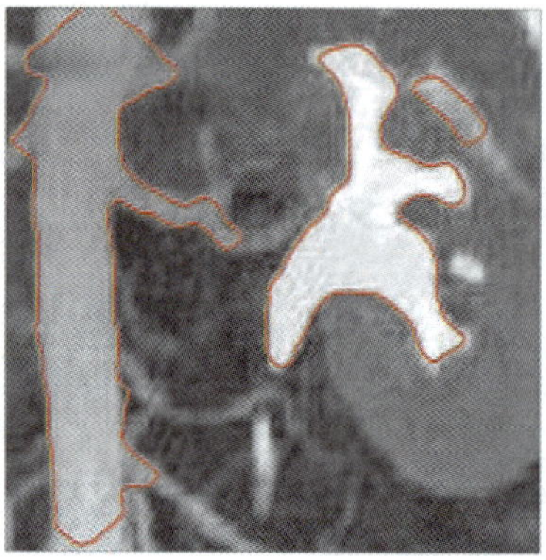

Figure 13.6. *Segmentation result obtained by phase-*II *with initialization by phase-*I*;* $\beta_1 = 1$, $\beta_2 = 1.0E4$, $\mu = 25$, $\theta = 0.45$.

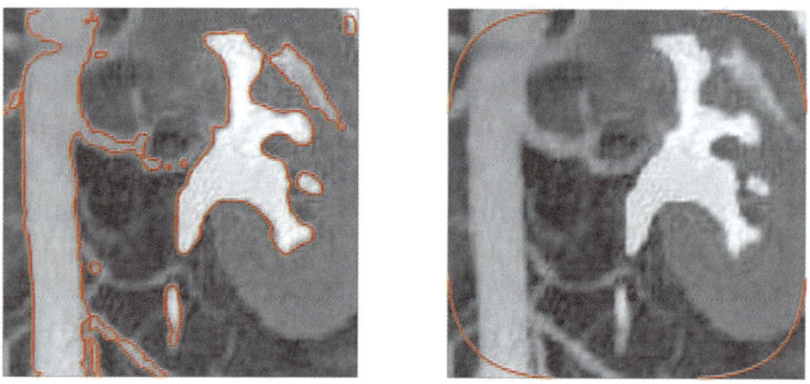

Figure 13.7. *Segmentation result (left) obtained by phase-*I *only and initial contour (right);* $\beta_1 = 1$, $\beta_2 = 1.0E4$, $\mu = 1.0E-2$, $\theta = 0.45$.

iteration count is in favor of our phase-II method. The optimal state and optimal adjoint state for the run documented in Figure 13.7 are shown in Figure 13.8.

In Figure 13.9 we demonstrate that the method is able to detect rather complicated topologies. The initial guess for our topology optimization algorithm is similar to the one shown in Figure 13.7. We ran phase-I with $\beta_1 = 1$, $\beta_2 = 1.0E4$, $\mu = 1.0E-2$, and $\theta = 0.525$. The algorithm stopped after 16 iterations (25s CPU time). The difference between u_*^h, the state upon termination, and u_d^h in the pointwise ℓ_∞-norm is shown in the left plot of Figure 13.10. The right plot depicts the behavior of $\{J^h(\Omega_k^h)\}$, the sequence of discrete objective values. Note that the scale on the vertical axis is a logarithmic one. Due to our line search procedure $\{J^h(\Omega_k^h)\}$ is monotonically decreasing.

Our final example is shown in Figure 13.11. The left plot is an MRI scan of a male

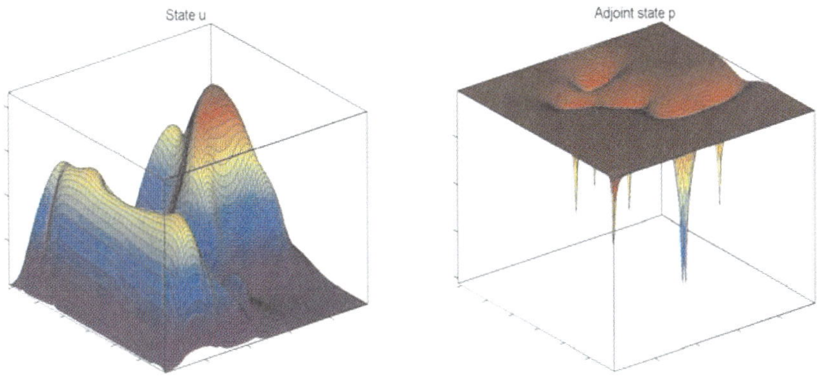

Figure 13.8. *State u (left) and adjoint state (right) for phase-*I*;* $\beta_1 = 1$, $\beta_2 = 1.0E4$, $\mu = 1.0E-2$, $\theta = 0.45$.

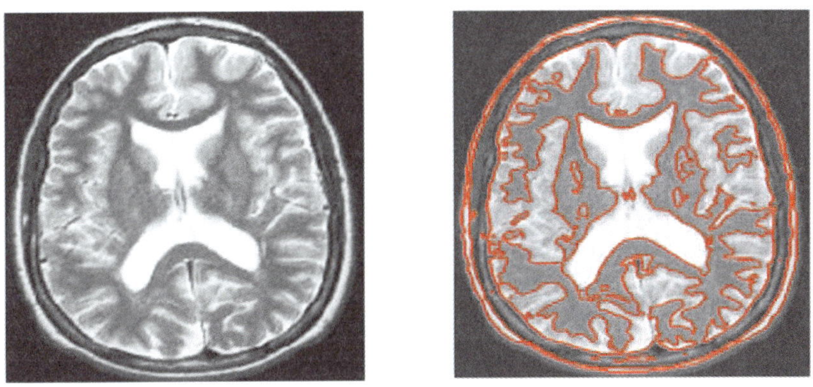

Figure 13.9. *Image and segmented image using phase-*I *only;* $\beta_1 = 1$, $\beta_2 = 1.0E4$, $\mu = 1.0E-2$, $\theta = 0.525$.

chest. The image contains many small components as well as comparatively large ones. Our simple initial guess is depicted in the right plot of Figure 13.11. It is completely different from the segmentation result shown in the left plot of Figure 13.12. The parameter values were the same ones as for the results in Figure 13.9. The algorithm stopped after 53 iterations. In the right plot of Figure 13.12 we show the zero-level set without the image in the background. As can be seen, our phase-I algorithm combined with our thresholding strategy allows us to detect complex topologies with components of quite different sizes. In Figure 13.13 we show the signed distance function whose zero-level set is displayed in the right plot of Figure 13.12. Also, we show the convergence behavior of $\{J^h(\Omega_k^h)\}$. Again, we can see that $\{J^h(\Omega_k^h)\}$ is monotonically decreasing due to our time-step size selection strategy.

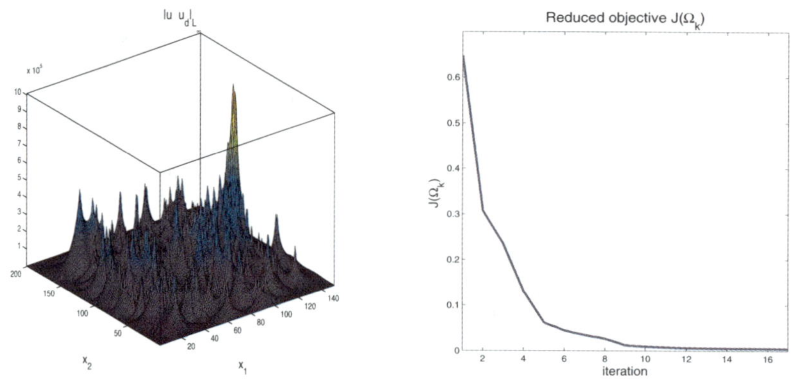

Figure 13.10. *Difference* $|u_*^h - u_d^h|_{\ell_\infty}$ *and behavior of* $\{J^h(\Omega_k^h)\}$*;* $\beta_1 = 1$, $\beta_2 = 1.0E4$, $\mu = 1.0E-2$, $\theta = 0.525$.

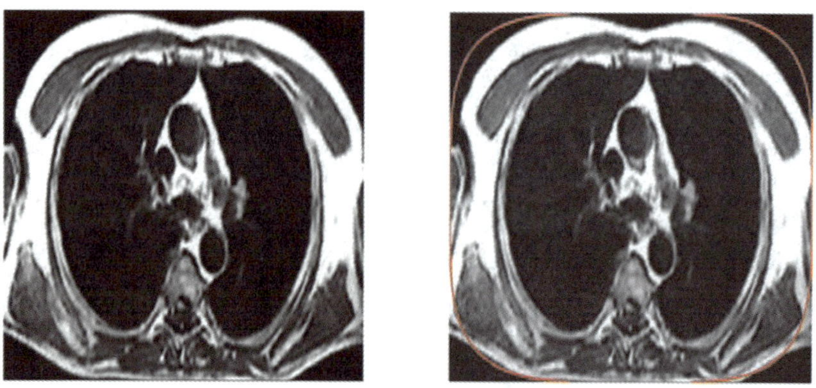

Figure 13.11. *Image (left) and initial guess (right) for phase*-I*;* $\beta_1 = 1$, $\beta_2 = 1.0E4$, $\mu = 1.0E-2$, $\theta = 0.55$.

13.5 Conclusion

We have presented a phase-I–phase-II concept for real-time image segmentation. Phase-I is based on topological sensitivity information in a given image, and phase-II uses shape sensitivity for computing descent flows for a shape functional. Both phases rely on geometrical implicit functions for representing the geometry and utilize line search techniques for computing adequate time-step sizes for updating the geometry. In addition, phase-II utilizes a level-set framework for updating the geometry.

From our numerical tests we find that phase-I can be used as an initialization method for phase-II and, thus, allowing us to dispense with manual initialization techniques. This feature is attractive in intraoperative imaging where the amount of required user interaction with the algorithm should be kept as small as possible. Another important consequence of

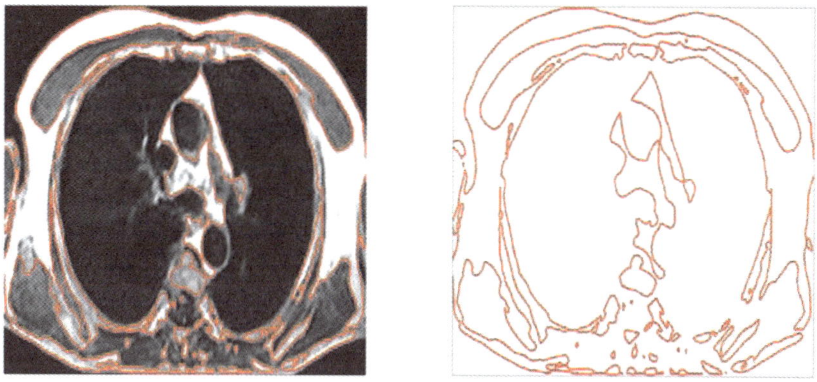

Figure 13.12. *Segmented image (left) and zero-level set without image background (right);* $\beta_1 = 1$, $\beta_2 = 1.0E4$, $\mu = 1.0E-2$, $\theta = 0.55$.

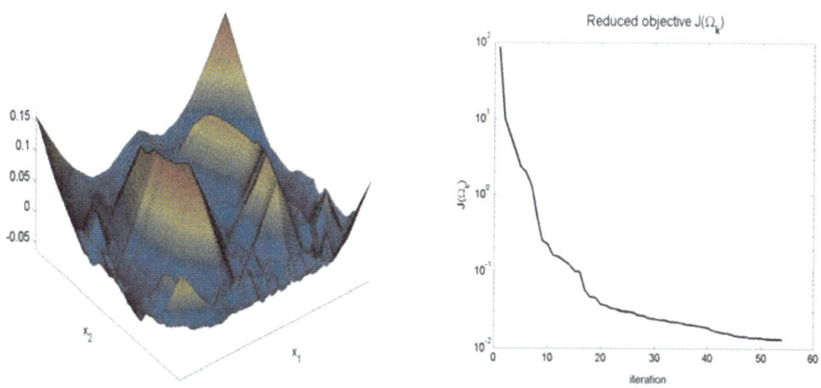

Figure 13.13. *Signed distance function at the solution (left) and behavior of* $\{J^h(\Omega_k^h)\}$ *(right);* $\beta_1 = 1$, $\beta_2 = 1.0E4$, $\mu = 1.0E-2$, $\theta = 0.55$.

our investigation is related to the fact that phase-I may serve as a segmentation method in its own right, rather than being an initialization method only. In our experience it provides excellent segmentation results. The latter efficiency is based on a preprocessing step for the intensity map. We use a simple thresholding technique which usually follows an image enhancement phase. However, it appears that in difficult situations where tumorous tissue and surrounding tissue are hard to distinguish further investigations for preprocessing the intensity map are necessary. Also more sophisticated techniques than convolution with a Gaussian-type kernel for obtaining a model problem for the topology optimization phase should be the subject of future investigations.

Acknowledgments

Partial support by the Austrian Sciences Fund FWF under SFB03 "Optimierung und Kontrolle" and support by Rice University, Department of Computational and Applied Mathematics, Houston, Texas, USA, where the author was visiting professor during this research project, are gratefully acknowledged.

Bibliography

[1] D. Adalsteinsson and J. A. Sethian, *The fast construction of extension velocities in level set methods*, J. Comput. Phys., 148 (1999), pp. 2–22.

[2] M. Burger, *Levenberg-Marquardt level set methods for inverse problems*, Inverse Problems, 20 (2004), pp. 259–282.

[3] M. Burger, B. Hackl, and W. Ring, *Incorporating topological derivatives into level set methods*, J. Comput. Phys., 194 (2004), pp. 344–362.

[4] V. Caselles, F. Catté, T. Coll, and F. Dibos, *A geometric model for active contours in image processing*, Numer. Math., 66 (1993), pp. 1–31.

[5] M. C. Delfour and J.-P. Zolésio, *Shapes and Geometries: Analysis, Differential Calculus, and Optimization*, SIAM, Philadelphia, 2001.

[6] G. Dziuk, *Finite elements for the Beltrami operator on arbitrary surfaces*, in Partial Differential Equations and Calculus of Variations, Lecture Notes in Math. 1357, Springer-Verlag, Berlin, 1988, pp. 142–155.

[7] S. Garreau, P. Guillaume, and M. Masmoudi, *The topological asymptotic for PDE systems: The elasticity case*, SIAM J. Control Optim., 39 (2001), pp. 1756–1778.

[8] D. T. Gering, A. Nabavi, R. Kikinis, N. Hata, L. J. O'Donnell, W. E. Grimson, F. A. Jolesz, P. M. Black, and W. M. Wells III, *An integrated visualization system for surgical planning and guidance using image fusion and an open MR*, J. Magnetic Resonance Imag., 13 (2001), pp. 967–975.

[9] M. Hintermüller and W. Ring, *A second order shape optimization approach for image segmentation*, SIAM J. Appl. Math., 64 (2003), pp. 442–467.

[10] M. Kass, A. Witkin, and D. Terzopoulos, *Snakes: Active contour models*, Int. J. Comput. Vision, 1 (1987), pp. 321–331.

[11] J. Nocedal and S. J. Wright, *Numerical Optimization*, Springer Series in Operations Research. Springer-Verlag, New York, 1999.

[12] S. J. Osher and R. P. Fedkiw, *Level Set Methods and Dynamic Implicit Surfaces*, Springer-Verlag, New York, 2002.

[13] S. J. Osher and J. A. Sethian, *Fronts propagating with curvature-dependent speed: Algorithms based on Hamilton-Jacobi formulations*, J. Comput. Phys., 79 (1988), pp. 12–49.

[14] J. A. Sethian, *Fast marching methods*, SIAM Rev., 41 (1999), pp. 199–235.

[15] J. A. Sethian, *Level Set Methods and Fast Marching Methods: Evolving Interfaces in Computational Geometry, Fluid Mechanics, Computer Vision, and Materials Science*, Cambridge University Press, Cambridge, UK, 1999.

[16] C.-W. Shu and S. J. Osher, *Efficient implementation of essentially non-oscillatory shock-capturing schemes*, J. Comput. Phys., 77 (1988), pp. 439–471.

[17] J. Sokolowski and A. Żochowski, *On the topological derivative in shape optimization*, SIAM J. Control Optim., 37 (1999), pp. 1251–1272.

[18] J. Sokolowski and J.-P. Zolésio, *Introduction to Shape Optimization: Shape Sensitivity Analysis*, Springer-Verlag, Berlin, 1992.

[19] A. Tsai, A. J. Yezzi, W. M. Wells III, C. Tempany, D. Tucker, A. Fan, W. E. L. Grimson, and A. S. Willsky, *A shape-based approach to the segmentation of medical imagery using level sets*, IEEE Trans. Med. Imag., 22 (2003), pp. 137–154.

[20] S. Warfield, F. Jolesz, and R. Kikinis, http://splweb.bwh.harvard.edu:8000/pages/papers/warfield/sc98/ (2004).

[21] A. Yezzi, S. Kichenassamy, A. Kumar, P. Olver, and A. Tannenbaum, *A geometric snake model for segmentation of medical imagery*, IEEE Trans. Med. Imag., 16 (1997), pp. 199–209.

Chapter 14

COFIR: Coarse and Fine Image Registration

Jan Modersitzki and Eldad Haber†*

14.1 Introduction

Image registration is one of today's challenging image processing problems. The problem can be formulated as follows: Given two images, find a "*reasonable*" transformation such that a transformed version of the so-called template image becomes "*similar*" to the so-called reference image. Image registration is applied whenever images resulting from different times, devices, and/or perspectives need to be compared or integrated; see, e.g., [20, 26] and the references therein.

A registration procedure is typically based on two main building blocks. The first one is a distance measure. The distance measure quantifies the meaning of similarity or proximity of images. A distance measure can be based on image features (e.g., moments [1], landmarks [20, 23], or markers [19]), on image intensities (e.g., L_2-norm or sum of squared differences [7], correlation, mutual information [10, 25]), on image surfaces [3, 4], on level sets [11], or on combinations thereof [8, 13]. For an overview and comparisons, see also [17, 20, 22].

The second building block is a regularizing term. Since image registration is an ill-posed problem, regularization is inevitable and becomes a central topic [20]. Typical regularization techniques are a restriction to a low-dimensional transformation space (e.g., the space of rigid or affine linear transformations) or an explicit regularization of the problem.

*Institut für Mathematik, Universität Lübeck, D-23560 Lübeck, Germany.
†Department of Mathematics and Computer Science, Emory University, Atlanta, GA 30322.

The regularizer of the latter approach is typically based on a bilinear form

$$\mathcal{S}[u] = \frac{1}{2}\int_{\Omega} \langle \mathcal{B}u, \mathcal{B}u \rangle \; dx, \tag{14.1}$$

where $\mathcal{B}$ is a partial differential operator. This includes well-known regularizers like, for example, the elastic [2, 6], diffusion [12, 24], or curvature [14], and, in a qualified sense, also the fluid registration [5, 9]; see also [20] for an overview.

In many applications one has to deal with large linear but also important nonlinear deformations of the displayed object. Typical examples arising from medical applications include the registration of a histological serial sectioning or the registration of two images from a joint taken at different positions; see also Figure 14.1. Therefore one would prefer a registration scheme which handles both the linear (more general: the *coarse*) and the remaining nonlinear (more general: the *fine*) part simultaneously. However, as it is well known to practitioners, existing registration schemes cannot handle this goal in an appropriate way, and in some cases such registration does not converge to an acceptable solution.

In the literature either of the following two ways is used to bypass this problem. The first one is to assume that the overall registration problem decouples into a linear and nonlinear part and to solve for the two parts separately and sequentially. However, though this approach seems to work quite well in practice, the decoupling assumption is artificial and it is an open question whether this assumption is true or not. The second remedy is to design a regularizer that does not penalize the affine linear part. In [14] it is suggested to use $\mathcal{B} = \Delta$, since a second-order derivative operator does not penalize affine linear transformation. However, not all applications allow for an arbitrary regularization. Higher-order derivative-based operators result in smoother transformations which may not be physically correct. Finally, from an implementation point of view, boundary conditions and the solution of the resulting system of equations become more troublesome.

Therefore we proposed a new way of dealing with these difficulties. We split the transformation space into two disjoint subspaces, a *coarse* space and a *fine* space. The coarse space is a low-dimensional linear space and the fine space is the remaining orthogonal complement. Only the components of the transformation belonging to the fine space are regularized. We call this novel COarse and Fine Image Registration scheme COFIR. In this new approach, the coarse and the fine spaces are guaranteed to be disjoint, and therefore the above-mentioned decoupling assumption holds by construction.

Different applications may demand different choices of the coarse space $\mathcal{C}$. For example, in studying tumor growth, one may want to consider only volume-preserving transformations and thus choose $\mathcal{C}$ to be the set of rigid transformations. In our mathematical formulation we therefore consider a general coarse space but concentrate on the space of affine linear transformations in our examples.

The chapter is organized as follows. In Section 14.2 we introduce the continuous mathematical framework which is of particular interest for our multilevel approach and in Section 14.3 we discuss discretization issues. The optimization schemes and the implementation of COFIR are discussed in Section 14.4. Finally, in Section 14.5 we present results for a highly nonlinear real life application; see also [18].

14.2 The Mathematical Setting

With $d \in \mathcal{N}$ we denote the spatial dimension of the given images $R, T : \mathbb{R}^d \to \mathbb{R}$ which are assumed to be sufficiently smooth. To be precise, we interpolate or approximate the given discrete data by smooth spline functions R and T, respectively. Thus, for any spatial position $\boldsymbol{x} = (x_1, \ldots, x_d)$, $T(\boldsymbol{x})$ gives the corresponding gray value. We assume that the supports of the images are contained in a bounded domain $\Omega :=]0, L[^d$, i.e., $R(\boldsymbol{x}) = T(\boldsymbol{x}) = 0$ for $\boldsymbol{x} \notin \Omega$.

Our goal is to find a "*reasonable*" deformation $\boldsymbol{u} = (u^1, \ldots, u^d)^T : \mathbb{R}^d \to \mathbb{R}^d$ such that the "*distance*" between the reference image R and the deformed template image $T(\boldsymbol{x}+\boldsymbol{u}(\boldsymbol{x}))$ becomes small. As explained in the introduction, we divide the space of feasible transformations into two disjoint subspaces, a coarse space $\mathcal{C}$ and a fine space $\mathcal{F}$. The space $\mathcal{C}$ is an m-dimensional linear space spanned by some basis functions c_j, $j = 1, \ldots, m$, such that a generic element reads

$$\boldsymbol{u}^{\mathcal{C}} = \sum_{j=1}^{m} z_j c_j(\boldsymbol{x}) =: \boldsymbol{C}(\boldsymbol{x})\boldsymbol{z}, \qquad \boldsymbol{z} \in \mathbb{R}^m. \tag{14.2}$$

Example: Probably the most important example is when $\mathcal{C}$ is the space of affine linear transformations. For spatial dimension $d = 2$ we have

$$\boldsymbol{u}^{\mathcal{C}} = \begin{pmatrix} z_1 + z_2 x_1 + z_3 x_2 \\ z_4 + z_5 x_1 + z_6 x_2 \end{pmatrix}$$

and for $d = 3$,

$$\boldsymbol{u}^{\mathcal{C}} = \begin{pmatrix} z_1 + z_2 x_1 + z_3 x_2 + z_4 x_3 \\ z_5 + z_6 x_1 + z_7 x_2 + z_8 x_3 \\ z_9 + z_{10} x_1 + z_{11} x_2 + z_{12} x_3 \end{pmatrix}.$$

To find $\boldsymbol{u} = \boldsymbol{u}^{\mathcal{C}} + \boldsymbol{u}^{\mathcal{F}}$, we formulate the following optimization problem:

$$\min \quad \mathcal{D}[R, T; \boldsymbol{u}^{\mathcal{C}} + \boldsymbol{u}^{\mathcal{F}}] + \alpha \mathcal{S}[\boldsymbol{u}^{\mathcal{F}}] \tag{14.3a}$$

$$\text{subject to} \quad \langle \boldsymbol{u}^{\mathcal{F}}, c_j \rangle = 0, \quad j = 1, \ldots, m, \tag{14.3b}$$

where $\mathcal{D}$ is some distance measure, $\mathcal{S}$ is some regularization term, and $\alpha > 0$ is a regularization parameter that compromises between similarity and regularity. In contrast to previous approaches the transformation has been divided into two parts and only the fine part is regularized. Moreover, and most importantly, the fine part is constrained to be orthogonal to the coarse part and therefore, unlike previous work, this splitting is disjoint.

Any differentiable choice of building blocks $\mathcal{D}$ and $\mathcal{S}$ is feasible. However, for ease of presentation, we restrict ourselves to the sum of squared differences (SSD) and the elastic regularizer,

$$\mathcal{D}[R, T; \boldsymbol{u}] = \frac{1}{2} \int_{\Omega} \Big(T(\boldsymbol{x} + \boldsymbol{u}(\boldsymbol{x})) - R(\boldsymbol{x}) \Big)^2 \, d\boldsymbol{x}, \tag{14.4}$$

$$\mathcal{B}[\boldsymbol{u}] = [\sqrt{\mu} \nabla u_1, \ldots, \sqrt{\mu} \nabla u_d, \sqrt{\lambda + \mu} \, \nabla \cdot \boldsymbol{u}]^{\top}, \tag{14.5}$$

$$\mathcal{A} := \mathcal{B}^* \mathcal{B} = \mu \Delta + (\lambda + \mu) \nabla \, \nabla \cdot , \tag{14.6}$$

where λ and μ are the so-called Lamé constants and natural boundary conditions are imposed; see, e.g., [20].

Remark: Let $\mathcal{P}$ denote the projector onto the orthogonal complement of $\mathcal{C}$. Hence, (14.3) can be replaced by

$$\min \quad \mathcal{D}[R, T; \boldsymbol{u}] + \alpha \mathcal{S}[\mathcal{P}\boldsymbol{u}]. \tag{14.7}$$

14.3 Discretization

The image registration problem (14.3) is discretized using staggered grids for the components of $\boldsymbol{u}$. Since the images may be noisy and derivatives are needed for efficient optimization procedures, we use a smoothing B-spline approximation R^{spline} and T^{spline} for the images R and T, respectively; see also [16]. In order to reduce the notational overhead, we denote discrete quantities with the same symbol as their continuous analogues. The precise meaning will be clear from the context. For ease of presentation, we also discuss only the two-dimensional case.

Let x_j denote the knots of the jth staggered grid, u_j the values of the jth component of $\boldsymbol{u}$ on x_j, and $\boldsymbol{u}$ the collecting of all these values. Our discretization of T is given by

$$T(\boldsymbol{u}) := T^{\text{spline}}(x_1 + P_1 u_1, x_2 + P_2 u_2),$$

where P_j denotes an averaging from the jth staggered to the cell centered grid and we use a B-spline interpolation or approximation scheme; see [16] for details. We denote the Jacobian of T by

$$T_{\boldsymbol{u}} := \frac{\partial T}{\partial \boldsymbol{u}}(\boldsymbol{u}) = \Big(\text{diag}(P_1^\top \partial_1 T), \text{diag}(P_2^\top \partial_2 T)\Big), \tag{14.8}$$

where the partial derivatives $\partial_j T$ have to be evaluated at $(x_1 + P_1 u_1, x_2 + P_2 u_2)$. For a discretization of the SSD distance measure (14.4), we use

$$\boldsymbol{D}(\boldsymbol{u}) := \tfrac{1}{2}\|T(\boldsymbol{u}) - R\|_2^2 \quad \text{and} \quad \boldsymbol{D}_{\boldsymbol{u}}(\boldsymbol{u}) = T_{\boldsymbol{u}}(\boldsymbol{u})^\top (T(\boldsymbol{u}) - R)$$

and for the regularizer (14.1) we take

$$\boldsymbol{S}(\boldsymbol{u}) = \tfrac{1}{2}\boldsymbol{u}^\top \boldsymbol{A}\boldsymbol{u} \quad \text{and} \quad \boldsymbol{S}_{\boldsymbol{u}}(\boldsymbol{u}) = \boldsymbol{A}\boldsymbol{u},$$

where $\boldsymbol{A} = \boldsymbol{B}^\top \boldsymbol{B}$ is based on a finite-difference-based discretization of $\mathcal{B}$. For details concerning the discretization of the elastic regularizer (14.5) see [16].

14.4 Optimization Scheme

In order to have an efficient optimization scheme we use a multilevel strategy (or grid continuation) where the problem is discretized and solved on a sequence of grids starting from a coarse grid . On level ℓ, the discretized analogue of the image registration problem (14.3) for $\boldsymbol{u}_\ell = \boldsymbol{u}_\ell^{\mathcal{F}} + \boldsymbol{C}_\ell z_\ell$ reads

$$\min \quad \boldsymbol{D}_\ell(\boldsymbol{u}_\ell^{\mathcal{F}} + \boldsymbol{C}_\ell z_\ell) + \alpha \boldsymbol{S}_\ell(\boldsymbol{u}_\ell^{\mathcal{F}}) \tag{14.9a}$$

$$\text{subject to} \quad \boldsymbol{C}_\ell^\top \boldsymbol{u}_\ell^{\mathcal{F}} = \boldsymbol{0}. \tag{14.9b}$$

The discrete version $\boldsymbol{C}_\ell$ on the ℓth level is computed directly by evaluating the basis functions c_j on the the underlying grids.

On each level, we use an iteration alternating between the minimization with respect to z and with respect to $\boldsymbol{u}^{\mathcal{F}}$. In the next subsections we describe the three parts of our algorithm: the minimization with respect to z, the minimization with respect to $\boldsymbol{u}^{\mathcal{F}}$, and the multilevel approach. The algorithm is summarized in Algorithm 14.1. For the discussion of the optimization schemes we consider a fixed level and therefore drop the level index in the description.

Algorithm 14.1.
Multilevel COarse and Fine Image Registration $[\boldsymbol{u}^{\mathcal{F}}, z] \leftarrow$ `COFIR`$(R, T, \boldsymbol{C})$.
for $\ell = 1, \dots, \max_\ell$
transfer the images to the level ℓ;
solve the coarse problem for z with $\boldsymbol{u}_\ell^{\mathcal{F}}$ fixed;
solve the fine problem for $\boldsymbol{u}_\ell^{\mathcal{F}}$ with z fixed;
prolongate $\boldsymbol{u}_\ell^{\mathcal{F}}$ to the finer level;
set $\ell \leftarrow \ell + 1$;
endfor

14.4.1 Optimizing the Coarse Part

Assuming we have an approximation to the fine part $\boldsymbol{u}^{\mathcal{F}}$ we compute a solution to the problem

$$\text{given} \quad \boldsymbol{u}^{\mathcal{F}}, \boldsymbol{C}, \quad \min \quad \boldsymbol{D}^{\mathcal{C}}(z) := \boldsymbol{D}(\boldsymbol{u}^{\mathcal{F}} + \boldsymbol{C}z), \tag{14.10}$$

we apply a Gauss–Newton scheme with an Armijo line search; see, e.g., [21]. Note that the computational expensive parts of this algorithm are the computation of $T(\boldsymbol{v})$ and $T_{\boldsymbol{u}}(\boldsymbol{v})$. Particularly the linear system in the Gauss–Newton approach are just m-by-m, where m is the dimension of the linear space $\mathcal{C}$; i.e., $m = 6$ for the two-dimensional linear space.

14.4.2 Optimizing the Fine Part

Assuming we have an approximation to the coarse part, i.e., z is given, we compute a solution to the problem

$$\begin{aligned} \min \quad & \boldsymbol{D}^{\mathcal{F}}(\boldsymbol{u}^{\mathcal{F}}) := bf\,D(\boldsymbol{u}^{\mathcal{F}} + \boldsymbol{C}z) + \alpha \boldsymbol{S}(\boldsymbol{u}^{\mathcal{F}}) \\ \text{subject to} \quad & \boldsymbol{C}^\top \boldsymbol{u}^{\mathcal{F}} = \boldsymbol{0}. \end{aligned} \tag{14.11}$$

To solve the problem we consider the Lagrangian

$$L(\boldsymbol{u}^{\mathcal{F}}, \boldsymbol{p}) := \boldsymbol{D}^{\mathcal{F}}(\boldsymbol{u}^{\mathcal{F}}) + \boldsymbol{p}^\top \boldsymbol{C}^\top \boldsymbol{u}^{\mathcal{F}},$$

where $\boldsymbol{p}$ denotes a vector of Lagrange multipliers. The necessary conditions for a minimizer $\boldsymbol{u}^{\mathcal{F}}$ of (14.11) read

$$\begin{pmatrix} \boldsymbol{D}_{\boldsymbol{u}}^{\mathcal{F}}(\boldsymbol{u}^{\mathcal{F}} + \boldsymbol{C}z) + \boldsymbol{C}\boldsymbol{p} \\ \boldsymbol{C}^\top \boldsymbol{u}^{\mathcal{F}} \end{pmatrix} = \boldsymbol{0}, \tag{14.12}$$

and a numerical solution of (14.11) is computed using a sequential quadratic programming (SQP) scheme (cf., e.g., [21]). Starting with a feasible initial guess $\boldsymbol{u}^{\mathcal{F}}$ (i.e., $\boldsymbol{C}^\top \boldsymbol{u}^{\mathcal{F}} = 0$), we solve

$$\begin{pmatrix} \boldsymbol{H} & \boldsymbol{C} \\ \boldsymbol{C}^\top & \boldsymbol{0} \end{pmatrix} \begin{pmatrix} \delta_u \\ \boldsymbol{p} \end{pmatrix} = - \begin{pmatrix} \boldsymbol{D}_u^{\mathcal{F}}(\boldsymbol{u}^{\mathcal{F}} + \boldsymbol{C}z) \\ \boldsymbol{0} \end{pmatrix} \tag{14.13}$$

and update $\boldsymbol{u} \leftarrow \boldsymbol{u} + \delta_{\boldsymbol{u}}$. Note that typical to SQP algorithms the Lagrange multipliers are computed directly. For the Hessian we use a Gauss–Newton approximation with respect to the data term

$$\boldsymbol{H} = T_u^\top T_u + \alpha \boldsymbol{A}.$$

For the computation of a numerical solution of the so-called KKT system (14.13) we use the multigrid-based solution scheme as proposed in [15]. The optimization procedure is summarized in Algorithm 14.2.

Note that in contrast to the linear part we now have to solve an $(n+m)$-by-$(n+m)$ linear system, where m is the dimension of the linear space and n is approximately the number of pixels/voxels.

Algorithm 14.2.
Fine Image Registration: $[\boldsymbol{u}_t^{\mathcal{F}}] \leftarrow \mathtt{FIR}(R, T, \boldsymbol{C}, \boldsymbol{u}^{\mathcal{F}}, z)$.
for $k = 0, \ldots$
Set $\boldsymbol{u} \leftarrow \boldsymbol{u}^{\mathcal{F}} + \boldsymbol{C}z$ and compute $T(\boldsymbol{u})$, $T_u(\boldsymbol{u})$, $\boldsymbol{D}^{\mathcal{F}}(\boldsymbol{u}^{\mathcal{F}})$, $\boldsymbol{D}_u^{\mathcal{F}}(\boldsymbol{u}^{\mathcal{F}})$.
Set $\boldsymbol{H} \leftarrow T_u^\top T_u + \alpha \boldsymbol{A}$ and solve
$\begin{pmatrix} \boldsymbol{H} & \boldsymbol{C} \\ \boldsymbol{C}^\top & \boldsymbol{0} \end{pmatrix} \begin{pmatrix} \delta_u \\ \boldsymbol{p} \end{pmatrix} = - \begin{pmatrix} \boldsymbol{D}_u^{\mathcal{F}} \\ \boldsymbol{0} \end{pmatrix}.$
set $\gamma \leftarrow 1$ **while** `line_search`
$\boldsymbol{u}_t^{\mathcal{F}} \leftarrow \boldsymbol{u}^{\mathcal{F}} + \gamma \delta_u$
if $\boldsymbol{D}^{\mathcal{F}}(\boldsymbol{u}_t^{\mathcal{F}}) < \boldsymbol{D}^{\mathcal{F}}(\boldsymbol{u}^{\mathcal{F}})$
STOP `line_search`
endif
$\gamma \leftarrow \gamma/2$
endwhile
if $\|\boldsymbol{u}_t^{\mathcal{F}} - \boldsymbol{u}^{\mathcal{F}}\| \leq \mathrm{tol}$
return
endif
$\boldsymbol{u}^{\mathcal{F}} \leftarrow \boldsymbol{u}_t^{\mathcal{F}}$.
endfor

14.4.3 The Multilevel Approach

For our multilevel approach we require two types of grid transfer operators. First, we need a grid transfer operator which coarsens the images, i.e., transfers an image from a fine level $\ell - 1$ to a coarser level ℓ. Second, we need two grid transfer operators which restrict and prolongate the transformation $\boldsymbol{u}^{\mathcal{F}}$.

For image transfer we use the down-sampled and convolved image

$$R_\ell = \mathrm{transfer}(R_{\ell-1}),$$

where the convolution stencil corresponds to a smoothing with $(1, 3, 3, 1)/8$ in each coordinate direction. For the prolongation of $\boldsymbol{u}^{\mathcal{F}}$ we use a linear interpolation scheme, as it is common for multigrid approaches. We denote the matrix representation of the prolongation by $\boldsymbol{P}_\ell$ and use $\boldsymbol{P}_\ell^\top$ as a restriction operator. We assemble $\boldsymbol{B}_1$ on the finest level and define $\boldsymbol{B}_\ell := \boldsymbol{B}_{\ell-1}\boldsymbol{P}_{\ell-1}$ for the coarser levels. Our numerical approach is summarized in Algorithm 14.1.

For the minimization on the coarsest level $\max_\ell$ we initialize $\boldsymbol{u}^{\mathcal{F}}_{\max_\ell} = \boldsymbol{0}$ and $z_{\max_\ell} = \boldsymbol{0}$. The numerical solutions of the optimization problems are transferred to the finer grid and used as starting values. For the linear part we set $z_\ell = z_{\ell+1}$ and for the nonlinear part we use the prolongation operator $\boldsymbol{P}_{\ell+1}$. However, since our prolongation scheme may not maintain the constraints, we apply an additional explicit orthogonalization step

$$\boldsymbol{u}_\ell = \operatorname{proj}(\boldsymbol{P}_\ell \boldsymbol{u}_{\ell+1}, \boldsymbol{C}_\ell), \quad \text{where} \quad \operatorname{proj}(\boldsymbol{u}, \boldsymbol{C}) = \boldsymbol{u} - \boldsymbol{C}(\boldsymbol{C}^\top\boldsymbol{C})^{-1}\boldsymbol{C}^\top\boldsymbol{u}.$$

The optimization scheme on the ℓth level is considered as converged if $\boldsymbol{u}^{\mathcal{F}}_\ell$ and/or z_ℓ converges.

14.5 Numerical Example

Figure 14.1 shows two two-dimensional sections of two three-dimensional magnetic resonance (MR) scans of a human knee; see also [18]. In this example, we choose $\mathcal{C}$ to be the space of affine linear transformations, such that the coarse and the linear registrations schemes coincides.

We start with some general remarks. Table 14.1 summarizes our results for four different multilevel registrations of the two 256×256 images, where five levels have been used such that the images on the coarsest level are 16×16. For comparison reasons, we used the same stopping criteria for the building blocks. The coarse registration on an ℓth level is considered as converged if $\left\|z^{(k)} - z^{(k-1)}\right\|_2 \leq 1\%$ or the number of iterations k exceeds 100. For the nonlinear/fine part, we stopped if $\left\|\boldsymbol{u}^{(k)} - \boldsymbol{u}^{(k-1)}\right\|_2 \leq 0.01\%$ or the number of iterations k exceeds 5.

In the following we discuss three registration approaches , the linear (or coarse), the common sequential linear/nonlinear, and the new COFIR registrations.

Plain coarse/linear. The most common approach is to use a linear preregistration. The results of the linear registration are depicted in Figure 14.1 ((d), (e), (f)). We observe that the overall result is fair in average, but considerable deformations are still observable particularly at the upper part of the Tibia. This is in accordance to the chosen integral-based distance measure; cf. (14.4). The computed deformation field is almost a pure rotation about 13 degrees.

Plain nonlinear. The results of a nonlinear registration without an affine linear preregistration are depicted in Figure 14.1 ((g), (h), (i)). Note that the results are meaningless (see, e.g., the deformation of the Tibia).

The new COFIR approach. Figure 14.1 ((j), (k), (ℓ)) shows the results of the new COFIR registration scheme. As is apparent from this figure, this registration gives reasonable results for the deformation as well as for the transformed image T^{COFIR}.

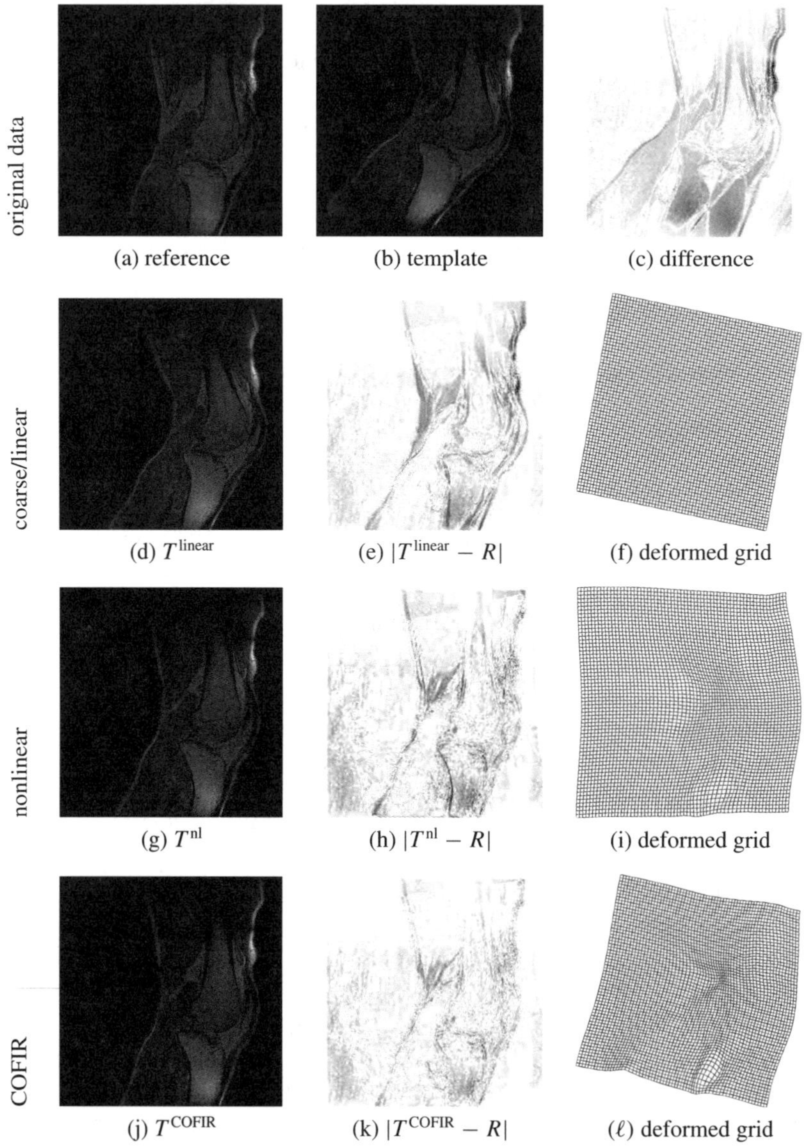

Figure 14.1. *Two MR images of a human knee* (a), (b) *and the difference image* (c) *as well as results for the linear or coarse* (d), (e), (f), *the nonlinear* (g), (h), (i), *and the new COFIR* (j), (k), (l) *registrations: final transformed images (first column), difference image (second column), and deformed grid (third column).*

Remark: The new COFIR approach combines two features, the clear segregation of the transformation space to coarse and fine parts and their treatment in a combined multilevel strategy.

Table 14.1. *SSD reduction and convergence results for the knee example. The tables show the image sizes and the reductions of the image difference versus the discretization level. The reduction in the difference is obtained from either the coarse grid correction, the coarse/linear, or the fine/nonlinear registration. The reduction is given in percent (%), where the initial difference on a particular level is considered as 100%. We also give the number of iterations spent for the coarse/linear and fine/nonlinear parts.*

Coarse/linear

Level	Size	Coarse	Red.	Iter.
1	16×16	–	46.4%	21
2	32×32	53.1%	52.0%	10
3	64×64	58.4%	58.0%	29
4	128×128	63.4%	62.8%	62
5	256×256	66.7%	66.6%	100

Sequential linear/nonlinear

Level	Size	Coarse	Red.	Iter.
1	16×16	52.1%	35.2%	5
2	32×32	38.5%	33.4%	5
3	64×64	38.2%	34.1%	5
4	128×128	39.1%	36.6%	5
5	256×256	42.9%	41.6%	5

COFIR

			Coarse		Fine	
Level	Size	Coarse	Red.	Iter.	Red.	Iter.
1	16×16	–	46.4%	21	28.2%	5
2	32×32	35.5%	33.6%	8	29.9%	5
3	64×64	37.2%	32.6%	5	29.9%	5
4	128×128	34.4%	34.0%	9	32.2%	5
5	256×256	38.8%	38.5%	17	37.0%	5

In all our examples, the new COFIR registration and the usual sequential linear/nonlinear registration lead to visually indistinguishable results. Therefore, one may conjecture that the separation assumption is true.

14.6 Conclusions and Further Discussion

In this chapter we present a novel and general framework for image registration based on a variational principle. As compared to other variational-based registration schemes, the new approach has two new features. First, a clean separation of the transformation space in a *coarse* and its orthogonal complement, the *fine* space, is performed. Only the fine part of the transformation is regularized. Second, rather than optimizing the parts separately using a multilevel scheme for each of the parts, we combine and solve for both parts simultaneously.

We present an implementation which is based on a multilevel sequential Gauss–

Newton scheme that allows for the introduction of coarse and fine transformations on coarse grids. For the fine part, we used a constrained optimization procedure to ensure the proper segregation of spaces.

For convex optimization problems, the optimization procedure is of minor importance. For nonconvex problems, like image registration, where one generally encounters many local minima, the procedure does play an important role. Therefore, the simultaneous treatment in the multilevel approach becomes essential.

Though the differences between the usual sequential linear/nonlinear and the COFIR registration appear to be very small, we prefer the COFIR scheme for two reasons. The first reason is that the COFIR scheme is based on a sound mathematical formulation. The second reason is that this clear approach comes almost for free: the computation times for a COFIR and a sequential linear/nonlinear registration are about the same.

Acknowledgments

Jan Modersitzki was supported by the US National Institutes of Health under grant NIHR01 HL 068904. We are indebted to Michele Benzi for suggestions and discussions.

Bibliography

[1] N. M. Alpert, J. F. Bradshaw, D. Kennedy, and J. A. Correia, *The principal axes transformation – A method for image registration*, J. Nuclear Medicine, 31 (1990), pp. 1717–1722.

[2] R. Bajcsy and S. Kovačič, *Toward an Individualized Brain Atlas Elastic Matching*, Technical report MS-CIS-86-71, Grasp Lap 76, Department of Computer and Information Science, Moore School, University of Philadelphia, Philadelphia, PA, 1986.

[3] J. Barequet and M. Sharir, *Partial surface and volume matching in three dimensions*, IEEE Trans. Pattern Anal. Mach. Intell., 19 (1997), pp. 929–948.

[4] P. J. Besl and N. D. McKay, *A method for registration of* 3-*d shapes*, IEEE Trans. Pattern Anal. Mach. Intell., 14 (1992), pp. 239–256.

[5] M. Bro-Nielsen, *Medical Image Registration and Surgery Simulation*, Ph.D. thesis, IMM, Technical University of Denmark, Copenhagen, Denmark, 1996.

[6] C. Broit, *Optimal Registration of Deformed Images*, Ph.D. thesis, Computer and Information Science, University of Pensylvania, Philadelphia, PA, 1981.

[7] L. Gottesfeld Brown, *A survey of image registration techniques*, ACM Computing Surveys, 24 (1992), pp. 325–376.

[8] G. E. Christensen and H. J. Johnson, *Consistent image registration*, IEEE Trans. Med. Imag., 20 (2001), pp. 568–582.

[9] G. E. Christensen, *Deformable Shape Models for Anatomy*, Ph.D. thesis, Sever Institute of Technology, Washington University, St. Louis, MO, 1994.

[10] A. Collignon, A. Vandermeulen, P. Suetens, and G. Marchal, *3d multi-modality medical image registration based on information theory*, Comput. Imaging Vision, 3 (1995), pp. 263–274.

[11] M. Droske, M. Rumpf, and C. Schaller, *Non-rigid morphological registration and its practical issues*, in Proceedings of the IEEE International Conference on Image Processing, Barcelona, Spain, 2003.

[12] B. Fischer and J. Modersitzki, *Fast diffusion registration*, in Inverse Problems, Image Analysis, and Medical Imaging, AMS, Providence, RI, 2002, pp. 117–127.

[13] B. Fischer and J. Modersitzki, *Combining landmark and intensity driven registrations*, PAMM, 3 (2003), pp. 32–35.

[14] B. Fischer and J. Modersitzki, *Curvature based image registration*, J. Math. Imaging Vision, 18 (2003), pp. 81–85.

[15] E. Haber and J. Modersitzki, *A multilevel method for image registration*, SIAM J. Sci. Comput., 27 (2006), pp. 1594–1607.

[16] E. Haber and J. Modersitzki, *Numerical methods for volume preserving image registration*, Inverse Problems, 20 (2004), pp. 1621–1638.

[17] G. Hermosillo, *Variational Methods for Multimodal Image Matching*, Ph.D. thesis, Université de Nice, Sophia-Antipolis, France, 2002.

[18] S. Kabus, T. Netsch, B. Fischer, and J. Modersitzki, *B-spline registration of 3d images with Levenberg-Marquardt optimization*, in Proceedings of SPIE 2004, Medical Imaging, San Diego, 2004, pp. 304–313.

[19] C. R. Maurer and J. M. Fitzpatrick, *Interactive image-guided neurosurgery*, in A Review of Medical Image Registration, American Association of Neurological Surgeons, Park Ridge, IL, 1993, pp. 17–44.

[20] J. Modersitzki, *Numerical Methods for Image Registration*, Oxford University Press, New York, 2004.

[21] J. Nocedal and S. J. Wright, *Numerical Optimization*, Springer-Verlag, New York, 1999.

[22] A. Roche, *Recalage d'images médicales par inférence statistique*, Ph.D. thesis, Université de Nice, Sophia-Antipolis, France, 2001.

[23] K. Rohr, *Landmark-Based Image Analysis*, Computational Imaging and Vision 21, Kluwer Academic Publishers, Dordrecht, The Netherlands, 2001.

[24] J.-P. Thirion, *Non-rigid matching using demons*, in Proceedings of the 1996 IEEE Conference on Computer Vision and Pattern Recognition, pp. 245–251.

[25] P. A. Viola, *Alignment by Maximization of Mutual Information*, Ph.D. thesis, MIT, Cambridge, MA, 1995.

[26] B. ZITOVÁ AND J. FLUSSER, *Image registration methods: A survey*, Image and Vision Computing, 21 (2003), pp. 977–1000.

Chapter 15

Real-Time, Large-Scale Optimization of Water Network Systems Using a Subdomain Approach

Carl D. Laird, Lorenz T. Biegler*, and Bart G. van Bloemen Waanders*†

15.1 Introduction

Certain classes of dynamic network problems can be modeled by a set of hyperbolic partial differential equations (PDEs) describing behavior along network edges and a set of differential and algebraic equations describing behavior at network nodes. In this chapter, we demonstrate real-time performance for optimization problems in drinking water networks. While optimization problems subject to partial differential, differential, and algebraic equations can be solved with a variety of techniques, efficient solutions are difficult for large network problems with many degrees of freedom and variable bounds. Sequential optimization strategies can be inefficient for this problem due to the high cost of computing derivatives with respect to many degrees of freedom [11]. Simultaneous techniques can be more efficient but are difficult because of the need to solve a large nonlinear program; a program that may be too large for current solver. This study describes a dynamic optimization formulation for estimating contaminant sources in drinking water networks, given concentration measurements at various network nodes. We achieve real-time performance by combining an efficient large-scale nonlinear programming algorithm with two problem reduction techniques. D'Alembert's principle can be applied to the PDEs governing behavior along the network edges (distribution pipes). This allows us to approximate the time-delay relationships between network nodes, removing the need to discretize along the

*Department of Chemical Engineering, Carnegie Mellon University, Pittsburgh, PA 15213.
†Optimization and Uncertainty Estimation Department, Sandia National Laboratories, Albuquerque, NM 87109.

length of the pipes. The efficiency of this approach alone, however, is still dependent on the size of the network and does not scale indefinitely to larger network models. We further reduce the problem size with a subdomain approach and solve smaller inversion problems using a geographic window around the area of contamination. We illustrate the effectiveness of this overall approach and these reduction techniques on an actual metropolitan water network model.

15.1.1 Contamination Source Determination

The vulnerability of municipal drinking water networks to intentional and accidental contaminations requires efficient systematic protection measures. Distribution networks cover a very large area, including private locations, making it impossible to use physical security alone to prevent drinking water contamination. If a contamination occurs, it is important to quickly identify the magnitude, time, and location of the contamination source to stop the spread of contaminant and devise a control strategy. Here we describe an approach for identifying contamination sources in water networks using concentration measurements from a set of installed contaminant sensors. Assuming that the cost of these sensors is high, the contaminant source must be effectively identified using few network sensors. It is also important to solve the inversion problem as quickly as possible. Since the inversion may be one component of an overall control strategy, it must be efficient and solve in real time for large water distribution networks.

Using known network flow rates and concentration measurements, we formulate a least-squares dynamic optimization problem subject to the constraints of the water quality model and unknown contaminant injection sources. Solving this dynamic optimization problem for the unknown, time-dependent source terms identifies both the time and location of possible contaminant injections.

A number of different techniques can be used to solve the dynamic optimization problem, progressing from minimal interfacing with an existing simulator to a more intrusive fully simultaneous technique. Although the fully simultaneous discretization technique produces a very large nonlinear program, in many cases it can be significantly more efficient than sequential techniques [2, 10, 11]. Because of the large number of degrees of freedom and the high cost of calculating derivatives for the time-discretized injection profiles, sequential techniques are not appropriate for the solution of the time-dependent network source determination problem. A straightforward application of the simultaneous approach is not possible, since a naive discretization of the network model in both time and space produces a nonlinear program that is too large for general nonlinear programming (NLP) tools. Indeed, given a network of 500 nodes, with 100 discretizations in time and an average of 1000 discretizations along the length of the pipes, the NLP would have on the order of $1 \cdot 10^8$ variables. Removing the need to discretize in space decreases this problem size by three orders of magnitude.

We recently developed a reformulation of the pipe constraints, converting the spatial PDEs into a set of algebraic time-delay expressions, allowing the use of a simultaneous technique for solving the time-dependent problem [4]. In that study, efficient numerical results were presented for a network model of approximately 500 nodes, with solution times under two minutes of wall clock time for all test scenarios. However, this approach performs the optimization over the space of the entire network and does not scale to significantly larger networks. The bottleneck in addressing very large systems is the need to find solutions for large linear systems at each iteration of the optimization algorithm. Since processing and memory requirements are not linear in the number of variables, increasing the problem size can cause a dramatic increase in solution times. Furthermore, with direct factorization techniques, memory limitations place restrictions on the overall size of the linear system, thus restricting the number of network nodes and the number of time steps that can be considered.

To consider larger metropolitan water networks, we apply a network subdomain approach to reduce the problem size further. Since the time delays associated with paths through water networks are typically long, a rapid source inversion method should identify contamination sources before the contaminant spreads significantly through the network. Therefore, the optimization problem can be formulated on a subset of the overall network. The required subdomain size is not dependent on the size of the entire network, and so this approach can deal with large municipal drinking water networks very efficiently. Moreover, the location and size of the subdomain need not be fixed, and tracking the spread of the contaminant and implementing a control strategy allows straightforward extensions of this approach to moving subdomains. While the primary purpose of the subdomain technique is to reduce computational effort, evidence suggests that this technique may also improve solution quality by allowing a finer discretization in time. To demonstrate this subdomain approach we consider the source inversion problem for a network model that describes the distribution system for a large city. We examine the quality of the inversion solutions under varying subdomain sizes. Our results indicate that, shortly following an injection, this subdomain approach is an effective problem reduction technique, and that the overall algorithm is an efficient method for contaminant source detection.

15.1.2 Outline of the Chapter

Section 15.2 presents the formulation of the contamination source determination problem and describes solution techniques for this problem. In particular, we consider an origin tracking algorithm that converts the PDE-constrained problem into a differential algebraic optimization problem. This problem can then be discretized in time and solved with large-scale NLP methods. We then develop an alternative formulation that applies to any selected subdomain of the network. Section 15.3 applies the subdomain approach on a real municipal water network model. This approach considers hydraulics and sensor measurements from the entire network, selects a subdomain for the optimization, and solves the optimization problem in the space of these selected network nodes only. A case study illustrates the effectiveness of this approach and the effect of subdomain size on computational effort and solution quality. Section 15.4 then concludes the chapter and offers a number of directions for future work.

15.2 Dynamic Optimization Formulation

Before developing the optimization formulation for the subdomain approach, we first describe the formulation for the full network model as given in [4]. Water distribution systems can be represented as networks with nodes and edges. The nodes represent zero-volume

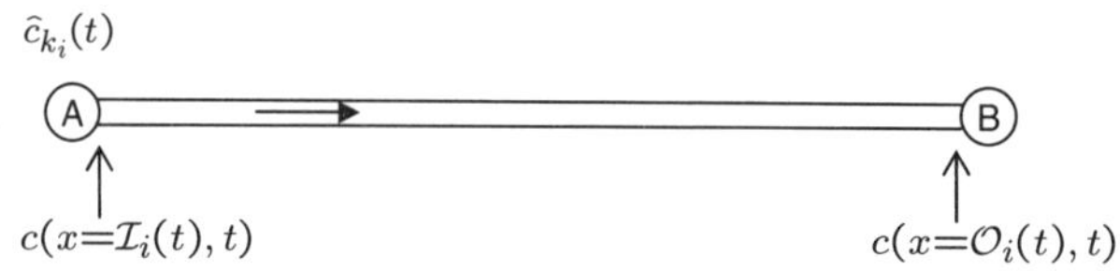

Figure 15.1. *Link boundary designation. $\mathcal{I}_i(t)$ indicates the inlet of the link, based on the current flow direction, while $\mathcal{O}_i(t)$ indicates the outlet of the link. The index $k_i(t)$ refers to the node connected at the inlet.*

junctions, storage tanks, or reservoirs (network hydraulic sources). The network edges represent pipes, valves, or pumps. We assume that all time-dependent pipe flows in the network can be determined in advance either from measurements or numerical simulation, and these become defined inputs to the water quality model [5, 7, 8, 9, 14]. Generally, low-flow residential areas are collapsed into single network nodes and only larger distribution pipes are modeled. Because the flow in these main distribution lines have high Reynolds numbers in the turbulent regime, we assume plug flow behavior in all pipe segments. Also, contaminant decay through reactions within the pipe segments is assumed to be negligible, although the model could be formulated with first-order decay kinetics without affecting the model properties.

The water quality model is developed using $\mathcal{P}$, $\mathcal{J}$, and $\mathcal{S}$ to refer to the index sets of all pipes, junctions, and storage tanks, respectively. In addition, $c_i(x,t), i \in \mathcal{P}$, represent the concentration in the pipes, and $\hat{c}_k(t), k \in \mathcal{N}$, represent the concentration at the nodes, where $\mathcal{N} = \mathcal{J} \cup \mathcal{S}$ is the index set of all nodes including junctions and storage tanks. We make no assumption about the time of the injection or the duration and introduce unknown, time-dependent mass injection terms, $m_k(t)$, for each node, $k \in \mathcal{N}$.

Here $t \in [0, t_f]$ defines the time horizon considered, and $x \geq 0$ is the displacement along a pipe. The notation for designating pipe boundaries is based on flow direction, as shown in Figure 15.1. Here $x{=}\mathcal{I}_i(t)$ refers to the boundary where fluid is entering pipe i and $x{=}\mathcal{O}_i(t)$ refers to the boundary where fluid is leaving pipe i. The index $k_i(t), i \in \mathcal{P}$, refers to the node connected at the inlet of pipe i at time t, again, determined by the flow direction. The source determination problem can therefore be stated as

$$\min_{\substack{m_k(t), c_i(x,t), \hat{c}_k(t) \\ \forall i \in \mathcal{P}, k \in \mathcal{N}}} \quad \Psi = \int_0^{t_f} \sum_{r \in \Theta^\star} \sum_{k \in \mathcal{N}^\star} w_k(t) \left(\hat{c}_k(t) - \hat{c}_k^\star(t)\right)^2 \delta(t - t_r) + \rho \sum_{k \in \mathcal{N}} m_k(t)^2 \, dt \quad (15.1)$$

subject to

$$\left.\begin{array}{l} \frac{\partial c_i(x,t)}{\partial t} + u_i(t) \frac{\partial c_i(x,t)}{\partial x} = 0, \\ c_i(x{=}\mathcal{I}_i(t), t) = \hat{c}_{k_i(t)}(t), \quad c_i(x,0) = 0 \end{array}\right\} \qquad \forall i \in \mathcal{P},$$

$$\hat{c}_k(t) = \frac{\left(\sum_{i \in \Gamma_k(t)} Q_i(t)\, c_i(x{=}\mathcal{O}_i(t), t)\right) + m_k(t)}{\left(\sum_{i \in \Gamma_k(t)} Q_i(t)\right) + Q_k^{ext}(t) + Q_k^{inj}(t)} \qquad \forall k \in \mathcal{J}, \quad (15.2)$$

$$V_k(t)\frac{d\hat{c}_k(t)}{dt} = \left(\sum_{i \in \Gamma_k(t)} Q_i(t)\, c_i(x{=}\mathcal{O}_i(t), t)\right) + m_k(t)$$
$$- \left[\left(\sum_{i \in \Gamma_k(t)} Q_i(t)\right) + Q_k^{ext}(t) + Q_k^{inj}(t)\right]\hat{c}_k(t), \quad \hat{c}_k(0) = 0 \quad \forall k \in \mathcal{S}, \quad (15.3)$$

$$m_k(t) \geq 0 \qquad \forall k \in \mathcal{N}. \quad (15.4)$$

Letting Θ be the complete set of discretized points in time, $\Theta^* \subseteq \Theta$ is the index set of sample times for the measured concentration data and $\mathcal{N}^* \subseteq \mathcal{N}$ is the index set of nodes with installed sensors. The objective function (15.1) minimizes the difference between measured nodal concentration data $\hat{c}_k^\star(t), k \in \mathcal{N}^*$, and corresponding values calculated from the network model, weighted by a flow-based function, $w_k(t), k \in \mathcal{N}^*$. The measured concentrations are known at the selected sampling times, $t_r, r \in \Theta_s$, and $\delta(t)$ is the Dirac delta function. The final term in (15.1) regularizes the objective and forces a unique solution to the inversion problem, where ρ is the regularization parameter.

Equation (15.2) models the network pipes and the junctions, (15.3) models the tanks, and (15.4) bounds the injection terms to be nonnegative. Here $\Gamma_k(t), k \in \mathcal{N}$, is the set of all pipes flowing into node k at time t. This designation is time dependent and changes with the flow direction. In addition, $u_i(t)$ is the known fluid velocity in pipe i, $Q_i(t), i \in \mathcal{P}$, is the known volumetric flow rate in pipe i at time t, $Q_k^{ext}(t), k \in \mathcal{N}$, is the volumetric flow rate for known external sources (e.g., reservoir flow), and $Q_k^{inj}(t), k \in \mathcal{N}$, is the volumetric flow rate of the unknown contaminant mass injection, $m_k(t)$. In practice the unknown value of Q_k^{inj} is set to a negligible quantity relative to other network flow rates. Finally, $V_k(t), k \in \mathcal{S}$, is the volume in tank k at time t.

Since the network flows are specified inputs, the only unknowns are the pipe and node concentrations, $c_i(x,t)$ and $\hat{c}_k(t)$ respectively, and the mass injections, $m_k(t)$. The traditional forward problem is that of setting the mass injections, $m_k(t)$, and solving (using (15.2)–(15.3)) for the network concentrations. In the inverse problem (15.1)–(15.4), the goal is to solve for the values of the injection terms, $m_k(t)$. Significantly positive values for the solution at a particular node indicate a potential contaminant source location. The optimization solution also gives the complete time profiles for the network concentrations. Problem (15.1)–(15.4) presents an infinite-dimensional optimization problem subject to algebraic, ordinary differential, and partial differential constraints.

Problems of this type can be solved with a *direct simultaneous* method, where we fully discretize all the state and control variables in the problem and solve the resulting system as a large-scale optimization problem with algebraic constraints. The forward problem is converged only once, along with the optimality conditions. Accurate analytical derivatives are often straightforward and efficient to calculate, and significant computational gains over alternative approaches are possible using this more intrusive technique. A review of this and other methods as applied to optimization of differential algebraic optimization problems can be found in [2]. However, straightforward application of the direct simultaneous method to problem (15.1)–(15.4) leads to a discretization in both time and space and a nonlinear program that is too large for current optimization tools. Real-time performance is a requirement for this application and many others, and while a Lagrangian viewpoint [5] of the pipe equations (15.2) provides a much more efficient simulation, a reformulation of the pipe equations is required within a simultaneous setting.

In the remainder of this section, we discuss two problem reduction techniques. The *origin tracking algorithm* [4] preprocesses the network flow patterns, derives a delay differential system of equations, and approximates the time delays. This reformulation removes the need to discretize in x. In addition, we briefly sketch a *network subdomain approach.* If the spread of contamination is slow, the relevant region is a small subdomain of the entire network and a much smaller NLP problem can be considered.

15.2.1 Origin Tracking Algorithm

The *origin tracking algorithm* approach has a physical interpretation for the water network inversion problem [4], but it is general in nature and applies to any problem where d'Alembert's principle can be applied to the PDEs governing behavior along network edges. Consider the discretization of the node concentrations, $\hat{c}_k(t)$, and the injection terms $m_k(t)$ in time $t_l, l \in \Theta$. To reformulate the pipe equations, we use the known flow profiles to calculate the time-delay relationships between pipe boundaries for all pipes and each time discretization. Specifically, we need to develop expressions for $c_i(x = \mathcal{O}_i(t_l), t_l)$ given $c_i(x = \mathcal{I}_i(t_l), t_l)$ for all $i \in \mathcal{P}$ and all points $t_l, l \in \Theta$. Assuming that the form of $u_i(t)$ is known, the relationship for the inlet pipe boundary is straightforward to write from the pipe boundary conditions as

$$c_i(x{=}\mathcal{I}_i(t_l), t_l) = \hat{c}_{k_i(t_l))}(t_l) \tag{15.5}$$

for nonzero $u_i(t)$.

Writing the relationship for $c(x{=}\mathcal{O}(t_l), t_l)$ requires further consideration of the pipe segment equations (15.2). Following elementary solutions of one-dimensional wave equations, we apply d'Alembert's principle and write

$$\frac{dc_i}{dt} = \frac{\partial c_i(x,t)}{\partial t} + \left(\frac{dx}{dt}\right)\frac{\partial c_i(x,t)}{\partial x} \tag{15.6}$$

$$= \frac{\partial c_i(x,t)}{\partial t} + u_i(t)\frac{\partial c_i(x,t)}{\partial x} = 0 \tag{15.7}$$

for each pipe segment, $i \in \mathcal{P}$. Equating (15.6) and (15.7) yields

$$\frac{dc_i}{dt} = 0, \qquad \frac{dx}{dt} = u_i(t), \tag{15.8}$$

and from the boundary conditions, we have

$$c_i(x,t) = c_i(\mathcal{I}_i(\bar{t}), \bar{t}) = \hat{c}_{k_i(\bar{t}))}(\bar{t}) \tag{15.9}$$

for values of x, t, and $\bar{t}$ that satisfy

$$x(t) = \mathcal{I}_i(\bar{t}) + \int_{\bar{t}}^{t} u_i(\tau)d\tau. \tag{15.10}$$

Since the form of $u_i(t)$ is known, we can write this equation for the outlet boundary as

$$\mathcal{O}_i(t_l) = \mathcal{I}_i(\bar{t}) + \int_{\bar{t}}^{t_l} u_i(\tau)d\tau, \tag{15.11}$$

where the only unknown is $\bar{t}$. Solving this equation for $\bar{t}$ then allows us to write an expression for the outlet concentration

$$c_i(\mathcal{O}_i(t_l), t_l) = c_i(\mathcal{I}_i(\bar{t}), \bar{t}). \tag{15.12}$$

Of course, it is unlikely that $\bar{t}$ will be in the set of discretized points $t_l, l \in \Theta$. Therefore, we accept an approximation to the time delay and exchange $\bar{t}$ with some nearby discretized point in time, $t_{\bar{l}}$,

$$c_i(\mathcal{O}_i(t_l), t_l) = c_i(\mathcal{I}_i(t_{\bar{l}}), t_{\bar{l}}). \tag{15.13}$$

The method for selecting this nearby point is problem specific and for physical reasons, we select the nearest point smaller than $\bar{t}$ (i.e., $t_{\bar{l}} \le \bar{t} < t_{\bar{l}+1}$).

In the drinking water network source inversion problem, $u_i(t)$ represents the known pipe velocities and is assumed piecewise constant over each interval $t_l < t \le t_{l+1}$. We write (15.11) as

$$\mathcal{O}_i(t_l) = \mathcal{I}_i(\bar{t}) + \int_{\bar{t}}^{t_l} u_i(\tau)d\tau \tag{15.14}$$

$$= \mathcal{I}(\bar{t}) + u_i(\bar{t}) \cdot (t_{\bar{l}+1} - \bar{t}) + \sum_{j=\bar{l}+1}^{l-1} u(t_j) \cdot (t_{j+1} - t_j). \tag{15.15}$$

In this expression we assume that the time discretization is sufficiently fine that neglecting the second term on the right-hand side does not lead to large errors. Determining the time delay then reduces to the problem of finding the index $\bar{l}$ such that we can satisfy the following inequalities:

$$\sum_{j=\bar{l}+1}^{l-1} u(t_j) \cdot (t_{j+1} - t_j) \le \mathcal{O}_i(t_l) - \mathcal{I}_i(\bar{t}) \le \sum_{j=\bar{l}}^{l-1} u(t_j) \cdot (t_{j+1} - t_j). \tag{15.16}$$

Here the quantity $\mathcal{O}_i(t_l) - \mathcal{I}_i(\bar{t})$ is the length of pipe segment i, or zero if the weighted sum of the flows is zero. In particular, if $\bar{l} = 0$ is reached before the solution is bracketed, $\mathcal{O}_i(t_l)$ would be set to the initial condition for the segment. This approach sets the time delay and hence the relationship between the concentrations at the inlet and the outlet by truncating the result at time discretizations. There is no guarantee that this truncated approach will preserve the mass traveling through the network. While the approach could be modified to include some interpolation of the individual packets, this leads to a smearing of the sharp

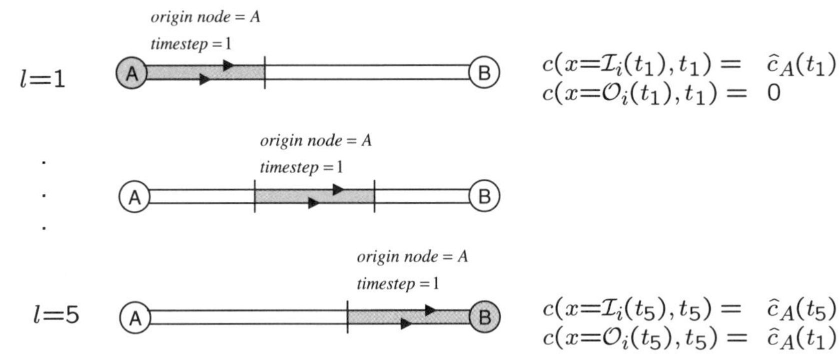

Figure 15.2. *Origin tracking algorithm. This figure illustrates the flow of a single volume element through a pipe. Tracking these elements allows us to determine relationships for the boundary concentrations at each point in time.*

front as it travels through the network, and this smearing is compounded as fluid travels through each node.

Furthermore, the simulated injections used in this study are all square pulse injections, and we feel it is as important to accurately identify the time and location of the injection as the true mass injected.

To determine the time delay that satisfies (15.16) we apply an origin tracking algorithm that calculates the position of past boundary conditions as they progess along the length of the pipe segment. This approach is equivalent to tracking the origin of discretized packets of fluid as they enter and travel through the pipe, allowing us to write an expression for $c(x{=}\mathcal{O}_i(t_l), t_l)$.

To illustrate this idea, Figure 15.2 shows an example for pipe i with flow conditions from left to right. At time step $l_1 = 1$, a fluid element enters the left boundary of the pipe, and we can record its originating node as "A" and its originating time step as $l_1 = 1$. The concentration at the left boundary of the pipe is equal to the immediate concentration from node A, that is, $c_i(\mathcal{I}_i(t_{l_1}), t_{l_1}) = \hat{c}_A(t_{l_1})$. As time progresses, the element advances through the pipe and at some future time step, say $l_2 = 5$, the concentration at the right boundary, however, is now the same as the concentration from node A at time step l_1, $c_i(\mathcal{O}_i(t_{l_2}), t_{l_2}) = \hat{c}_A(t_{l_1})$. The linear time relationships in Figure 15.2 describe the concentration of the pipe boundaries as a function of the concentrations of connected nodes.

While this simple example illustrates the basic idea, the more detailed origin tracking algorithm developed in [4] formalizes the approach and handles situations of changing pipe velocity and flow reversals. This algorithm defines expressions for all network pipes under varying flow conditions. We denote these linear linking equations (15.13) by $p_i(c_i(\mathcal{O}_i, t_l), \hat{c}_{k_i(t_{\bar{l}})}(t_{\bar{l}})) = 0$ for all $l \in \Theta$. These linking equations replace (15.2) in the optimization problem, thus removing the spatial dependence. From [4], our Lagrangian-based algorithm has two advantages over previous work. First, it allows application of the simultaneous approach where the time discretization is fixed a priori by some scheme selected for the differential tank equations. Second, the analysis is performed on a pipe by pipe basis, not on the network as a whole. The processing and memory requirements are then linear with the number of pipes and efficient for large networks. While these simplifications introduce estimation errors in the time delays associated with paths through the network, they provide efficient scaling and sparsity.

15.2.2 Network Subdomain Approach

The above algorithm has successfully inverted for the magnitude, time, and location of contamination sources on a relatively complex network model of approximately 500 nodes. Due to the limitations of direct linear solvers, it becomes significantly more expensive to solve the full network problem for increasingly larger networks. The challenge in this current work is to invert for contamination sources in much larger networks, where solving the optimization problem on the entire network is not possible. Instead, hydraulics and sensor measurements are considered from the entire network, a subdomain of the entire network is selected, and the optimization problem is formulated and solved for this subdomain only.

The subdomain is identified by choosing some subset of *selected* nodes and edges from the full network and is defined by the sets $\mathcal{SN} \subseteq \mathcal{N}$, $\mathcal{SJ} \subseteq \mathcal{J}$, $\mathcal{SP} \subseteq \mathcal{P}$, and $\mathcal{SS} \subseteq \mathcal{S}$. We then rewrite problem (15.1)–(15.4) for this subset of selected nodes. The node concentrations, $\hat{c}_k(t)$, injection terms, $m_k(t)$, and node expressions, (15.2)–(15.4), are only necessary for selected nodes. Given a set of selected nodes, it is straightforward to determine which pipes need to be considered in $\mathcal{SP}$. If pipe i is connected to a selected node at each end, the pipe itself is *selected* and must be modeled. On the other hand, if pipe i is not connected to any selected nodes, it is deselected and can be completely excluded from the model. When pipe i is connected to a selected node at one end and a deselected node at the other, the pipe is still classified as deselected, but it represents a boundary to the subdomain. Since the flow conditions for the node at this boundary must be rectified, the flow into the subnetwork from this pipe is treated as a known external source and does not contribute any contaminant to the node.

The optimization problem in the space of the selected subdomain is given by

$$\min_{\substack{m_k(t), c_i(x,t), \hat{c}_k(t),\\ \forall i \in \mathcal{SP}, k \in \mathcal{SN}}} \quad \Psi = \int_0^{t_f} \sum_{r \in \Theta^\star} \sum_{k \in \mathcal{SN}^\star} w_k(t) \left(\hat{c}_k(t) - \hat{c}_k^\star(t)\right)^2 \, \delta(t - t_r) + \rho \sum_{k \in \mathcal{SN}} (m_k(t))^2 dt \tag{15.17}$$

subject to

$$p_i(c_i(\mathcal{O}_i, t_l), \hat{c}_{k_i(t_{\bar{l}})}(t_{\bar{l}})) = 0 \quad \forall l \in \Theta \quad \forall i \in \mathcal{SP}, \tag{15.18}$$

$$\hat{c}_k(t) = \frac{\left(\sum_{i \in \mathcal{SP} \cup \Gamma_k(t)} Q_i(t)\, c_i(x{=}\mathcal{O}_i(t), t)\right) + m_k(t)}{\left(\sum_{i \in \Gamma_k(t)} Q_i(t)\right) + Q_k^{ext}(t) + Q_k^{inj}(t)} \quad \forall k \in \mathcal{SJ}, \tag{15.19}$$

$$V_k(t) \frac{d\hat{c}_k(t)}{dt} = \left(\sum_{i \in \mathcal{SP} \cup \Gamma_k(t)} Q_i(t)\, c_i(x{=}\mathcal{O}_i(t), t)\right) + m_k(t)$$

$$-\left[\left(\sum_{i\in\Gamma_k(t)} Q_i(t)\right) + Q_k^{ext}(t) + Q_k^{inj}(t)\right]\hat{c}_k(t), \quad \hat{c}_k(0) = 0 \quad \forall k \in \mathcal{SS}, \quad (15.20)$$

$$m_k(t) \geq 0 \qquad \forall k \in \mathcal{SN}. \quad (15.21)$$

Note that the objective has been modified to sum the concentration errors for the selected sensor nodes in $\mathcal{SN}^\star$ and to regularize over the injection terms in $\mathcal{SN}$. Similarly, the mass balances for the junctions and tanks have been modified to include nonzero concentrations from selected pipes only. In the event of a contamination, a subdomain of the entire network can be selected and the solution techniques described in Section 15.2 can be used to solve the smaller, reduced network problem (15.17)–(15.21). While the formulation presented is general enough for any set of selected network nodes, it remains to select a suitable subdomain. Any subdomain selection technique must satisfy the following goals:

1. The resulting nonlinear program must be small enough for current optimization techniques; therefore the selection technique must choose a reasonable number of nodes from the network.

2. This source inversion problem uses real-time sensor measurements from the network and needs to be solvable in an online setting. Therefore, any preprocessing to select the network subdomain must be efficient for large networks.

3. The algorithm must select a subdomain that sufficiently represents the area of interest to the source inversion problem. This can be difficult, since the actual source is, of course, not known.

In this study, a simple topological window is used to select the network subdomain. The geographical location of the first sensor to detect contaminant is identified. Then, the n closest nodes to this location are selected for the subdomain, where the value of n is chosen large enough to represent the area of interest but small enough to produce a solvable nonlinear program. Of course, there is no explicit guarantee that this subdomain will include the actual contamination locations, but if a contamination source is identified on the boundary of the subdomain, the window could be expanded or moved.

15.3 Numerical Results

Previous work has illustrated the effectiveness of formulation (15.1)–(15.4) and the origin tracking algorithm for identifying contamination sources for network models with approximately 500 nodes [4]. Nevertheless, we observe that problem sizes cannot increase indefinitely for the source inversion problem. This is particularly true because, at every iteration, the interior point NLP solver used in this study requires the solution of a symmetric indefinite linear system; solver reliability requires the inertia of this linear system to ensure the generation of descent directions. Moreover, the arbitrary structure of the network and the linear system makes preconditioned iterative methods difficult to apply; direct methods are needed instead. This limits us to a particular class of sparse direct linear solvers, and we currently use MA27 from the Harwell Subroutine Library. With arbitrary network structures, we do not have an easily defined complexity with respect to problem size; however, our experience has shown scaling greater than linear for these problems.

As a result, the primary purpose of the subdomain approach is to sufficiently reduce the size of the resulting nonlinear program, thus allowing solutions of large network inversion problems with optimization packages that use direct linear solvers. As the size of the subdomain is varied, we are interested in both the computational requirements (time and memory) and the quality of solutions. We first examine these effects by decreasing the size of the subdomain while keeping the time discretization fixed. The size of the nonlinear program decreases with the size of the subdomain, showing the efficiency gains that can be realized with the subdomain approach. Of course, the problem size cannot be reduced indefinitely, and we are more interested in determining the best subdomain size given a fixed budget of computational effort. A second set of results examines the effect of decreasing the subdomain size while keeping the size of the nonlinear program constant, by increasing the number of time discretizations. This result shows how the subdomain approach can be used to produce higher quality solutions by allowing a finer discretization for the same computational effort.

Results are demonstrated using a real municipal water network model, shown in Figure 15.3, with 3500 nodes with 350 randomly placed sensors.[10] Figure 15.4 shows the circled portion of the network in more detail, where shaded nodes indicate nodes with installed sensors. The attack scenario is simulated using EPANET [6], where an hour long injection is simulated at hour 7 from node 2921.

We have developed a formulation tool that produces the nonlinear program for the subdomain problem. Specifying the optimization horizon, the subdomain size, and the time discretization, the software tool reads the simulated hydraulic information for the entire network and the concentration measurements for sensor nodes. The tool then forms the subdomain by identifying the first sensor to detect contaminant and selecting the nearest n nodes.

For a given time discretization, the formulation tool performs the origin tracking algorithm and determines the time delays for all the selected network pipes. It then writes the time-discretized nonlinear program (15.17)–(15.21) using a simple backward Euler technique for the tank equations (15.20). Although more elaborate discretization techniques could be used (like collocation on finite elements [1]), the stepsize tends to be governed by accuracy requirements on the time delays and a simple technique is sufficient for the tank equations. The discretized nonlinear program is written in AMPL [3] format. AMPL is an optimization modeling language that provides first- and second-order derivatives using automatic differentiation. Although the optimization problem (15.17)–(15.21) is a large-scale quadratic program, the addition of general nonlinear reaction terms would destroy the linearity of the constraints. As well, (15.21), once discretized, produces a large number of variable bounds, and interior point algorithms provide an efficient alternative to active-set strategies for these problems. For these reasons, the nonlinear program is solved with IPOPT [12, 13], a large-scale, interior point, optimization tool (see http://www.coin-or.org). All optimization results were formulated with a 6-hour time horizon from the 5th to the 11th hour and were solved on a 1.8 GHz Pentium 4 machine.

[10]This image has been cropped and does not show the full network model.

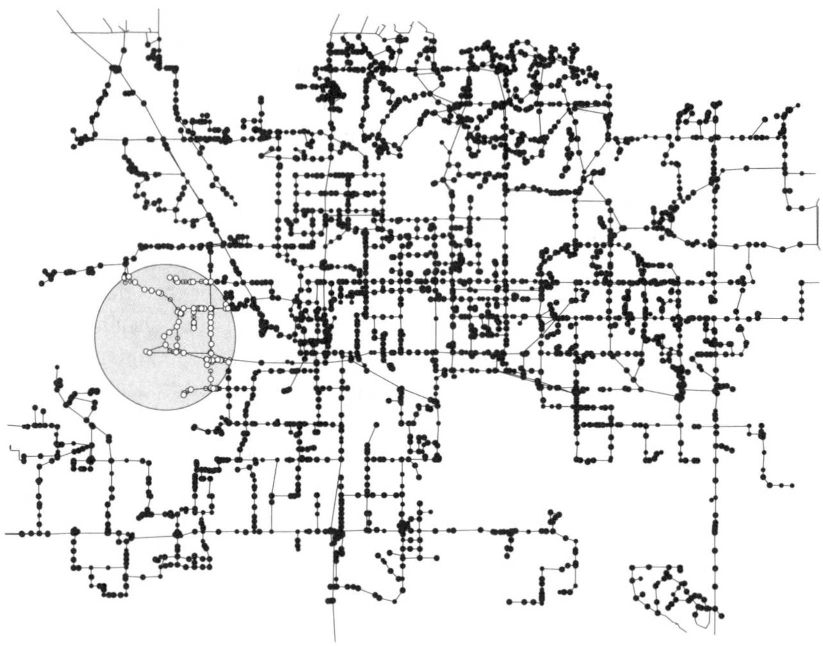

Figure 15.3. *Municipal water network. The circled area is shown in more detail in Figure* 15.4.

15.3.1 Results: Fixed Discretization, Variable Problem Size

In this set of results, the size of the subdomain is varied, but the time step is kept fixed at 5 minutes, giving 72 discretizations over the 6-hour optimization horizon. Figure 15.5 shows the inversion solution for a subdomain size of 900 nodes, plotting all injection profiles that, at some point, have significant values. Here we see that the inversion solution is nonunique and has indicated four possible injection locations at nodes 2820, 2879, 2921, and 3612. Nevertheless, we can see two significant profiles for nodes 2921 (the actual injection location) and 2879 (a neighboring node), and the method has reasonably identified a small region containing the injection location (2921) and also correctly identified the injection time (7–8 hours) for this location.

The solution nonuniqueness is primarily due to the sensor locations and network topology [4]. Since the nonlinear program makes no assumption about the location, time, or number of injections, any linear combination of injections that produces the observed data is a valid solution and hence the need for the regularization term in the objective. Errors in the time-delay approximation (the use of $t_{\bar{l}}$ instead of $\bar{t}$) can also contribute to poor solution quality. Since these errors propagate, they are more pronounced when optimizing over larger networks (or larger subdomains) and longer time horizons. We can see the evidence of this effect by examining the solution profile for a smaller subdomain. Smaller subdomains provide less information for the optimization problem, and it is expected that the solution

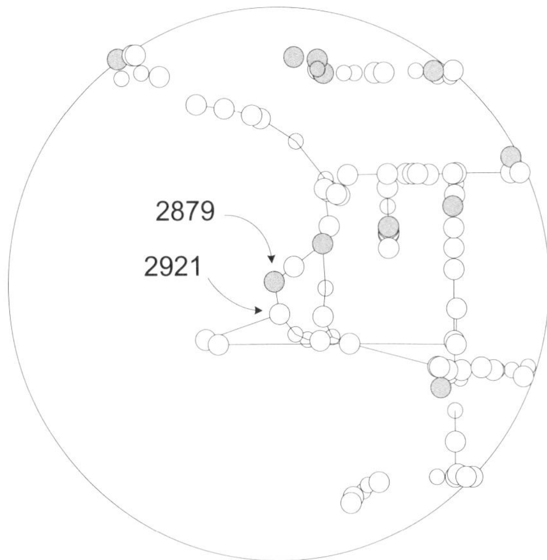

Figure 15.4. *Municipal water network. This figure shows increased detail near the injection location, where shaded nodes indicate sensor locations.*

quality would deteriorate as the subdomain size decreased. This is not necessarily the case. Figure 15.6 shows the inversion solution using a subdomain size of 100 nodes. Here we can see that the two suspect locations (2820 and 3612) are no longer a part of the solution. For window sizes smaller than 400, these two nodes are no longer included within the window. While it may seem that this approach has excluded possible solutions, the injection time for these suspect nodes was later than the time that the first sensor detected contaminant. This, coupled with the fact that they are far from the center of the window, leads us to believe that they are a result of model errors alone.

Table 15.1 summarizes the inversion results for subdomain sizes ranging from 900 to 50 nodes. The first column shows the number of nodes selected for the subdomain, the second column is the number of variables in the resulting nonlinear program, the third column is the approximate computational time required to perform the optimization, and the last column indicates the number of possible injection nodes identified in the solution. In all cases, the actual injection node was also identified as a possible injection location.

As expected, the computational time is dramatically reduced as the size of the subdomain is decreased. Interestingly, for this example, the solution quality does not deteriorate with smaller subdomains but actually improves. This result is not general and using increasingly smaller subdomains may result in loss of necessary sensor information. On the other hand, our experience with smaller subdomains shows that they can remove network nonuniqueness and reduce the effect of errors in the time-delay approximation.

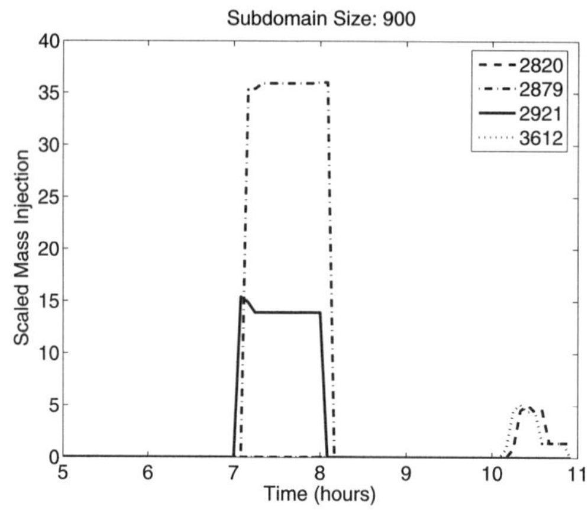

Figure 15.5. *Solution with* 900 *node subdomain. This figure shows the solution of the inversion problem for a* 5*-minute time discretization and a subdomain size of* 900 *nodes. Although the solution is nonunique, showing* 4 *possible injection locations, the significant profiles* 2921 *and* 2879 *are the actual injection location and its neighbor. The solution has also correctly identified the injection time.*

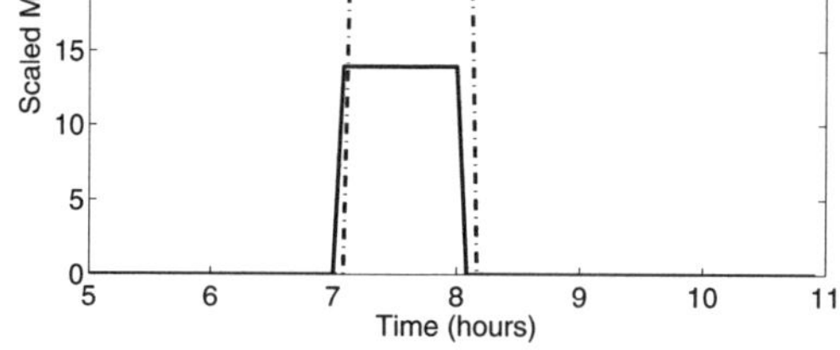

Figure 15.6. *Solution with* 100 *node subdomain. This figure shows the solution of the inversion problem for a* 5*-minute time discretization and a subdomain size of* 100 *nodes. The solution quality has improved over the larger* 900 *node subdomain.*

Table 15.1. *Optimization results with fixed time discretization.*

Subdomain size	Number of variables	Execution time (sec)	Identified nodes
900	267698	70	4
800	238329	62	4
700	208656	56	4
600	178848	47	4
500	148176	38	4
400	117072	29	4
300	88128	18	2
200	57888	10	2
100	28800	5	2
50	14256	2	2

Table 15.2. *Formulation parameters for different subdomain sizes.*

Subdomain size	Number of discretizations	Timestep size (min)
300	24	15
200	36	10
100	72	5
40	180	2

15.3.2 Results: Fixed Problem Size, Variable Discretization

In this section, we examine the effect of changing the size of the subdomain while keeping the size of the nonlinear program constant. We are still interested in solutions using the 6-hour time horizon. Therefore, as the size of the subdomain is decreased, the number of time discretizations are increased proportionally, keeping the computational effort approximately constant. The quality of solutions is examined as the subdomain size decreases and the discretizations are refined.

Table 15.2 shows the parameters used to generate the inversion results. Here we have kept the size of the nonlinear program approximately constant with $|\mathcal{N}||\Theta| = 7200$, where $|\cdot|$ denotes the cardinality of the set. For example, this corresponds to a subdomain of 300 nodes, and 24 time discretizations, requiring a 15-minute time step to cover the complete 6-hour time horizon. Using this expression, we calculate the number of allowed time discretizations (and hence the length of each time step) for each subdomain size. Therefore, finer discretizations are used at the expense of smaller subdomain sizes.

Using the formulation tool with the contamination scenario described previously, we generate the inversion solutions shown in Figure 15.7. The solution quality is very poor for large subdomains with the coarse discretizations. This can be attributed to the approximation errors resulting from the problem discretization and time delay calculation, which increase with the discretization size. Using a smaller subdomain reduces the modeling errors by decreasing the span of the network considered and by allowing finer discretization. Figure 15.7 shows how higher quality solutions can result as the subdomain size is decreased. Of course, if the subdomain size is decreased further, however, we may exclude nodes with important sensor information.

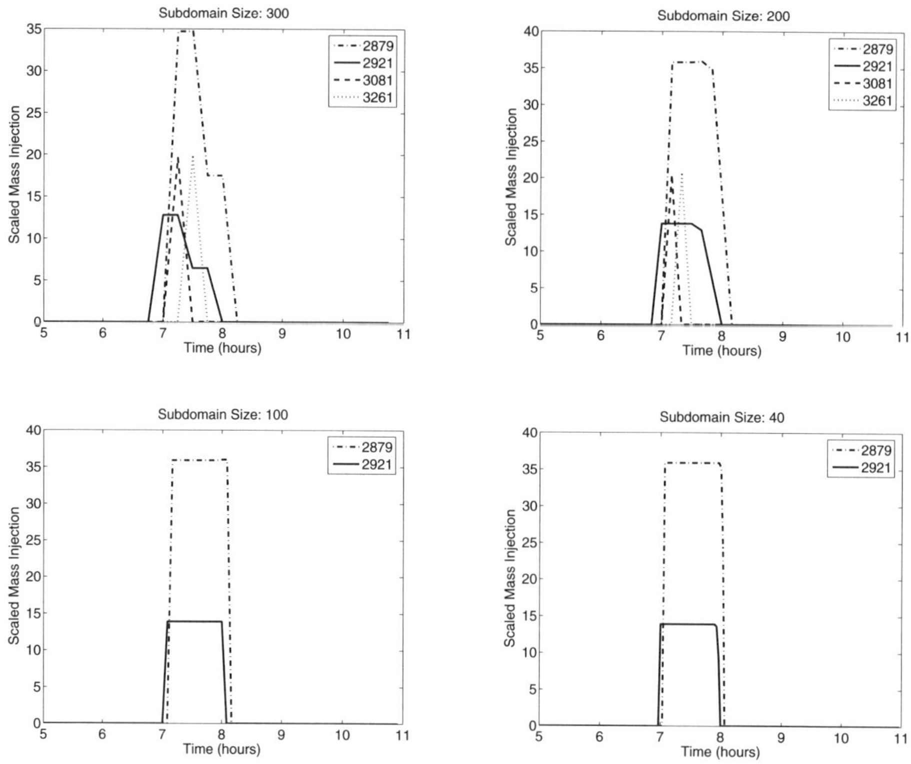

Figure 15.7. *Optimization solutions for fixed problem sizes. This figure shows the optimization solution for different subdomain sizes, where the size of the nonlinear problem was kept constant using the parameters specified in Table* 15.2.

15.4 Conclusions and Future Work

The origin tracking algorithm successfully reduces the size of the optimization problem, allowing the use of simultaneous techniques on medium-sized network models in a real-time setting. Previous results have demonstrated the effectiveness of the dynamic optimization formulation on over 1500 test scenarios with real municipal network models. The use of the subdomain approach further reduces the problem size allowing real-time solutions for large network models.

The subdomain approach can also be an effective technique for improving the quality of the inversion solution. The cause for this is twofold. While decreasing the subdomain size reduces the amount of information available to the optimization, it also allows a finer time discretization for the same computational effort. Furthermore, errors in the time-delay approximations are less pronounced in a smaller network, helping to improve solution quality. Care must be taken, of course, to select a subdomain that does not exclude necessary information for the inversion problem.

In these examples, we found solutions with randomly placed sensors. It is important to identify optimal sensor locations to reduce solution nonuniqueness and decrease identification time. Also, statistical analysis is necessary to determine the effectiveness of the formulation when there are errors in flow and sensor data.

The formulation presented here assumes that the network flow rates are known inputs to the water quality model. In actuality, the network hydraulics may not be known exactly due to uncertain demands and few flow measurements. We propose to formulate a multiperiod optimization problem that will account for the demand uncertainty in the inversion results by solving a single-estimation problem that includes a large number of realizable demand scenarios. The problem grows linearly in the number of samples but has a specific structure that can be exploited to provide efficient solutions.

This chapter presents a rudimentary algorithm for selecting the subdomain of interest based on geographical location. Future work includes the investigation of more advanced techniques for selecting this subdomain. If contamination sources are found on the boundary of the subdomain, this indicates that the subdomain may need to be expanded or moved. The subdomain can be expanded up to acceptable computational limits of the direct linear solver. Also, the window itself can easily be moved, solving the formulation again, until the identified contamination sources are interior to the subdomain. Moreover, multiple windows could also be used in a single optimization, each over a different period of time. This requires a multistage approach, where the first window is used for $t = [0, t_1]$, the second for $t = [t_1, t_2]$, etc. Multiple windows could also be solved in parallel. Using distributed parallel processing, each window can be solved independently, with an external algorithm to resolve the connected boundaries.

In conclusion, these preliminary results indicate the potential of simultaneous optimization techniques for efficient, effective identification of contamination sources in municipal drinking water networks. The subdomain approach allows consideration of large water networks, and reasonable solution times allow its use in a real-time setting. Finally, although the simultaneous optimization formulation using the origin tracking algorithm has only been demonstrated on the contamination source inversion problem, its importance extends to other applications, including real-time optimization for contaminant control of pipe networks.

Acknowledgments

Financial support from Sandia National Laboratories and the National Science Foundation (under ITR grant ACI-021667) is gratefully acknowledged.

Bibliography

[1] U. M. Ascher and L. R. Petzold, *Computer Methods for Ordinary Differential Equations and Differential Algebraic Equations*, SIAM, Philadephia, 1998.

[2] A. Cervantes and L. T. Biegler, *Optimization strategies for dynamic systems*, in Encyclopedia of Optimization, C. Floudas and P. Pardalos, eds., Kluwer Academic Publishers, Dordrecht, The Netherlands, 2003, pp. 216–227.

[3] R. Fourer, D. M. Gay, and B. W. Kernighan, *AMPL: A Modeling Language for Mathematical Programming*, The Scientific Press, Danvers, MA, 1993.

[4] C. D. Laird, L. T. Biegler, B. van Bloemen Waanders, and R. A. Bartlett, *Time dependent contaminant source determination for municipal water networks using large scale optimization*, J. Water Resour. Plng. and Mgmt., 131 (2005), pp. 125–134.

[5] C. P. Liou and J. R. Kroon, *Modeling the propogation of water-borne substances in distribution networks*, J. American Water Works Assoc., 79 (1987), pp. 54–58.

[6] L. A. Rossman, *EPANET User's Manual*, Risk Reduction Engineering Laboratory, U.S. EPA, Cincinnati, OH, 2000.

[7] L. A. Rossman and P. F. Boulos, *Numerical methods for modeling water quality in distribution systems: A comparison*, J. Water Resour. Plng. and Mgmt., 122 (1996), pp. 137–146.

[8] L. A. Rossman, P. F. Boulos, and T. Altman, *Discrete volume-element method for network water-quality models*, J. Water Resour. Plng. and Mgmt., 119 (1993), pp. 505–517.

[9] F. Shang, J. G. Uber, and M. M. Polycarpou, *Particle backtracking algorithm for water distribution systems analysis*, J. Envir. Engrg., 128 (2002), pp. 441–450.

[10] B. van Bloemen Waanders, R. A. Bartlett, K. Long, P. T. Boggs, and A. Salinger, *Large Scale Nonlinear Programming for PDE Constrained Optimization*, SAND Report 2002-3198, Sandia National Laboratories, Albuquerque, NM, 2002.

[11] B. G. van Bloemen Waanders, R. A. Bartlett, L. T. Biegler, and C. D. Laird, *Nonlinear programming strategies for source detection of municipal water networks*, in Proceedings of theASCE EWR/World Water and Environmental Resources Congress, EWRI 2003, Philadelphia, 2003.

[12] A. Wächter and L. T. Biegler, *On the implementation of an interior-point filter line-search algorithm for large-scale nonlinear programming*, Math. Program., 106 (2006), pp. 25–57.

[13] A. Wächter, *An Interior Point Algorithm for Large-Scale Nonlinear Optimization with Applications in Process Engineering*, Ph.D. thesis, Carnegie Mellon University, Pittsburgh, PA, 2002.

[14] M. L. Zierolf, M. M. Polycarpou, and J. G. Uber, *Development and autocalibration of an input-output model of chlorine transport in drinking water distribution systems*, IEEE Trans. Control Sytstems Technology, 6 (1998), pp. 543–553.

Index